KNOWLEDGE, INDUSTRY AND ENVIRONMENT

Knowledge, Industry and Environment

Institutions and innovation in territorial perspective

Edited by
ROGER HAYTER
Simon Fraser University, Canada

RICHARD LE HERON
University of Auckland, New Zealand

Routledge
Taylor & Francis Group

LONDON AND NEW YORK

First published 2002 by Ashgate Publishing

Reissued 2018 by Routledge
2 Park Square, Milton Park, Abingdon, Oxon OX14 4RN
711 Third Avenue, New York, NY 10017, USA

Routledge is an imprint of the Taylor & Francis Group, an informa business

Publisher's Note
The publisher has gone to great lengths to ensure the quality of this reprint but points out that some imperfections in the original copies may be apparent.

Disclaimer
The publisher has made every effort to trace copyright holders and welcomes correspondence from those they have been unable to contact.

A Library of Congress record exists under LC control number: 2002101563

ISBN 13: 978-1-138-72747-2 (hbk)
ISBN 13: 978-1-138-72745-8 (pbk)
ISBN 13: 978-1-315-19032-7 (ebk)

Contents

List of Figures   viii
List of Tables   xi
List of Contributors   xiii

1   Introduction to Knowledge, Industry and Environment   1
     Richard Le Heron and Roger Hayter

2   Industrialization, Techno-economic Paradigms and the
     Environment   11
     Roger Hayter and Richard Le Heron

3   Paths of Sustainable Industrialization in the
     Knowledge-based Economy   31
     Sam Ock Park

4   Building up Competence, Institutions and Networks: A
     Perspective on 'Catch-up' in the Knowledge Economy   49
     Ayda Eraydin

5   Global-local Networking of PC Manufacturing in
     Dongguan, China   67
     Tong Xin and Wang Jici

6   Development of the Internet in China and Its Spatial
     Characteristics   87
     Weidong Liu

7   Characteristics and Development of Industrial
       Districts: The Case of Software Clusters in
       Seoul, South Korea   107
       Joo-Seong Hwang

8   Qualified Labour Migration and Regional Knowledge
       Economies   125
       Martina Fromhold-Eisebith

9   Improving Embedded Knowledge in Old Industrial
       Districts: Case Studies, from Valencia, Spain   145
       Javier Alfonso Gil, Antonia Sáez Cala,
       Antonio Vázquez-Barquero and Ana Isabel Viñas Apaolaza

10   The Internalization of Knowledge in Manufacturing
       Value Added: The Geography of Ericsson's Mobile
       Systems, and the China Connection   165
       Claes G Alvstam

11   Environmental Knowledge, the Power of Framing
       and Industrial Change   187
       Dietrich Soyez

12   Environmental Service-providers, Knowledge
       Transfer, and the Greening of Industry   209
       Christian Schulz

13   Competitive and Green? Determinants of
       Successful Environmental Management in
       the Manufacturing Sector   227
       Boris Braun

14   Globalization of the New Zealand Forestry Industry:
       The New Impact of Japanese Linkages   253
       Christina Stringer

15   Cultural Embeddness, Corporate Strategy and
      Foreign Investment in Poland: A Tale of Two
      Firms   273
      Jane Hardy

16   Paths of Industrial Transformation in Poland
      and the Role of Knowledge-based Industries   289
      Tadeusz Stryjakiewicz

17   Knowledge-based Regions in the Global Periphery:
      The Case of South Africa   313
      Chris Rogerson

18   The Sustainable Renovation of the Industrial
      Complex in Inner Tokyo: The Case of the
      Japanese Machinery Industry   337
      Atsuhiko Takeuchi and Hideo Mori

19   Globalization and the Reorganization of a
      Metropolitan Knowledge System: The Case of
      Research and Development in
      Frankfurt/Rhein-Main, Germany   355
      Eike W. Schamp

20   Enhancing Competencies, Networking and
      Institutions for Knowledge Economy:
      Singapore's Garment Industry   377
      Leo van Grunsven

21   Conclusion: Institutions and Innovation in Territorial
      Perspective   399
      Roger Hayter and Richard Le Heron

Index   411

List of Figures

Figure 2.1   Eden's model of the 'greening' of corporate behaviour   22

Figure 5.1   First stage export processing activities in Dongguan: Plants keep global linkages through Hong Kong while local linkages are weak   72

Figure 5.2   Local PC-related hardware processing activities and products in Dongguan   78

Figure 5.3   Dongguan: Interplant linkage and export institutions   82

Figure 6.1   Number of Internet users and domains under 'CN' (1997-2000)   93

Figure 6.2   Number of computer hosts and bandwidth of leased international connections in China (1997-2000)   93

Figure 7.1   Basis for industrial districts   109

Figure 7.2   Five core dimensions of industrial districts in the evolutionary process   111

Figure 8.1   Qualified labour migration as a supporting factor of regional knowledge economies – a conceptual model   137

Figure 10.1   Ericsson's sales by geographic regions, 2000   173

Figure 10.2   The gradual externalization process of physical manufacturing within Ericsson   175

Figure 10.3   The gradual internalization of the creation of knowledge within Ericsson   176

Figure 10.4   The development towards third generation of mobile communications   178

Figure 10.5   Direct and indirect research and development
               at Ericsson, 1990-2000   180
Figure 10.6   Spatial dynamics of research and development
               creation within the telecommunications
               industry   183
Figure 11.1   The 'greening' of industry 1960-2000 in
               Western industrial countries – a tentative
               synopsis   190
Figure 12.1   Factors in the 'greening' of industry   211
Figure 12.2   Providers and client types for knowledge
               intensive environmental services   213
Figure 12.3   Types of relationships between environmental
               service providers and clients   216
Figure 12.4   Networking between service providers, clients
               and research institutions   219
Figure 13.1   Determinants and dimensions of successful
               environmental management in manufacturing   235
Figure 13.2   Firm-related strategies to achieve central
               goals of environmental management   248
Figure 14.1   Fletcher Challenge Forests: Production
               chain components   261
Figure 14.2   Juken Nissho: Production chain components   264
Figure 16.1   Outlays for research and development in
               selected countries   292
Figure 16.2a   Forms of adaptation of industrial enterprises
               to the new industrial system: Globalized type   299
Figure 16.2b   Forms of adaptation of industrial enterprises
               to the new economic system: Home-market
               oriented type   300
Figure 16.2c   Forms of adaptation of industrial enterprises
               to the new industrial system: De-industrialized
               type   301
Figure 16.2d   Forms of adaptation of industrial enterprises
               to the new industrial system: Paternalistic type   302
Figure 17.1   Spatial pattern of employment in high-
               technology manufacturing in South Africa,
               1999   318

Figure 17.2   Spatial patterns of enterprises in high-
technology manufacturing in South Africa,
1999   319
Figure 17.3   Spatial pattern of employment in information
technology service activities in South Africa, 1999   323
Figure 17.4   Spatial pattern of enterprises in the information
technology service in South Africa, 1999   324
Figure 18.1   Concentration of industry in Ota-ku:
Distribution of plants, 1993   339
Figure 18.2   Concentration of industry in Ota-ku: Change
in land use in Shimomaruko district, 1968-1997   340
Figure 18.3   Transition from vertical to horizontal inter-firm
relationships   348
Figure 18.4   Participation of new generation of entrepreneurs
and evolution of community spirit   350
Figure 19.1   A Triple Helix model of innovation system   358
Figure 19.2   Research and development employment in
the districts of metropolitan region,
Frankfurt/Rhein-Main   362
Figure 19.3   A portfolio of the regional knowledge system
of Frankfurt/Rhein-Main   372
Figure 20.1   Configuration of production and distribution
networks of OEM producers   388
Figure 20.2   Configuration of production and distribution
networks of OBM producers   389

List of Tables

Table 2.1   Selected features of techno-economic
              paradigms   13
Table 2.2   Techno-economic paradigms and the
              environment   15
Table 3.1   Environmental impacts of waste emissions   36
Table 3.2   Selected aspects of ECOFIT-Parks   43
Table 5.1   Origins of PC-related firms invested in
              Dongguan   75
Table 5.2   Top ten largest PC-related firms in Qingxi,
              Dongguan   77
Table 6.1   Geography of international connection band
              width of the Internet in China   96
Table 6.2   Results of the multiple regression: model
              summary   97
Appendix 6.1 Statistical basis of model summarized in
              Table 6.2   104
Table 7.1   Perceived locational advantages by areas
              and dimensions of industrial districts   118
Table 7.2   Comparative characteristics of the
              Kangnam-Seochu industrial district   120
Table 8.1   Statistics of the 'brain drain': Highly
              educated immigrants (tertiary level) from Asian
              less developed countries to the USA and all
              OECD states: Numbers and shares in 1990   131
Table 9.1   Research and development (R&D) indicators
              for EU, Spain and Valencia, 1994-1997   149

Table 9.2   National patents applications in Spain by some
                regions of origin, 1994-1998   150
Table 9.3   Services offered by the technological institutes
                in the region of Valencia   157
Table 9.4   Services provided by technological institutes:
                INESCOP, AICE, AITEX and AIJU, 1996   159
Table 13.1   Selected indicators for successful implementation
                of EMS   231
Table 13.2   Variables of logistic regression models   238
Table 13.3   Logistic regression results   242
Table 15.1   Selected features of Enginco's and Vehico's
                operations in Poland   277
Table 16.1   Per capita research and development
                outlays, 1998   293
Table 16.2   Patents granted in Poland in the
                years 1980-1998   294
Table 16.3   Sources of information considered by enterprises
                to be basic for innovations, by firm size and
                ownership sector   297
Table 18.1   Composition of manufacturing in Ota-ku:
                Composition of firms, 1998   342
Table 18.2   Composition of manufacturing in Ota-ku:
                Size-structure of the plants (%)   343
Table 18.3   Categories of renewed enterprises   347
Table 19.1   Research and development (R&D) employment in
                manufacturing sectors in the region
                Frankfurt/Rhein-Main, 1993   364

List of Contributors

Claes G Alvstam, Professor, Department of Human and Economic Geography, School of Economics and Commercial Law, Goteborg Univiersity, P.O. Box 630, SE-405 30, Goteborg, Sweden. E-mail: claes. alvstam@geography.gu.se Professor Alvstam's research interests focus on direct foreign investment, trade, industrial transformation and location dynamics of services.

Ana Isabel Viñas Apaolaza, Dr., Departmento de Estructura Económica y Economía del Desarrollo, Facultad de Ciencias Económicas y Empresariales,Universidad Autónoma de Madrid, Campus de Cantoblanco, Ctra. Colmenar km. 15, 28049 Madrid, Spain. E-mail: ana.vinas@uam.es Dr. Apaolaza's research interests focus on regional industrial policy.

Antonio Vázquez-Barquero, Professor, Departmento de Estructura Económica y Economía del Desarrollo, Facultad de Ciencias Económicas y Empresariales,Universidad Autónoma de Madrid, Campus de Cantoblanco, Ctra. Colmenar km. 15, 28049 Madrid, Spain. E-mail: vazquez-barquero@uam.es Professor Barquero's research interests focus on regional and local economic development, and industrial districts.

Boris Braun, Dr., Geographisches Institut, der Universität Bonn, Meckenheimer Allee 166, 53115 Bonn, Germany. E-mail: braun@giub.uni-bonn.de Dr. Braun's research interests focus on economic geography, environmental economics, regional and urban planning, Europe and Asian economies.

Antonia Sáez Cala, Professor, Departmento de Estructura Económica y Economía del Desarrollo, Facultad de Ciencias Económicas y Empresariales, Universidad Autónoma de Madrid, Campus de Cantoblanco, Ctra. Colmenar km. 15, 28049 Madrid, Spain. E-mail: antonia.saez@uam.es Professor Cala's research interests focus on the innovation patterns of small firms, and local industrial systems.

Ayda Eraydin, Professor, Department of Urban and Regional Planning, Middle East Technical University, 06530 Ankara, Turkey. E-mail: ayda@metu.edu.tr Professor Eraydin's research interests focus on the economic geography of industrial organization, industrial districts, local economic development, small firms, innovation networks and regional policy.

Martina Fromhold-Eisebith, Dr., ARC Seibersdorf Research GmbH, Systems Research Technology-Economy-Environment, A-2444 Seibersdorf, Austria. E-mail: martina.fromhold@arcs.ac.at Dr. Fromhold-Eisebeth's research interests focus on technology oriented regionaldevelopment, innovation systems in Europe and Asia, the economic geography of qualified labour migration, MNCs.

Javier Alfonso Gil, Professor,Departmento de Estructura Económica y Economía del Desarrollo, Facultad de Ciencias Económicas y Empresariales, Universidad Autónoma de Madrid, Campus de Cantoblanco, Ctra. Colmenar km. 15, 28049 Madrid, Spain. E-mail: javier.alfonso@uam.es Professor Gil's research interests focus on institutions and economic change.

Leo van Grunsven, Dr., Department of International Economics and Economic Geography, Faculty of Geographical Sciences, Utrecht University, P.O. Box 80.115, 3508 TC Utrecht, The Netherlands. E-mail: L.Grunsven@geog.uu.nl Dr. Grunsven's research interests focus on global commodity chains, industrial upgrading and local innovation systems in S.E. Asia

Jane Hardy, Dr., Business School, University of Hertfordshire, Mangrove Road, Hertford SG138QF, UK. E-mail: J.A.hardy@herts.ac.uk Dr. Hardy's research interests focus on foreign investment in transforming economies, firms as institutions.

Roger Hayter, Professor, Department of Geography, Simon Fraser University, Burnaby, British Columbia V5A 1S6, Canada. E-mail: hayter@sfu.ca Professor Hayter's research interests focus on industrial location dynamics and regional industrial transformation, the forest sector, Japanese MNCs.

Joo-Seong Hwang, Senior Research Fellow, Korea Information Society Development Institute, Seoul, Korea. E-mail: jshwang@kisdi.re.kr Dr. Hwang's research interests focus on industrial districts and the Korean economy.

Richard Le Heron, Professor, Department of Geography, University of Auckland, New Zealand, Private Bag 92019, New Zealand. E-mail: r.leheron@auckland.ac.nz Professor Le Heron's research interests focus on industrial restructuring, internationalization processes, resource-based commodity chains, the New Zealand economy.

Weidong Liu, Dr., Department of Regional Studies, Institute of Geographical Sciences and Natural Resources Research, Chinese Academy of Sciences, Beijing, 100101 China. E-mail: liuhequn@public.east.cn.net Dr. Liu's research interests focus on the economic geography of the Internet, the auto industry, regional disparity in China.

Hideo Mori, Mr. Department of Geography, Nippon Institute of Technology, Gakuendai, Miyashiro, Saitama, 345-8501, Japan. E-mail: atsuhiko@mbe.sphere.ne.jp Mr. Mori's research interests focus on the mapping of economic activities in Tokyo.

Sam Ock Park, Professor, Department of Geography, Seoul National University, Seoul 151-742, Korea. E-mail: parkso@snu.ac.kr Professor Park's research interests focus on industrial restructuring, industrial districts, regional innovation systems, and regional policy.

Chris Rogerson, Professor, School of Geography, Archaeology and Environmental Studies, University of Witswaterand, Private Bag 3, P.O. Wits 2050, South Africa. E-mail: 017cmr@cosmos.wits.ac.za Professor Rogerson's research interests focus on small enterprise development and local economic development in South Africa.

Eike W. Schamp, Professor, Institute fuer Wirtschafts- und Sozialgeographie, Johann Wolfgang Goethe-Universitaet, Dantestr. 9, 60054 Frankfurt/Main, Germany. E-mail: schamp@em.uni-frankfurt.de Professor Schamp's research interests focus on the economic geography of globalization, low and medium-tech industries, knowledge intensive services, and urban economic change.

Christian Schulz, Dr., Geographisches Institut, Universität zu Köln, Albertus-Magnus-Platz, 50923 Köln, Germany. E-mail: ch.schulz@ uni-koeln.de Dr. Schulz's research interests focus on economic geography, border regions, urban and environmental planning, and France.

Dietrich Soyez, Professor, Geographisches Institut, Universität zu Köln, Albertus-Magnus-Platz, 50923 Köln, Germany. E-mail: d.soyez@uni-koeln.de Professor Soyez's research interests focus on the resources/environment interface of economic geography, industrial tourism and the transnationalization of civil society actors.

Tadeusz Stryjakiewicz, Dr., Institute of Socio-Economic Geography and Space Economy, Adam Mickiewicz University, Fredry 10, PL 61-701 Poznan, Poland. E-mail: tadek@amu.edu.pl Dr. Stryjakiewicz's research interests focus on industrial geography, innovation systems and the socio-economic transition of East-Central Europe.

Christina Stringer, Dr., Department of Geography, University of Auckland, New Zealand, Private Bag 92019, New Zealand. E-mail: c.stringer@auckland.ac.nz Dr. Stringer's research interests focus on forestry and other resource-based commodity chains in the Pacific Rim, and New Zealand.

Atsuhiko Takeuchi, Professor, Department of Geography, Nippon Institute of Technology, Gakuendai, Miyashiro, Saitama, 345-8501, Japan. E-mail: atsuhiko@mbe.sphere.ne.jp Professor Takeuchi's research interests focus on the regional dynamics of industry in Japan.

Tong Xin, PhD candidate, Department of Geography, Peking University, Beijing, 100871, P.R. China. E-mail: 1th@pku.edu.cn Ms. Tong's research interests focus on the geography of information technology industry in China.

Wang Jici, Professor, Department of Geography, Peking University, Beijing, 100871, P.R. China. E-mail: wjc@pku.edu.cn Professor Wang's research interests focus on industrial districts, high tech parks, innovation systems and China's economy.

1 Introduction to Knowledge, Industry and Environment

Richard Le Heron and Roger Hayter

For 30 years, local, national and global development has been challenged by the imperatives of economic restructuring and environmental sustainability. This challenge has proven to be a formidable one. Each imperative represents profoundly complex changes while debate and policy responses have often been complicated by the framing of the relationships between them in the jarring, antagonistic terms of a negative sum trade-off summarized as 'more economy and less environment', or vice versa. Even in the affluent west (and Japan), economy and environment have posed lasting, politically charged conundrums. Indeed, in the early 1970s, as governments in these countries sought to give priority to environmental concerns, their economies simultaneously became significantly more volatile while unemployment rates returned to levels not witnessed since the 1930s. Beyond the affluent west (and Japan), in countries with fewer resources and less democratic forms of governance, the priorities of economy and environment have appeared to be even more daunting, and the negative sum trade-off between the two even more stark.

Fortunately, there have been insistent, coherent voices, particularly those associated with institutional economics in the Veblen and Schumpeterian traditions, led by Freeman (1992) and Kuttner (1984), that have sought to reframe economic and environmental imperatives from negative- to positive sum games. In their view, the policy challenge is precisely to facilitate such outcomes, that is for public and private sector behaviour and policies to realize both economic and environmental goals. A key theme in their prescriptions is the role of education, learning and innovation, institutional as well as technological, in the creation and implementation of ideas and policies that will underpin economically and ecologically sustainable societies. *Knowledge, Industry and Environment* explores this theme from an economic-geography perspective.

References to knowledge-driven industries, and to the 'bigger' idea of the knowledge economy, have gained widespread currency in recent years in academic circles and policy debates. Admittedly, this contemplation has been largely driven by concerns for economic restructuring. Indeed, in our view in many countries during the 1990s the policy spotlight on national and regional development beamed away from environmental issues and increasingly on the growth and employment prospects associated with knowledge-based industrial development. This vacillation between economic and environmental imperatives is a least symptomatic of the tendency to fragment these imperatives inside different policy and theoretical boxes. Yet, if economy and environment imperatives represent distinct spheres of interest with distinct ideological roots, they need not – and must not – be divorced from one another. The underlying assumption of a positive sum game mindset is that sound economic performance is vital to ensuring environmental sustainability, and vice versa.

Knowledge, Industry and Environment grapples with the nexus of relationships formed by knowledge, industry and environment, and how these relationships are shaped in different parts of the world. These relationships remain poorly understood. In particular, there are few attempts to explicitly link the idea of the knowledge economy with the wider adjustments towards what Freeman (1992) terms a new green techno-economic paradigm of capitalist production-consumption, relations that are rooted in increasingly reflexive knowledge about environmental and industrial processes. Yet, such reflexivity is happening, as is underlined by Florida's (1996) pioneering study in the USA that demonstrates that best practice firms in terms of product innovation, labour relations and technology also tend to be best practice in terms of environmental performance (see also Florida et al 2001). *Knowledge, Industry and Environment* seeks a better understanding of the geographical manifestations of the knowledge economy and of the growing reflexivity between industrial and environmental processes. We are especially interested in how institutional innovations in different regional and local economies around the world are seeking knowledge-based development and positive sum outcomes between economy and environment. While our conceptualizations of these dimensions address global dynamics, the empirical base to our territorial perspective is not truly global, for example, we lack case studies from the USA or South America. Nevertheless, *Knowledge, Industry and Environment* discusses a globally diverse range of social experimentation drawn from the west and east of Europe, South Africa and around the Pacific Rim.

All the chapters in this book were originally presented at a conference organized by the Commission on the Organization of Industrial Space of the International Geographical Union at Dongguan, China in August of 1999. As such, *Knowledge, Industry and Environment* builds on previous volumes of the Commission, and predecessor Commissions. The book particularly integrates and enlarges upon two previous volumes, namely those edited by Taylor (1995) and Malecki and Oinas (1999), that respectively dealt with environmental and space and the technological connections underlying contemporary economies. However, the structure and themes of, and the list of authors contributing to, *Knowledge, Industry and Environment* are distinctive, characteristics we suggest that point to the vitality of the Commission. The book's distinctiveness in this regard, relates to its conceptualization and policy implications, as well as for its substantive illustrations.

We argue that the three domains of industry, environment and knowledge should be reset into new discourses and conceptualizations. In particular the contemporary course of world industrialization and the myriad of local industrialization experiences should be re-approached by a view that is sensitive to the interactions between industrial, environmental and knowledge processes. As the chapters of this book indicate, a case can be made that the nexus of knowledge, industry and environment deserves urgent and ongoing theoretical, empirical and policy attention.

The starting point for *Knowledge, Industry and Environment* is that a fruitful and necessary direction is to consider the interest in environment and knowledge as integral to changes in capitalist accumulation and regulatory processes. In particular, the book offers critical perspective on the geographic study of the knowledge economy and the industry-environment interface. By critical we mean at least two things. In the first place, the book takes a strong position over the association of knowledge, industry and environment. It seeks, by both drawing on and critiquing current research on industrial change, to present an improved understanding of how increasingly reflexive knowledge about environmental and industrial processes might feed a new green techno-economic paradigm of capitalism. This intervention into the contested arenas of academic interpretation and policy formulation is necessarily selective. *Knowledge, Industry and Environment* contemplates the nexus between industry-environment-knowledge in a number of territorial contexts. But the selectivity is based on a particular intellectual-political position. Academics have a responsibility to reveal the workings of the world. In this book an attempt is made to bring *Knowledge, Industry and Environment* together rather than keeping them in separate domains, as they are so often conceived in the literature, in policy and in industrial practice.

Two important goals of *Knowledge, Industry and Environment* are thus to extend, in modest ways, understanding of the territorial 'workings out' of the knowledge economy and to integrate the processes of regional industrial development and environmentalism. This leads to the second sense in which the book is a critical engagement with the literature and with the policy. The book frames an argument that it is potentially helpful to explore the green-knowledge economy in terms of a fusion between technological and economic innovation *and* innovation in social, political, cultural and related institutions. Innovation, whether in the spheres of the economy (and therefore immediately tied into the strategies of actors and the wider processes of accumulation) or in the institutional sphere (which can constrain or enable accumulation efforts), implies learning processes of many kinds. Readers of this book are also likely to bring their own critical perspectives, and challenge both the general argument and the illustrative insight of the case studies outlined by contributors. But in questioning why *Knowledge, Industry and Environment* might be bundled together and in reflecting on the contingent and spatially specific conditions of the international case studies, a fuller and clearer sense of the conditions under which policy is formulated and implemented may be reached.

Themes

There are four themes (highlighted in italics in the following paragraphs) evident within *Knowledge, Industry and Environment* that we believe are part of any articulation of a geographical perspective on the emerging knowledge economy constituted through industrial and environmental processes. First, the theoretical literature on techno-economic paradigms and industrialization is insistent that *institution-building strategies*, especially for innovation, are an important dimension of sustainable development. Our contention, derived from an institutionalist and regulationist reading of accumulation and regulatory regimes, is that institutional innovation is as important as technological change. Institutional change, and re-regulation, can facilitate technological change, and can be an independent source of productivity improvement. That institutional change can be misdirected reinforces the need to place institutions in analytical focus. The chapters by Hayter and Le Heron, Park and Eraydin examine the knowledge, industry and environment nexus primarily through the lens of institutional innovation, collectively establishing a broad framework for the rest of the book. Hayter and Le Heron specifically take up Freeman's (1990) challenge that profound changes in attitudes towards the environment must be followed by paradigmatic techno-economic changes *and* a matching of supportive institutional initiatives.

This call for social responsibility on the part of Freeman is both admirable, as it stresses conscious changes in technological, economic and institutional strategies. However, there are considerable obstacles to giving effect to new trajectories, with differing temporalities and spatialities, from context to context. Hayter and Le Heron review the theoretical ideas behind the extension of the basic techno-economic paradigm into a 'green' techno-economic paradigm and explore some of the contradictory dimensions associated with the new model. Their chapter poses a series of questions about the wider implications of a green techno-economic paradigm, some of which are revisited in the book's concluding chapter. Park considers various dimensions of sustainability in relation to changing production systems and environmental conditions, a theme underplayed in the geographical literature and then examines possible directions that sustainable development might take in an era where the influence of knowledge economy thinking is potentially high. He identifies in particular the formative role of networking in developing and maintaining sustainability initiatives, at whatever spatial scale. The chapter leads directly to discussions of learning processes that are covered by Eraydin in a discussion that deals with what might be needed by lagging economies to effect a development 'catch-up' to core regions on the basis of achieving a knowledge economy. The chapter can be read in at least two ways. First, the chapter represents a search for a formula to achieve regional and national competitiveness through capacity building, innovation and networking. Second, it starkly reveals the difficulties connected with making the shift called for by Freeman, towards creating new institutions to accompany product and process innovation.

A second theme features the role of *agglomerations as territorial context*. In the contemporary world, an increasing portion of the world's population lives in metropolitan areas and knowledge economy workers (and many researchers) have re-enshrined agglomerations as both outcomes and mechanisms of capitalist expansion. Agglomerations are thus likely settings for new waves of economic activities and the basis for environmental contestation. Seven chapters provide very different territorial windows on contemporary knowledge economy 'experiments'. Tong and Wang provide critical insights into the global-local dynamics that have shaped the remarkable export-oriented growth of the electronics industry in Dongguan, South China, noting how further value added developments are now being held back by the very regulations that made this production possible. This realization represents a strategic opportunity in regulation and governance – to explore alternative policy configurations that might enhance more local know-how while ensuring a green techno-economic future. Huang also focuses on the electronic sector to assess the extent to which several emerging software clusters in Seoul, South Korea can be said to constitute industrial districts. He shows that these clusters vary from one another and have a dynamic that is geared to technological but not institutional innovation. Liu through his documentation of the expansion of the Internet in China challenges geographers to explicitly study the spatial and temporal pulses of industrialization. He argues that without the recognition of how new industrial spaces are being created, the capacity to fuse new patterns of thought into industrial and environmental thinking is considerably reduced. In the Valencia region of Spain, Gil et al reveal the inherent tension in present innovation processes, and the difficult choice facing small firms between the internal generation of technology or dependence on the embodied knowledge of others. Most of these firms, they note, favour the latter, 'dependent' strategy while simultaneously gaining a degree of technological improvement by accessing newly created technological institutes.

In the South African case, Rogerson reveals that the rapid emergence of new high tech manufacturing activities and information technology service activities has been strongly clustered, not in new industrial spaces, but in existing agglomerations. He argues that this reinforcement of established agglomerations by a new layer of activities illustrates the latter's powerful advantages and principles of cumulative causation. Schamp's analysis, that focuses on Frankfurt/Rhein-Main, Germany, explores the anatomy of the idea of the regional innovation system, revealing the importance of the metropole for its evolution and organization. He also shows that metropolitan knowledge systems are fragmented and locally incomplete systems often dominated by national and international influences through actors and institutions. Moreover, he suggests that trajectories of metropolitan knowledge systems are place-specific, depending on both historical structures and contradictory actions of local and non-local actors.

Thus, the idea of *agglomerations as territorial contexts* for economic and institutional innovation must be thoroughly prefaced by the contingent nature of developments. This is an often acknowledged fact but one which, in terms of the argument of the book, is an invitation to insert freshly imagined research and political agendas, springing from local and non-local inspiration. Takeuchi and Mori add to this theme. They provide a salutary reminder that in situ geographical industrialization can be rapid, though invariably with strong elements of continuity alongside and associated with emerging patterns. Their micro-analysis of inner Tokyo describes the role of successive generations of ('father-son') entrepreneurs to achieve sustainable renovation of a core area in Japan's machinery industry. The case study illustrates subtle changes in the behaviour of both firms and government over a number of decades that cumulatively have resulted in significant re-vitalization of the industrial area, much guided by Japanese norms. This culturally specific yet open renovation involves a symbiotic relationship among industry, the community, land use and the physical environment.

A third theme considers the *regulation and governance of industrial-environmental processes* towards sustainable interactions. Chapters by Soyez, Schulz and Braun examine the interrelations between environmental imperatives and business behaviour and probe the difficulties connected with developing, putting in place and maintaining effective environmental regulations and governing arrangements. In a broad review, Soyez analyzes the highly discursive nature of environmental knowledge and explores how such knowledge shapes political and corporate thinking and action. His analysis reveals, with case studies from British Columbia and Washington D.C., the global reach of civil society actors, most notably environmental non-government organizations, as they influence each part of the production chain. He concludes his chapter with a vigorous call for industrial and economic geographers to address the greening of industry, in order to help ensure that environmental concerns become the motor of industrialization. Two empirical chapters explore efforts to date by German businesses to modify new environmental behaviours. Schulz reports on a survey of the transfer of environmental knowledge in German firms. His arguments indicate the growing (and strategic) role of specialized producer service companies which impact upon corporate environmental performance through knowledge transfer and inter-organizational learning. Braun asks whether it is possible to be both competitive and green. Drawing on the findings of a large-scale survey regarding the implementation of environmental monitoring systems, he systematically tests various hypotheses about environmental performance. A significant conclusion is the positive correlation between economic and environmental performance, underlying that positive sum games are a legitimate policy hope.

A fourth theme developed in the book is that of *learning and location dynamics*. Thus several chapters examine how the incorporation of new knowledge (at plant, firms and production system levels) shapes change, including the perception of locations as sites for investment, the forms of technological transfer, the organization of production and distribution activities and the nature of institutional responses and initiatives. Stringer's chapter on the globalization of New Zealand's forestry industry, explored from the perspective of Japanese linkages, reveals re-positioning of forestry operations in New Zealand, within the broader realm of Asia Pacific Rim forestry. This dynamic underscores the need to develop national and international regulatory and governance frameworks that are commensurate with the geographic scales at which firms and industries are operating. Hardy examines the role of foreign investment in transferring know-how to Poland. Her case studies reveal that there is no one best way or blueprint to maximize this transfer which is strongly shaped by intra-corporate cultures, as well as local conditions. In her view, a more nuanced approach is needed if academics and policy makers are to understand how firms transfer new knowledge and managerial practices, be it across national boundaries, from knowledge surplus to knowledge deficit areas.

Stryjakiewicz assesses the innovation potential of the Polish economy and classifies types of firm-based adjustments that define present day transformations in Polish industry. Two case studies illustrate two main forms of adjustment, each with alternative lessons for local development. As in several other chapters, the question is whether and how peripheral economies might join the knowledge economy set. Van Grunsven's investigation of Singapore's garment industry suggests that mature industries in Asian Newly Industrialized Economies are not necessarily rendered unsustainable, as resources are diverted to new knowledge-driven industries and activities within the framework of knowledge-based economic policies. His general optimism rests on the fact that knowledge-based economic initiatives provide a framework that enables firms to build competences and capacities. He sees attitudinal changes as a key in any alteration in strategic direction by firms.

Two chapters explore the interconnection of the knowledge economy with globalization processes, specifically linking developed and developing countries, albeit in different ways. Thus, Fromhold-Eisebith contends that migration patterns of highly qualified people induces internal connections of networked regions leading to mutually supportive technological developments in home and host countries. Her analysis is a timely reminder of the importance of individuals in the technology creation and transfer process and serves to reveal a significant 'spread' effect generated by such global high tech growth poles as Silicon Valley. Finally, Alvstam links the developed and developing world in the very creation of knowledge within knowledge intensive MNCs, specifically the telecommunication giant, Ericcson. He notes the little known geography of the value added within MNCs whose accounting of the location of value added may not coincide with its creation. In the case of Ericcson, for example, intellectual capital is not only developed in formal research and development organizations in Sweden but also in its distant manufacturing and service operations in China.

These four themes run through *Knowledge, Industry and Environment* but, since they are not mutually exclusive, they do not compartmentalize it. In one way or another, every chapter contemplates institutional innovation, while varying emphasis is given to the other themes. In addition, the subsequent chapters give a territorial perspective to economic and institutional innovation. As the chapters in the book foreshadow, constructing industrial policy knowledgeably, at different spatial scales, is a major challenge for policy makers. Geographers, however, are in a position to offer some concrete suggestions about how to insert 'steering practices' that give a broader and more penetrating beam to the industrial policy spotlight. To this end, the book's final chapter explores what might go into a research agenda aimed at fostering a reflexive green techno-economic paradigm.

Acknowledgements

We are especially grateful for the support we received in the Geography Departments of Simon Fraser University and the University of Auckland. At Simon Fraser, Marcia Crease and Ray Squirrell provided tremendous help in organizing the tables and graphics, and in preparing a camera ready manuscript. In Auckland we are very appreciative for the research assistance of Christina Kaiser and the secretarial support of Firoza Cooper throughout the preparation of this book.

References

Florida, R. (1996), 'Lean and Green: The Move to Environmentally Conscious Manufacturing', *California Management Review*, vol. 39, pp. 80-105.

Florida, R., Atlas, M. and Cline, C. (2001), 'What Makes Companies Green? Organizational and Geographic Factors in the Adoption of Environmental Practices', *Economic Geography*, vol. 77, pp. 209-24.

Freeman, C. (1992), *The Economics of Hope*, London: Pinter Publishers.

Kuttner, R. (1984), *The Economic Illusion: False Choices Between Prosperity and Social Justice*, Houghton Mifflin, Boston.

Malecki, E.J. and Oinas, P. (eds) (1999), *Making Connections: Technological Learning and Regional Economic Change*, Ashgate, Aldershot.

Taylor, M. (ed) (1995), *Environmental Change: Industry, Power and Policy*, Averbury Press, Aldershot.

2 Industrialization, Techno-economic Paradigms and the Environment

Roger Hayter and Richard Le Heron

Introduction

As Soyez (chapter 11) reminds us, the research agenda of Industrial Geography has been slow to respond to the public and corporate priorities given to 'environmental imperatives'. In recent years, however, there are signs of a willingness to build upon earlier pioneering studies (Ullman, 1958; Thomas, 1968; Soyez, 1985) in a more serious contemplation of these imperatives for industrial location dynamics (Eden, 1996; Florida, 1996; Taylor, 1995; see also chapters 12-14). In this chapter, we contribute towards this contemplation within the framework of a model of the long run processes of industrialization, specifically the model of techno-economic paradigms (TEPs) pioneered in three papers by Perez (1985), Freeman (1987) and Freeman and Perez (1988).

The chapter is in two main parts. The first part, in an admittedly heuristic way, incorporates an environmental dimension within the TEP model, including Freeman's (1992) suggestion that the next paradigm for economic development will be a green one. In Freeman's (1974, 1992) view, profound changes in attitudes towards the environment require paradigmatic technological changes, as well as associated ('matching') institutional initiatives. The second part explores the idea of the 'green paradigm' including the emergence of the 'green corporation' and the implications raised for 'green geographies', especially in relation to industrial location dynamics. In turn, these implications pose research challenges for Industrial Geography.

Techno-economic Paradigms and the Environment

The TEP model provides an explanation of industrial growth and change since the Industrial Revolution (Perez, 1985; Freeman and Perez, 1988). The model argues that the driving force underlying Kondratiev long waves or cycles is provided by technological changes that are buttressed and shaped by institutional innovations. Technological innovation can range from small scale 'incremental' changes to 'paradigmatic' shifts while between these polar types 'major' and 'radical' innovations are possible. Incremental change involves minor productivity improvements that are nevertheless collectively significant; indeed, they are vital to the full exploitation of major and radical innovations as well as to new TEPs. TEPs, stimulated by deep crises that cannot be solved by the old paradigm, provide the basis for new Kondratiev waves and exert 'pervasive' effects throughout the economy (Table 2.1). In general, each TEP features new lead industries that dominate the economy, new rapidly emerging industries, new forms of infrastructure, new key factor industries that provide abundant, low cost and widely applicable inputs, new ideas about productivity and engineering common sense, and new 'geographic spaces'. Simultaneously, massive technological re-thinking requires a set of 'matching' institutional innovations, for example, regarding forms of business organization, labour relations, and R&D structures as well as international forms of regulation.

Clearly, given the importance of 'institutional matching', the TEP model, does not constitute technological determinism, as is sometimes supposed (Sabel and Zeitlin, 1985). Indeed, Freeman and colleagues have been pioneers in understanding the nature of national innovation systems, consistently stressing the importance of culture, history and the institutional linkages of technological trajectories (Freeman, 1995). Geographically, the TEP model is admittedly limited to national scales. On the other hand, two important alternative models of long run industrialization, namely regulation theory (Jessop, 1995) and flexible specialization (Piore and Sabel, 1984), tend to underplay technology. Moreover, the geographical tension between the forces of agglomeration and dispersal that is at the heart of the flexible specialization thesis, is readily incorporated in the TEP model.

Table 2.1 Selected features of techno-economic paradigms

Techno-economic Paradigm	Key Factor Industry	Main Carrier Industry	Newly Emerging Industry	Productivity Base	Forms of Organization Innovation
Mechanical, 1760s-1820s	cotton	textiles	steam engines	scale econ's, factories	entrepreneurs (drive system)
Steam, 1820s-1870s	coal	steam engines	heavy engineering	deepened scale economies, factory	limited liability
Electrical, 1870s-1920s	steel	electrical engineering	autos	deepened scale economies, factory, firm	captains of industry (Taylorism)
Fordism, 1920s-1970s	oil	autos	electronics	scale economies factory, MNC product line	techno-structure (collective bargains)
ICT, 1970s-2020s	chips	electronics	biotech	economies of scope	collaborative networks

Sources: Perez (1985) and Freeman and Perez (1987). See also Hayter (1997).

As a formal expression of the principle of creative destruction, the TEP model envisages that deepening crisis, exploitation, job loss and deskilling associated with the late stages of the Kondratiev wave are ultimately countered by new rounds of innovative investments generating new skills and jobs. From this perspective, industrialization is interpreted as a social learning process, the technological and institutional conditions of which change with each TEP. That is, industrialization is also presented as adaptive and open-ended, with no particular 'end-state' in mind, nor based on one (or more) universal truth(s). Admittedly, the original formulation of the TEP model did not address environmental issues, but then neither does flexible specialization nor regulation theory. Moreover, a major problem for the latter models is the lack of attention to existing and emerging technology, a neglect that inevitably compromises a full comprehension of environmental problems and solutions.

Indeed, Lipietz (1992), a founding guru of (marxian) regulation theory, and a noted 'green', expresses 'disillusionment' with that theory's thinking for coping with environmental problems, even if his ideas for a 'green' paradigm are rooted in a strongly socialist political ecology. At the same time, the potential for technological innovation in resolving environmental dilemmas is perfunctorily dismissed as a 'productivist' position that is the basic cause of environmental problems. In contrast, Freeman (1992) offers a more optimistic, socialist prescription which recognizes that the basic motivations for and impacts of technological changes historically. In his hopes for a 'green paradigm' technology has a vital role.

To further explore this line of inquiry, we argue that each previous TEP is characterized by a distinct environmental phase, each associated with a specific scale of geographic impact and sets of public policies and industry attitudes and strategies (Table 2.2). Thus, in the first TEP the principles of laissez-faire imposed a new interpretation of nature as freely available commodities to be used and abused as economic calculus dictated. In this environmental phase, which may be defined as 'frontier economics' (Rejeski, 1997, p. 51), deleterious impacts on air, land and water were simply treated as 'negative externalities'. The impacts were strongly localized in this period, but the living and health conditions of workers and their families were often awful. There were some 'utopian' (and private sector) – responses by entrepreneurs to create healthier housing conditions, such as by Robert Owen in Lancashire. But these initiatives were isolated; generally, disregard for the environment and human welfare went hand in hand.

Frontier economics (frontierism) remained the dominant environmental phase during the steam and electrical TEPs as the scale of impacts became national and international in scope. If environmental exploitation remained the credo of these TEPs, awareness of environmental impacts did increase and there were important policy initiatives. For example, in the UK, the Alkali Acts of the 1860s signalled government recognition of pollution as a public bad. There were also important initiatives that sought to conserve environmental values, including the introduction of sustained yield principles in forestry. In the USA, the conservation movement was impressively

Table 2.2 Techno-economic paradigms and the environment

Techno-economic Paradigm	Environ-mental Phase	Scale of Impacts	Enviro-Policy/Industry Strategies
Mechanical, 1760s-1820s	Frontier Economics	Local	Laissez-faire, environment as free good; pollution as externality. Robert Owen's social reforms, and improved housing for workers, (also evident in New England), are isolated, 'utopian' initiatives.
Steam, 1820s-1870s	Frontierism, Awareness	Regional	Ditto – but Alkali Acts (UK) of 1860s signalled government recognition of pollution as public bad.
Electrical, 1870s-1920s	Frontierism, Awareness	National	Policy initiatives to: a) preserve enviro-values (Yellowstone National Park, 1872, and b) improve living conditions of workers, notably garden city movement. Industry attitudes remain same.
Fordism, 1920s-1970s	Amenity and Protection	Global	Amenity (clean air, gardens, parks) is systemic feature of urban/regional planning, including bans on home coal fires. 'Environment' as location factor for firms. But mass production and consumption creates fears for limits to 'growth' and 'Silent Spring' becomes metaphor of environmental catastrophe.
ICT, 1970s-2020s	Resource Management	Global	Rapid escalation of legislation to drastically reduce pollution, biodiversity becomes a priority and Montreal/Kyoto protocols seek global responses. Green consumer appears. Recycling is widespread, industry develops technologies to reduce environmental impacts (eg. EVs), ecocertification spreads.
Green	Eco-development	Local	Dematerialization of the economy, industry internalizes environmental values (extends ecocertification, take back policies, sell services rather than goods), and sustainability is defined. Environment is R&D priority.

Sources: Perez 1985, Freeman and Perez 1987 and Freeman 1992 in relation to the 'green paradigm'. Terminology for the 'environmental phases' is based on Colby as reported in Rejeski 1997: 51-2.

signalled by the creation of Yellowstone National Park in 1872. In addition, the garden city movement in the UK, beginning with the creation of Letchworth Garden City in 1902, explicitly acknowledged the suffering of workers and their families from living in close juxtaposition to works freely discharging pollutants that were not only damaging to the environment but to people's health. Garden cities, which recognized the importance of green-space and 'unpolluted' environmental values as positive externalities in turn anticipated the development of 'council housing' (subsidized single family housing, with gardens, rented from local governments) in the UK in the 1920s, at the beginning of the Fordist TEP.

Indeed, Fordism heralded a new environmental phase of 'amenity and protection' as 'attempts to protect people from environmental pollution and to provide environmental amenity became a systemic feature of urban and regional planning throughout the OECD. In addition, the formal recognition of environmental values by planners as an important dimension of the standard of living of people was paralleled by recognition of environmental amenity as a location factor for increasingly footloose segments of the population, including retirees and entrepreneurs of high tech firms (Ullman, 1958).

During Fordism there were also attempts to reduce environmental impacts generated by industry and by households, for example, laws requiring reductions in the use of coal as a source of heat. Yet, environmental regulations remained broad and ill defined, and any positive environmental performances by industry were overwhelmed by the massive, unprecedented escalation in resource and industrial production that occurred in support of the mass production and mass consumption ethic at the heart of Fordism. Natural resources were seen as abundant and cheap, solely useful as inputs to industrial processing. Tremendous assumptions were routinely made about the 'resilience' of the environment, and its ability to absorb and dissipate effluents.

As Meadows et al (1972) argue, however, Fordism was threatened by resource scarcity, while itself threatening the surviving vestiges of the non-industrial values of the environment. Indeed, Fordist production created truly global pollution problems, soon to be seen most forcefully in damages to the earth's ozone layer and changes in global climates. If these problems are rooted in the rise of capitalism, and its related growth ethic, material culture, and technological dynamism, it might be noted that the centrally planned communist countries were both less successful at growth and preserving the environment. This unhappy achievement at least gives pause to an overly simple relationship between capitalism and environmental disaster. Indeed, the transformation from Fordism to the Information and Communication techno-economic paradigm (ICT) revealed the weakness of the centrally planned economies in coping with change, thus hastening their collapse or at least ideological re-think. Simultaneously, the transformation emphasized the (admittedly geographically uneven) adaptability of capitalism and its periodic tendency to radically alter its trajectory. Moreover, the restructuring of capitalist economies featured a significant new environmental phase of 'resource management'.

The ICT and Resource Management

During the 1970s, Fordism was undermined by technological changes led by micro-electronics, the freeing of exchange rates, energy crises, and stagflation. Simultaneously, environmental legislation became a public policy priority among OECD countries, reinforced by widespread public support. The literal boom in environmental legislation is illustrated by the USA, where, of the 42 environmental laws passed in the 87 years between 1899 and 1986, 28 were passed in the 17 year period between 1970 and 1986 and 34 since 1964 (Richards and Forsch, 1997, p. 6). Such an exponential expansion of environmental legislation is evident throughout the OECD, imposing increasingly strict controls on permissible levels of pollution, in general. In industry specific ways, more acknowledgment has also been given to the protection of (non-industrial) environmental values (for example, endangered species legislation in the USA). These initiatives underpin widespread environmental policy experimentation that is further evidenced by a variety of environmental taxes, outright bans, compliance programmes which encourage voluntary efforts, the introduction of eco-certification and green labelling initiatives. In addition, there are increasingly strong commitments to the 'recycling society' that in Japan are reinforced by an 'eco factory' research effort and in Germany by 'take back policies' which require manufacturers to take back products discarded by consumers (Richards and Frosch, 1997, pp. 7-8).

In the 1990s, local and national policies were supplemented by tentative international efforts to cooperate to combat global pollution, notably the Montreal and Kyoto protocols that respectively seek agreements to reduce CFCs (that destroy the ozone layer) and carbon emissions (that cause green house effects). A recent, rapidly expanding international scale effort, principally involving non-governmental organizations (NGOs), focuses on eco-certification, green labelling and market reforms that attempt to inform consumers about products which are judged to have been manufactured throughout the production chain according to defined standards of environmental acceptability. In 1996, for example, The International Standards Organization (ISO) introduced the ISO 14000 series, beginning with 14001, which offers certification for compliance with comprehensive environmental management systems that specify procedures but not discharge standards, the latter being the responsibility of governments (Kuhre, 1995). However, the ISO is also in competition with alternative organizations, most notably environmental NGOs, that wish to establish international standards for environmental behaviour.

Indeed, during the ICT environmental NGOs have become key actors, in one way or another, encouraging and threatening governments, industry and related organizations such as the World Bank, to radically change their behaviour in favour of sustaining environmental values (Soyez, chapter 11). Business practice and strategy has been jolted by NGOs that, often with anti-capitalist and anti-growth rhetoric, have aggressively implemented anti-logging, anti-whaling, anti-sealing, anti-nuclear energy, anti-mining, and anti-dam campaigns. These activities have been further reinforced by highly publicized consumer boycotts and lobbies of 'globalization' meetings, as well as by rather quieter initiatives to pressure financial organizations. As a globally powerful institution group, environmental NGOs are posing questions not only for MNCs but also for local development, and their actions should not be automatically assumed to be socially beneficial (Hayter and Soyez, 1996). Yet, environmental NGOs have ensured that grappling with environmental imperatives has been sustained, leaving no place for temerity by governments or 'dynasaurism' by business.

Consequently, concern for environmental amenity and protection during Fordism has been extended and re-focused during the ICT to a deeper concern with resource management and sustainability. Thus, there has been a major shift from non-renewable to renewable resource utilization; in the USA, for example, the latter became more important than the former for the first time in the 20th century in the mid-1970s (Rejeski, 1997, p. 54). In general, leading edge corporations have shifted from reactionary (responding to regulations) to anticipatory behaviour (establishing standards for or even pre-empting regulation) environmental strategies (Richards and Frosch, 1997, p. 9). Thus, end of pipe responses are being replaced by beginning of pipe responses; there is a growing 'environmental industry; and firms are investing massively in R&D to create processes and products to eliminate or reduce environmental impacts, such as the creation of electric vehicles (Patchell, 1999).

Freeman (1992, pp. 199-211) argues that ICT has made substantial contributions to meeting environmental imperatives. He notes that the ICT: a) has allowed opportunities for more accurate monitoring and control of multiple processes in industries, buildings and households that reduce energy needs and material use; b) has led to the elimination or reduction of defective products, thus greatly reducing wastes; and c) is a miniaturizing technology based around semi-conductor technology which has significantly reduced energy and material consumption in electronics products and many engineering products, by reducing the number of components required. Despite these contributions, and his belief that the potential for ICT to create environmentally friendly technologies is far from realized, Freeman suggests "a change of techno-economic paradigm beyond ICT will be needed" (p. 202) to reduce energy needs and intensify material use to establish a path of sustainable development. Others agree that industrial response to environmental imperatives remains insufficient (Freeman, 1992; Lipietz, 1992).

The Green Techno-economic Paradigm

From an environmental perspective, the problem of the ICT is that environmental imperatives are not paramount – they do not drive the thinking underlying productivity. Rather, the ICT is based on achieving productivity breakthroughs by the systemic flexibilization of the economy, featuring an increasingly important role for economies of scope as well as economies of scale. In practice, this flexibilization of industrial systems has realized important positive implications for the environment. As Freeman emphasizes, however, not all the potentials of the ICT for mitigating environmental issues have been achieved while other problems have been created. As an example of unrealized potential, he refers to home-based work. Such work, although made increasingly possible by the ICT thus reducing the environmental costs of commuting, has been limited by slow (organizational) adjustments to employment practices, working hours, management control systems and technical developments in video phones and teleconferencing (Freeman, 1992, pp. 204-5). As an example of environmental problems created by the ICT, he refers to world tourism which has been significantly affected by the ICT, including replacing the standardized packaged holidays of Fordism with a more flexible range of options (Freeman, 1992, pp. 205-6). But this change has been associated with more, longer journeys, and longer holidays creating a huge increase in air travel damaging to the environment.

In a green TEP, environmental imperatives become explicit motives for systemic change. In a green paradigm, innovation priorities are oriented to radically reducing the use of energy and materials in transportation, construction and manufacturing systems. In terms of environmental attitude, the resource management phase of the ICT is replaced by the phase of eco-development as ecological definitions of productivity recognize the non-industrial values of nature and seek to provide practical definitions for the idea of sustained development. Moreover, just as the activities that damage the environment are themselves destroyed, along with the jobs they supported, activities that maintain, replenish and enhance the environment are created, along with new jobs and skills required. If the green paradigm implies further restructuring, it is not based on a simple trade-off between jobs and the environment, just as paradigmatic technological change in the past never involved a simple trade-off between jobs and capital (Freeman, 1992, pp. 175-89).

TEP theory predicts that the seeds of a new paradigm are already sown in the existing one. Thus, Henry Ford and other auto manufacturers had established mass production decades before the 'age of mass production' while the ICT paradigm had its roots in radar, computer and other electronic developments in the 1940s, much inspired by World War 2. Presently, nanotechnologies and biochemistry-based technologies are rapidly developing and could be the engine of a Green TEP that displaces the ICT paradigm in a couple of decades or so (Freeman, 1992, p. 203; see Rejeski, 1997, p. 68). These technologies have tremendous potential for 'dematerializing' the economy by reducing material use and non-renewable resources (Table 2.2).

In developing technological policies that aspire to principles of eco-development, Freeman argues that support for the new technologies can be combined with new developments in ICT. In the context of transportation, for example, he argues that the atmospheric pollution and energy demands associated with air travel can be reduced by switching more passengers to sea (and other water-based) travel, by developments in computer-controlled sailing

technology to build wind-assisted ships, along the lines of Japanese experimentation, and, more radically, by the creation of airships. On land, electric
vehicles are based partly on ICT technology and partly on new principles of
energy creation that did not originate in auto or electronic R&D, while
magnetism and superconductivity promise radical developments in rail
systems, neither originating outside the transportation sector. Inevitably, such
trends will require financial support for R&D reinforced by appropriate pricing
to help stimulate fundamental changes in consumer transportation behaviour.

Industrial ecologists are also detailing the changes required in the
energy, construction and manufacturing sectors to meet the principles of eco-
development. As England and Cope (1997) note, changes in energy use are
fundamental to a Green paradigm because all sectors use energy and energy
use is a major contributor to environmental problems. Again, while significant
improvements in energy efficiency are possible with existing knowledge,
for example, through adjustments in the building stock, the diffusion of non-
conventional and new technologies, for example, solar energy and especially
hydrogen systems will be vital in the future. Indeed, in the Green TEP,
hydrogen may complement the 'chip' as the 'key factor industry' for the
paradigm as a whole, that is as an input that is abundant, cheap and with
widespread application (Table 2.1).

In the case of manufacturing, the Green TEP promises enormous
changes as new industries become dominant, new ones emerge rapidly and
old ones adjust or fade away. In this context, 'dematerialization' does not
mean less consumption or that manufacturing activities are less important to
people's standards of living. Thus, in Freeman's Green TEP, consumption
and jobs remain important, and technological (and matching institutional)
innovations are seen as a source of creativity by which to solve environmental
problems in the long run rather than as a relentless threat to environmental
values. Consequently, Freeman's Green TEP does not automatically dismiss
the corporation as the enemy of the environment, or the environment as the
enemy of development.

The Green Corporation

Concern for the natural environment has re-defined business environments
profoundly during the ICT. Eden (1996), for example, systematically
documents the external forces shaping the greening of corporate behaviour
(Figure 2.1).

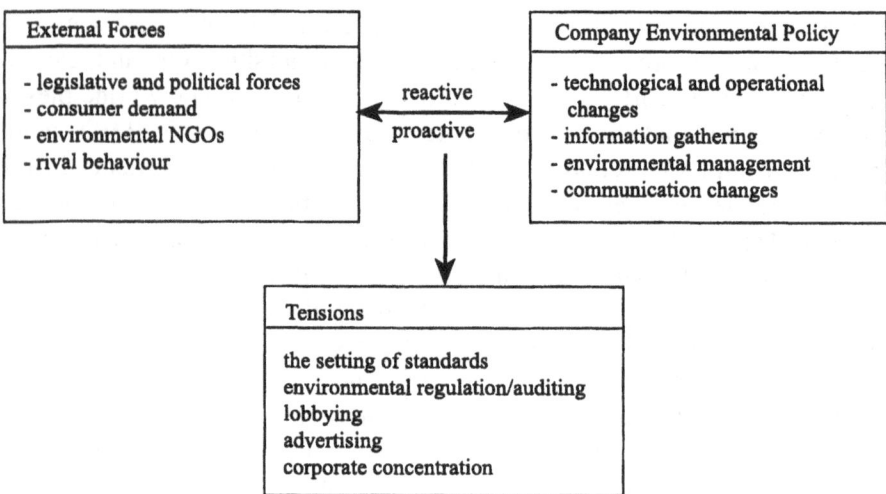

Figure 2.1 Eden's model of the 'greening' of corporate behaviour
Source: Based on Eden (1996).

As she notes, these forces (legislative and political factors, consumer demand, environmental NGOs and rivals) have forced extensive reactions by corporations in terms of technological operations, information gathering, environmental management and communication. In addition, corporations have become increasingly proactive in dealing with environmental forces, thereby creating tensions between policy makers and corporations, for example, in terms of the setting of standards and forms of regulation. In Richards' and Frosch's (1997, p. 9) terms, corporate responses to environmental imperatives are no longer 'unprepared' but have successfully developed 'reactive', 'anticipatory' and 'high integration' strategies.

There is no simple linear development in corporate attitudes towards the environment, however, as Eden's tensions reveal. An important battleground concerns regulation. There are two basic, related issues. First, there is a debate as to who should set environmental standards and monitor them. Thus, environmental NGOs have been aggressive in questioning the role of governments and industry regulations in setting and monitoring environmental standards, roles for which they see themselves fitted because of their 'independence' and single focus. In the context of eco-certification for timber products, for example, NGOs are developing rival standards to the procedures proposed by ISO and by governments and industry (Wallis et al 1997). Second, especially as (some) corporations have become more pro-active, a debate has arisen as to whether regulation actually stifles innovation. Richards and Forsch (1997, p. 6), for example, note evidence that supports the negotiation of performance-based measures as leading to less pollution at less cost. In this view, environmental regulations potentially impose a bureaucratic quagmire adding to costs, even making agreements 'unworkable' while the site specific nature of many environmental problems invites a role for distinctive approaches.

This issue poses dilemmas for NGOs to the extent that regulations reinforced by a lack of trust of corporations may be seen as dampening innovative behaviour, not to mention the associated jobs. More generally, the need to promote socially desirable innovations as a way of both resolving environmental problems and achieving environmentally superior performance, as advised by TEP theory (Freeman, 1992), may be contrasted with Lipietz's (1992) regulation heavy (innovation light) approach. In this regard, the institutional innovations advocated by Freeman (1992, pp. 208-9) focus on how R&D on environmental priorities might be organized. A related question relates to pricing. While there is a widespread view that prices should reflect environmental costs and benefits, evident in Lipietz (1992) as well as Freeman (1992), for the latter, appropriate pricing reinforces the innovations required to meet environmental imperatives, rather than constituting the solution itself.

In TEP theory therefore, the green corporation is an innovative firm that has internalized environmental responsibility. As Allenby (1997), notes such internalization of a 'social responsibility' raises interesting questions as whether it is the boundary of the firm that is being changed, or the nature of the firm itself. There are also interesting questions as to whether the green corporation is more likely to be a MNC or a SME. The answer is not straightforward. On the one hand, environmental rhetoric is strongly critical of capitalism's growth ethic and the size and power of giant corporations. Indeed, the thesis that MNCs are motivated to seek jurisdictions that not only offer low wages but lax environmental standards, reinforces such a view. The 'small is beautiful' ethic underlies much environmental polemics, based on the assumption that SMEs are more localized in their activities (demanding less in terms of global inputs and sometimes even markets). This view also assumes that SMEs are inherently more caring of their local environments, perhaps in part because they have to be – they only operate in one location – and in part because of their embeddedness in community values. Recently,

Patchell (2000) developed this argument, noting how SMEs can indeed form 'convivial communities' that can successfully compete, add value and achieve external economies of scale and scope without the contradiction posed by the emergence of dominant firms and internalization tendencies.

On the other hand, the technological imperatives that underlie the emergence of the modern industrial corporation provide impetus to its further growth even as it turns green. Some solutions to environmental problems are massively expensive, demanding not only the resources of the largest corporations, but also support from governments. In his analysis of the R&D needed to create electric vehicles in Japan, Patchell (1999) underlines this argument in an industry and country where SMEs are significant. Similarly, although his survey focused on larger manufacturers, Florida's (1996) analysis of USA firms concludes that the most innovative in relation to environmental performance were typically the most innovative in other respects of behaviour as well, such as labour relations, marketing and production technology. Stahel's (1997) proposal for a 'functional economy', in which firms lease or rent the functions of goods and services, rather than the goods themselves, so that firms 'take-back' their products to stimulate efficient resource use, further implies a strong role for giant firms, as his reference to Xerox attests.

In TEP theory, the giant corporation is seen as a key institution of modern economies, since the third (electrical) TEP. During the evolution of each paradigm, the relative roles of SMEs and MNCs may vary somewhat, with concentration tendencies becoming more powerful as industries mature. Thus, Freeman (1992, p. 225) did not anticipate that flexible specialization would replace MNCs during the ICT. The same observation can be applied to the forthcoming Green TEP. In the early stages, innovative possibilities for SMEs will be at their maximum even as the need for expensive, radical innovations will ensure that MNCs will remain the dominant business segment, over time slightly increasing their power. MNCs will not be disadvantaged by environmentalism. Indeed, since environmental regulation typically adds greatly to costs and bureaucratic burdens, SMEs may well be at a relative disadvantage (Eden, 1996; Wallis et al 1996; Allenby, 1997).

Green Geographies

Theoretically, each new TEP is orchestrated by a lead nation (or nations). During Fordism, the USA was the leading exemplar while in the ICT Japan, along with Germany, became the model for restructuring, although neither country replaced the USA as the globe's hegemonic power. In the Green TEP, environmental leadership, in terms of technological and institutional innovation, may well originate in a group of established advanced countries, including the USA, Japan and Germany. In broad terms, the geography of the Green TEP raises interesting questions about: the locus of creation of leading edge environmental innovations; their diffusion within and among countries (as well as within and among industrial sectors); regional development implications, for example, with respect to most rapidly industrializing countries and resource regions; and, in more micro-terms, the nature of location dynamics.

Just how green geographies will unfold is hard to predict with any precision. Here, we will simply touch on the contours of the debate with respect to implications for location dynamics and for development (in the 'south') to at least contribute towards setting a research agenda. With respect to industrial location dynamics, whether trends in industrial organization favour MNCs or SMCs, environmental arguments support spatial concentrations of activity, albeit concentrations that are globally and nationally dispersed. Thus, industrial ecologists favour (regionally distributed) agglomeration tendencies to facilitate recycling, energy transfers, take-back policies, servicing, and to minimize transportation costs and waste streams (Richards and Forsch, 1997, pp. 8-10; England and Cope, 1997, p. 75; Stahel, 1997, p. 99). In the Green TEP, the (potential) benefits of agglomeration rooted in skill formation, flexible labour, just-in-time, quality control, differentiated production, and above all encouragement to innovate are extended to incorporate ecological mandates.

Such arguments readily associate with the idea of flexible specialization, especially in industrial districts dominated by SMEs. Indeed, Patchell (2000) has explored this view by articulating the nature of 'convivial communities' that are locally autonomous, creative, feature high levels of horizontally disintegration in search of a differentiated range of high quality goods produced in small communities, and which explicitly equate environmental and community sustainability. As he demonstrates in the context of Bordeaux wine growers and Kyoto house contractors, sustainability requires creativity (innovation) and "The greatest potential for convivial organization is enhancing localization of production" (Patchell, 2000, p. 29). This argument also reveals the practicality of SME-based flexible specialization, without drawing on romanticized 'back to nature' localized forms of economic activity.

But how far can conviviality be taken in a world in which corporate concentration remains a defining characteristic? Admittedly, regionalized production has become a feature of the strategy of several MNCs such as Honda who are seeking localized networks in different parts of the world, primarily motivated by desires to avoid political barriers and to differentiate production to reflect distinct local tastes (Mair, 1997). In such strategies, however, local autonomy (and associated implications for creativity) remains problematic while branch plant location choices may or may not contribute to environmentally sensible patterns of dispersed concentration. This issue links into the development debate.

In general, environmental arguments council against dispersed forms of mass production serving global markets based on a cost minimizing calculus, especially if the environment is dismissed as an externality. This criticism directly clashes with development aspirations of those countries that see large scale export driven industrialization, initially based on low costs, as the recipe for economic growth. From this perspective, 'environmental imperatives' readily translate into accusations of first world protectionism or even as another form of imperialism.

The development-environment debate, with its north-south green-brown geography, poses an extraordinary difficult dilemma. Indeed, for socialist observers, such as Lipietz (1992) who closely link environmental crisis with consumerism, technology is a marginal solution, providing only short term 'fixes' at best. In this view, mass consumerism must be mitigated to reduce mass production. But with less production, how can wealth be more equitably distributed, especially since the problems of distribution are difficult enough under conditions of growth? It is perhaps not surprising that Lipietz (1992) should combine a gloomy view of the future with some rather vague critical musings about regulation theory which he helped inspire. Freeman (1992), on the other hand, offers more hope.

Indeed, Freeman (1992) is adamant that the Green TEP cannot be at the expense of development. Rather, he insists on development as a moral, economic, political and even environmental necessity. Development is an environmental necessity, for example, because broadly based environmental opposition to development will be comprehensively rejected in a way that will ensure continued environmental damage. Freeman is able to be both green and pro-development because innovation permits positive sum games. While a framework detailing the nature of the positive sum game between the environment and development still awaits articulation, Freeman (1992) offers numerous suggestions to promote education, R&D and overall technical capacity within developing countries that will help promote local innovativeness and improved environmental and industrial performance. The development challenge is often been presented as one of technological catch-up; environmental know-how can readily be incorporated within this admonition.

Conclusion

Industrial geographers are rightly suspicious of universal models of development and predictions. If local economic variations seem infinite, however, local economies are also open to outside influences that link their fortunes together. In this context, we suggest that a strength of the TEP model is its sensitivity to local variation, explicitly in terms of national variations in innovation systems and implicitly by its ability to incorporate the debate over flexible specialization. Moreover, if the idea of a green TEP is highly speculative, it is given legitimacy by the remarkable body of empirical evidence Freeman and colleagues have amassed in support of the TEP framework. At least, the incorporation of the environment within the TEP model raises an interesting research agenda.

Is the idea of a 'green TEP' a legitimate one? If so, what and how will green geographies evolve? For example, what are the implications of the greening of the economy for core-periphery relations? How will the global role of resource regions be modified? What are the possibilities for a positive sum game between the development of the south and the environment in different places? Can environmental industries be created in the south as well as the north? How innovative will these industries be? In a green economy what are the implications for industrial organization and for location dynamics? In a green economy what does concentration, dispersal and dispersed concentration of economic and other activities really mean? What is the geography of environmental regulation? How does environmental know-how diffuse? We contend that these and related questions need to be addressed by industrial geography if it is to retain its relevance.

Acknowledgements

Roger Hayter gratefully acknowledges stimulating conversations with Dietrich Soyez and the hospitality provided by the Geography Department, University of Cologne where a draft of this paper was written.

References

Allenby, B.R. (1997), 'Environmental Constraints and the Evolution of the Private Firm', in D.J. Richards (ed), *The Industrial Green Game: Implications for Environmental Design and Management*, National Academy Press, Washington DC, pp. 101-113.

Eden, S. (1996), *Environmental Issues and Business: Implications of a Changing Agenda*, John Wiley and Sons, Chichester.

England, G. and Cope, D.R. (1997), 'Energetic Concepts Drawn from Electricity Production and Consumption', in D.J. Richards (ed), *The Industrial Green Game: Implications for Environmental Design and Management*, National Academy Press, Washington DC, pp. 73-90.

Florida, R. (1996), 'Lean and Green: The Move to Environmentally Conscious Manufacturing', *Califiornia Management Review*, vol. 39, pp. 80-105.

Forsch, R. A. (1997), 'Closing the Loop on Waste Materials', in D.J. Richards (ed), *The Industrial Green Game: Implications for Environmental Design and Management*, National Academy Press, Washington DC, pp. 37-47.

Freeman, C. (1974), *The Economics of Industrial Innovation*, Penguin, Harmondsworth.

Freeman, C. (1987), *Technology Policy and Economic Performance: Lessons from Japan*, Pinter, London.

Freeman, C. (1992), *The Economics of Hope*, Pinter, London.

Freeman, C. (1995), 'The 'National Innovation System of Innovation' in Historical Perspective', *Cambridge Journal of Economics*, vol. 19, pp. 5-24.

Freeman, C. and Perez, C. (1988), 'Structural Crises of Adjustment: Business Cycles and Investment Behaviour', in G. Dosi, C. Freeman, R. Belson and G. Silverberg (eds), *Technological Change and Economic Theory*, Pinter, London, pp. 38-66.

Hayter, R. and Soyez, D. (1996), 'Clearcut Issues: German Environmental Pressure and the British Columbia Forest Sector', *Geographische Zeitschrift*, vol. 84, pp. 143-56.

Jessop, R. (1995), 'The Regulation Approach, Governance and Post-Fordism: Alternative Perspectives on Economic and Political Change?', *Economy and Society*, vol. 24, pp. 307-33.

Lipiet, A. (1992), *Green Hopes: The Future of Political Ecology*, translated by Malcolm Slater, Polity Press, London.

Mair, A. (1997), 'Strategic Localization: The Myth of the Postnational Enterprise', in K. R. Cox (ed), *Spaces of Globalization: Reasserting the Power of the Local*, Guilford, New York, pp. 64-88.

Meadows, D.H., Meadows, D.L., Randers, D.L., and Behrens, W.W. (1972), *The Limits to Growth*, Universe Books, New York.

Patchell, J. (1999), 'Creating the Japanese EV Industry: The Challenges of Uncertainty and Cooperation', *Environment and Planning A*, vol. 31, pp. 997-1016.

Patchell, J. (2000), 'Organization for Conviviality', paper presented at a Workshop on: The Firm in Economic Geography, University of Portsmouth, March.

Perez, C. (1985), 'Micro-electronic, Long Waves and Structural Change', *World Development*, vol. 13, pp. 441-63.

Rejeski, D. (1997), 'Metrics, Systems, and Technological Choices', in D.J. Richards (ed), *The Industrial Green Game: Implications for Environmental Design and Management*, National Academy Press, Washington DC, pp. 48-72.

Richards, D.J. and Forsch, R.A. (1997), 'The Industrial Green Game: Overview and Perspectives', in D.J. Richards (ed), *The Industrial Green Game: Implications for Environmental Design and Management*, National Academy Press, Washington DC, pp. 1-34.

Sabel, C. F. and Zeitlin, J. (1985), 'Historical Alternatives to Mass Production: Politics, Markets and Technology in Nineteenth Century Industrialization', *Past and Present*, vol. 108, pp. 13-76.

Soyez, D. (1985), *Ressourcenverknappung und Konflikt: Entstehung und Raumwirksamkeit mit Beispielen aus dem Mittelschwedischen Industriegebiet*, Arbeiten aus dem GeographischenInstitut der

Universtät des Saarlandes, Band 35, Saarbrücken (English summary, pp. 329-32).

Stahel, W.R. (1997), 'The Functional Economy: Cultural and Organizational Change', in D.J. Richards (ed), *The Industrial Green Game: Implications for Environmental Design and Management*, National Academy Press, Washington DC, pp. 91-100.

Taylor, M. (ed) (1995), *Environmental Change: Industry, Power and Policy*, Averbury Press, Aldershot.

Thomas, M.D. (1969), "Regional Economic Growth: Some Conceptual Aspects', *Land Economics*, vol. 46, pp. 43-51.

Ullman, E.L. (1958), 'Amenities as a Factor in Regional Growth', *Geographical Review*, vol. 44, pp. 119-32.

Wallis, E., Stokes, D., Wescott, G. and McGee, T. (1997), 'Certification and Labelling as a New Management Tool for Sustainable Forest Management', *Australian Journal of Environmental Management*, vol. 4, pp. 24-38.

3 Paths of Sustainable Industrialization in the Knowledge-Based Economy

Sam Ock Park

Introduction

One of the most important challenges to confront industrial society is how to devise ecologically and environmentally sustainable forms of production. This challenge is rooted in growing concerns about serious damage to environment and health from industrial growth (see chapter 2 by Hayter and Le Heron, and chapter 11 by Soyez). It is only in recent years, especially since the Rio Earth Summit in 1992, that concerns about the sustainability of current industrial development patterns have spread globally (UNIDO, 1997). During the 1980s, diverse environmental problems such as global carbon pollution, ozone depletion, and loss of species, forests, and fertile soils suggested that environmental damage was more widespread and serious than expected (Schmidheiny and Zorraquin, 1996). The importance of sustainable development policies has been widely recognized, including developing countries where rapid industrialization and economic growth, especially in China, has led to severe environmental degradation and pollution (UNIDO, 1996; Zhang et al 1999).

Despite the worldwide movement towards more sustainable forms of development during the last decade, most companies still regard reductions in environmental damage solely in terms of rising costs, decreasing competitiveness and increasing legal challenges. From this perspective, environmental regulations are regarded as an unfavourable factor for industrial location and competitiveness. Nevertheless industrial attitudes are changing, and some leading companies are now environmentally proactive. Moreover, empirical studies not only dispute evidence of the negative impacts of environmental control on industrial competitiveness, but suggest that strong environmental regulations contribute to innovativeness and the international competitiveness of industry (Jaffe et al 1995; Robinson, 1995; Wallace, 1995).

Following challenges to the business sector at the UN Conference on Environment and Development in Rio in 1992, the World Business Council for Sustainable Development (WBCSD) was formed in 1995 through a merger of the Business Council for Sustainable Development and the World Industry Council for Environment in Paris. The WBCSD plays a significant role in policy development in relation to sustainable development through closer cooperation between business and government (Schmidheiny and Zorraquin, 1996). The United Nations Industrial Development Organization (UNIDO, 1991; 1996) performs a similar role. Diverse strategies for sustainable industrial development in developing countries have also been suggested by several scholars (for example, Schmidheiny and Zorraquin, 1996; Wallace, 1995, 1996; Wallner, 1999). However, as Soyez's chapter 11 notes, there have been few attempts in Regional Science and Economic Geography to conceptualize the links between environment, economy, and society and to suggest appropriate strategies and policies for sustainable development (see Angel and Rock, 2000; Taylor, 1995).

Most research on sustainable industrial development emphasizes the need for new production systems in order to make the process of industrialization itself an agent for sustainable development. To contribute towards this need this chapter examines, first, relationships between production systems and the environment, including with respect to various dimensions of sustainability and, second, possible directions of sustainable industrialization in the (present) era of the knowledge-based economy.

Changes of Production Systems and Environment

Changing production systems are a feature of industrial history. In the Industrial Revolution in the United Kingdom, craft production techniques were developed by the integration of traditional skills of artisans with new sources of power such as steam engines powered by coal (Womack et al

1990). In the early twentieth century, led by the technical and managerial innovations in the USA, a new mass production system was developed. From the second half of the 1960s, as advanced capitalist countries experienced a crisis of profitability, firms sought alternatives to mass production technologies. Epitomized by industrial districts in the Third Italy, and the rediscovery of craft production (Hudson, 1995), small firms and 'flexible production' again became widely regarded as an alternative to mass production. This development is based on the networks and cooperation between capital and labour as well as between firms linked with horizontal division of labour in a supportive social environment. However, mass production has itself become more flexible, notably in a form known as the lean production system that has been most associated with developments in Japan (Womack et al 1990).

The changes towards more flexible production systems have emerged particularly rapidly in industrialized countries where alternative types frequently co-exist. Nevertheless, based on an historical argument, Wallace (1996) argues that those developing countries where rapid industrialization is now beginning are the natural location for the evolution of any new production system based on principles of sustainable development. In his view, among industrialized countries, past production systems developed around the logic of profitability of capital (Hudson, 1995), and this logic has carried over into alternative production systems, where sustainability is viewed from the perspective of capital rather than from the perspective of the environment. Whether or not satisfactory options for sustainable industrialization, in which environmental protection and management are emphasized, can be achieved in developing countries remains a moot point, however.

Production systems have different implications for the environment in terms of the management of materials (Wallace, 1996). The craft production system had no critical impact on the sustainability of environment because there was little waste of resources caused by scrapping and unwanted products. However, craft production is inappropriate for modern societies based around a mass production culture. While mass production allows great efficiencies in the per capita use of resources, based on principles of standardization and economies of scale, long production runs also result in large inventories. In addition, there is waste of physical resources due to the scrapping or dumping of goods at below cost, resulting in widespread environmental degradation. In the lean production systems that adopt flexible manufacturing techniques and just-in-time production problems of large inventories and the scrapping of finished goods are not serious. That is, the relationship between environment and production systems, around the management of physical resources, has evolved effectively with changes in production systems. However, with increasing pressures of population growth and rapid industrialization in developing countries, the conventional management focus on production processes (which include primary processing, design, manufacture, and distribution) is not enough to evaluate the sustainability of environment. Beyond these boundaries, material extraction, consumer's use, and disposal should be included in the evaluation of environmental impacts (Wallace, 1996).

Impacts on environment occur at every stage of the production chain. Increasingly firms are being asked to consider life-cycle analysis for their products, with a view to understanding the final fate of products and product use by consumers (Wallace, 1996). There are also increasing pressures on firms to reduce their consumption of materials or to shift to less environmentally damaging materials. Pressures on firms and international organizations for the rational use of raw materials and improved manufacturing processes are leading to development of low waste technologies in engineering industries (ECE, 1994). The development of environmentally sound and innovative technologies and processes is now critical for environmental sustainability and firm competitiveness in the industrial world (Porter and Linde, 1995). Furthermore, firms should adopt techniques such as eco-auditing and should consider consumers and local communities in their decision-making processes. The role of knowledge-based engineering technologies and services is more important in the development of sustainable production systems in this new century (see chapter 12 by Schulz).

Dimensions of Sustainability

In the report of World Commission on Environment and Development (WCED, 1987, p. 43), sustainable development is described as development that meets "the needs of the present, without compromising the ability of future generations to meet their own needs". Sustainability in this definition encompasses the relationships between environment and economy, and is characterized by commitment to inter-generational equity as well as to intra-generational equity across space. The 'sustainability' in the term of 'sustainable industrialization' is related to firms' use of natural resources, human resources, and the society around them (Wallace, 1996). However, 'sustainability, in sustainable industrial development is a slippery concept to capture and pin down because 'sustainability' can be defined from different perspectives or dimensions (Hudson, 1995).

Doryan (1993, p. 452) identified four dimensions of sustainability underlying a competitive view of development. First, productivity sustainability is defined by market, macroeconomic and financial dynamism, infrastructure development, industrial efficiency, and globalization of the domestic economy. Second, environmental sustainability is defined by pollution indicators, rules and procedures to control waste, institution building to manage natural resources, and behaviour of citizens toward the environment. Third, socio-political sustainability is defined by institutions that allow freedom of expression, human rights, and confidence toward the judiciary system and the government. Finally, 'humanware', or human resources sustainability, is defined by educational and training opportunities and by openness of the population and the labour force toward technology and managerial changes and science and technology infrastructure.

Given that current industrial production systems cause negative environmental and social impacts, an environmental point of view has been emphasized for the improvement of the industrial systems in recent studies (Angel and Rock, 2000; Erkman, 1997; Wallner, 1999). That is, shared industrial growth and eco-efficiency are considered a priority for sustainable development. Similarly, Hudson (1995, p. 39) considers sustainability in terms of the "social sustainability of the level and distribution of employment and of income, and of the ecological sustainability of the level and composition of output". Even if ecological sustainability is emphasized in recent studies, industry cannot be viable without economic sustainability or profitability. Therefore, three dimensions - economic, social, and ecological sustainabilities – should all be considered in sustainable industrialization paths.

Problems for Sustainable Industrialization

Rapid industrialization of developing countries in recent years has resulted in rapid depletion of natural resources and, pollution and health problems. It is expected that most of the increase in world atmospheric carbon emissions in the next 20 years will originate from the developing countries (UNIDO, 1996). Waste emissions have serious environmental impacts at global, regional, and local levels (Table 3.1). The environmental degradation and problems resulting from the current industrial production systems suggest that a new paradigm for industrial development is needed for sustainable development in the global society.

Table 3.1 Environmental impacts of waste emissions

Global Impacts:
* The build up of greenhouse gases in the atmosphere and the consequent warming of world climate
* A rise in sea level with climate warming
* Damage to the stratopheric ozone layer
* Loss of biodiversity

Regional Impacts:
* Acidification and toxification of soils and surface waters
* Deforestation, desertification and erosion
* Loss of natural diversity

Local Impacts:
* Exposure to toxic chemicals, especially pesticides
* Urban air pollution
* Contamination of soils, rivers, streams and ground water with untreated sewage, pesticide and other toxic industrial wastes

Source: UNIDO, 1996.

The United Nations Industrial Development Organization (UNIDO) is actively developing a new model of sustainable industrial development. UNIDO (1996, p. 92) defines ecologically sustainable industrial development as "patterns of industrialization that enhance the contribution of industry to economic and social benefits for present and future generations without impairing basic ecological processes". The patterns might differ by country depending on resource endowments and stage of economic development. In this definition, developments of appropriate technology, institutions, policy framework and incentive structures are assumed to be pre-requisites for a complementary relationship between industrial development and environmental sustainability.

Wallace (1996) argues that the ways of achieving sustainable industrialization are more likely to emerge from developing countries, though there are several issues to be considered in the evolution of new production systems in this context. First, even though environmental degradation results from rapid industrialization in developing countries, industrialization cannot be stopped because economic growth and sustainability are critical for these countries. Second, 'dirty' industry may move to developing countries to avoid strict environmental regulation and control in the advanced industrialized countries. Third, economic globalization and liberalization of markets, in association with tight control from the parent country of multinational corporations (MNCs) and intensified spatial divisions of labour, may have negative impacts on the sustainability of developing countries. Fourth, financial problems for the development of environmentally sound technologies such as cleaner production processes and waste minimization technologies exist in the developing countries. Lastly, the innovation potential of developing countries is relatively weak for a variety of institutional and human resource problems.

Considering the above problems and the dimensions of sustainability, eco-efficiency in industrial development, the role of MNCs in globalization, knowledge creation and innovation networks for environmentally sound technologies, and financing change are the key elements for sustainable industrialization. These four directions of sustainable development are now discussed.

Directions Towards Sustainable Industrialization

Industrial Ecology and Sustainability

Achieving ecologically and environmentally sustainable industrial development is a major challenge of the new century. The related fields of industrial ecology and industrial metabolism allow some insights into the links between production system and environment for sustainable industrialization (Chertow, 1998; Erkman, 1997; Hudson, 1995; Wallner, 1999). Industrial metabolism is an approach to sustainable production that features the construction of a balance sheet of the physical and chemical inputs to and output from production. It is basically an application of the material-balance principle, "aimed at understanding the circulation of the material and energy flows linked to human activity, from their initial extraction to their inevitable reintegration, sooner or later, into the overall biochemical cycles" (Erkman, 1997, p. 1). Industrial ecology is a systems view of industrial operations that seeks to optimize the total material cycles from virgin materials, to finished materials, components, products, obsolete products, and ultimate disposal (Graedel and Allenby, 1995; Chertow, 1998). In the industrial ecology perspective, the industrial system is regarded as a kind of ecosystem, which is different from the conventional view which considers the industrial system as separate from the biosphere (Erkman, 1997).

Although there are some methodological and historical distinctions, most scholars do not make a clear difference between industrial ecology and industrial metabolism. Thus, Erkman (1997, pp. 1-2) identifies three key elements of these approaches. First, they provide systemic, comprehensive, integrated views of all the components of the industrial economy and their relations with the biosphere. Second, they emphasize the biophysical substratum of human activities, that is, the complex pattern of material flows within and outside the industrial system, in contrast to current approaches which mostly consider the economy in terms of abstract monetary units, or alternatively energy flows. Finally, they consider technological dynamics, that is, the long-term evolution (technological trajectories) of clusters of key technologies as a crucial (but not exclusive) element for the transition from unsustainable industrial systems to viable industrial ecosystems.

One of the earliest applications of industrial ecology can be found in Japan. Problems of pollution and industrial waste were serious in the 1960s in Japan. Because of these problems, the Ministry of International Trade and Industry (MITI) set up the Industry - Ecology Working Group to develop the idea of a reinterpretation of the industrial system in terms of scientific ecology (Erkman, 1997). Since the 1970s, Japan has taken seriously the idea of industrial ecology and attempted to replace material resources with technology.

Since the late 1990s, many industrialized countries have attempted to introduce the idea of industrial ecology to production systems with regard to waste management. For example, within firms, there are desires to "design out waste" at the beginning of product life cycles and to "design for environment" while between and among firms, across the many stages of production waste, water, and energy can be shared and traded (Chertow, 1998). In addition, the flow of materials and energy can be traced across regions, economies, and the globe, so that the places of the greatest harm can be identified and targeted for policy attention. In recent years two major directions have been suggested for the application of the concept of industrial ecology, namely eco-industrial parks and systemic dematerialization (Erkman, 1997). Thus, in an eco-industrial park, firms and community participants share waste, water, and energy across firm and town boundaries, and the waste or by-products of a firm are used as resources by another firm. A dematerialization strategy intends to optimize the flows of materials within the economy by the increase in resource productivity, which is largely based on technological evolution. Systemic dematerialization is related to the strategy of service economy evolution. In particular, it refers to "the fact of increasing the resource productivity not only at the level of the product, but at the level of global infrastructures, but also, most importantly, to decrease its speed within the industrial system, thus minimizing the problem of dissipative emissions during normal use" (Erkman, 1997, p. 7).

Even though the industrial ecology approach can contribute to ecological sustainability, it is based on a biological analogy and ignores the fact that production is a social process and has manifold social implications (Hudson, 1995). Accordingly, industrial ecology has been called "the science of sustainability", which is critically necessary but not wholly sufficient, for achieving sustainable development (Chertow, 1998). Economic and social principles also have to be considered.

Role of Multinational Corporations

Multinational corporations have significantly contributed to the diffusion of industry and industrialization of developing countries since the 1970s. Prior to the 1970s, environmental regulation for industrial location and public awareness of the problem of industrial pollution in the advanced countries had some impacts on the diffusion of specific type of industries to peripheral or developing countries. Nowadays, however, there is little evidence that environmental regulations have significant impacts on industrial location (Robinson, 1995). Rather, it is argued that environmental regulation has positive effects on industrial innovation and firm competitive advantage in some industries (Wallace, 1995). Due to the public awareness of the importance of environmental protection, and environmental regulation under the WTO, developing countries are much less attractive for polluting factories.

Most developing countries have difficulties developing environmental technologies due to the lack of capital and knowledge. MNCs in the industrialized countries have relative advantages in accessing the capital and environmental technology needed for environmental sustainability. MNCs can thus contribute to the sustainable industrialization in two ways. Wallace (1996) suggests that MNCs, in all their diversity, have critical knowledge and technologies required for sustainability and that foreign direct investment (FDI) is a key mechanism for diffusion of the technology internationally. MNCs have capacity to create new environmental knowledge and through their internal processes can transfer knowledge to host countries through customer and supplier chains in local areas. Furthermore, the interactions of tacit and codified knowledge along supplier chains in host countries can provide opportunities for the creation of new environmental knowledge that is appropriate to host countries. Managerial expertise, technological competence, and tacit knowledge can be absorbed more effectively and cheaply through employment than through aid or training programmes.

Second, many MNCs are in the vanguard of developing corporate responses to environmental problems and are contributing to sustainable industrialization in host countries. Evidence in developing countries suggest that many MNCs are applying environmental standards far in excess of those required by host governments by bringing their own donor country environmental standards with them (Wallace, 1996; Schmidheiny and Zorraquin, 1996). MNCs are increasingly sensitive to the possibility of consumer's rejection or other image-damaging outcomes of being affiliated with environmentally irresponsible local operations (Zhang et al 1999). Some MNCs have adopted proactive approaches to environmental issues and are "exploring new concepts such as design for environment, eco-audits, life-cycle analysis, cleaner and zero-emission production, industrial metabolism and industrial ecology" (Wallace, 1996, p. 68).

Some developing countries are now undergoing rapid industrialization with serious environmental degradation. MNCs and other forms of FDI can help to break the cycle in which environmental degradation follows on from industrial development, so improving sustainability (Spofford Jr.,1996). For sustainable industrialization in the developing countries, however, more cooperative networking among international organizations, MNCs, domestic firms, and host governments is necessary, beyond the positive role of MNCs.

Networks for Sustainability

Most economic geography studies of intra-organizational and inter-organizational networking have emphasized economic aspects such as innovation potential, savings in transaction costs and sunk costs, regional development, and other effects. Meanwhile, insufficient attention has been paid to environmental issues (Camagni, 1991; Cooke and Morgan, 1993; Malecki and Oinas, 1999; Park, 1996, 1998). The network concept is critical, however, to the industrial ecology/metabolism approaches. Since the general importance of the economic and social aspects of networking is well recognized by economic geographers, discussion in this section focuses on environmental aspects.

The importance of networking for environmental sustainability can be summarized as follows. First, cooperative networks with suppliers, customers, and stakeholders including competitors can provide opportunities for knowledge creation process and contribute to developing environmentally sound new products. Collective learning through strong regional networking can facilitate the interaction of tacit and codified knowledge and increase the possibility of innovation as a part of the process of knowledge creation.

Second, close networks with environmental service providers can contribute to the creation of environmental knowledge through interactive learning process. Such networking is critical for development of environmental management systems. Third, industrial recycling networks, by substituting primary raw materials with external recycled materials, bring about ecological advantages due to reduced consumption of resources and decreasing industrial wastes. The concept of eco-industrial park, in which waste or by-products of one company are used as resources by another company, draws on the network concept.

Fourth, cooperation and networks of specialists from different sectors at both local and global levels can contribute to the efficient use of material and development of low-waste technologies in engineering industries. The development of new environmental technologies in the engineering industries are the result of inter-organizational networks as well as inhouse R&D

activities. Fifth, networks of local government, non-profit organizations (NPOs), non-government organizations (NGOs), private firms, and individuals are a useful procedure for achieving regional industrial development harmonized with natural environment and livelihood (Imaizumi et al 2000). Networks of NPOs and NGOs have had significant impacts on the investigation and solution of environmental issues in recent years. Lastly, global networks as well as local or regional networks are important for the transfer of environmental technologies and development of new environmental products. Strategic alliances for innovative research at the international level is increasing. The development of information and communication technologies facilitates global networks with regard to environmental issues and can enhance the interaction of codified knowledge with local knowledge.

In recent years, network concepts have been applied to create an eco-cluster for sustainable development, which is called the ECOFIT-Park (Wallner, 1999). Selected aspects of the ECOFIT-Parks are summarized in Table 3.2. The possible implementation of the ECOFIT-Park project is now under discussion in Styria, Austria. The concept of the eco-cluster is based on two aspects: first, an industrial ecology approach with the idea of recycling and business networking and; second, construction of a centre of innovation and activity for a region. Accordingly, the ECOFIT-Park aims to promote ecological sustainability as well as socio-economic sustainability.

Financing Environmental Change

Financial markets used to encourage short-term goals and undervalue environmental resources and accordingly did not sufficiently support sustainable development. Commercial banks, insurers, and others in the financial community have been far from the front line of environmental protection and sustainability. However, in recent years, with growing awareness on the importance of sustainable development from the public and business sectors, it is expected that financial markets will change the ways in which they value the business and will start rewarding eco-efficiency for purely financial reasons (Schmidheiny and Zorraquin, 1996). The concept of eco-efficiency is to increase value added while minimizing resource consumption, waste, and pollution.

Table 3.2 Selected aspects of ECOFIT-Parks

Function of the ECOFIT-Parks
- Focus not on foundation of an industrial park but an activity center for sustainable development of the region
- Selected existing enterprises are integrated into the network concept and new, selected enterprises are located (new enterprises must FIT into the ECO-concept)
- Disadvantages for farming population should be cushioned (regional marketing and processing of agricultural products)
- New standards must be set up with innovative integrated technology

Alternative sectors in the ECOFIT-Parks
- Centre for initiative (incubation center)
- Local exchange trading systems (LETS)
- Centre for regional services
- Quality centre – local supply markets
- Centre for creative reuse and recycling (CCRR)
- Social and culture centre

Work in the ECOFIT-Parks
- Culture of cooperation and trust within and between firms
- Trust in workers and their skills
- Man-machine relationship towards machine as a tool
- Safeguarding and expanding know-how
- Active involvement of workers in decision-making processes
- Worker motivation through further training

Source: Wallner, 1999.

Commercial banks in the industrialized countries are beginning to take an interest in environmental issues. Before the early 1990s, most commercial banks did not see environmental degradation and pollution as part of their agenda and had no interest in the sustainable development.

However, under the US Comprehensive Environmental Response, Compensation and Liability Act (known as the "Superfund"), liability for cleanup is imposed on owners of contaminated sites. Companies threatened with such costs have tried to find banks to share the costs (Schmidheiny and Zorraquin, 1996). With growing environmental liabilities, in 1992, about 30 leading banks, working with UN Environment Programme (UNEP), signed a "Statement by Banks on the Environment and Sustainable Development", and the number of banks signed has continuously increased since then. The Statement identified that ecological protection and sustainable development are collective responsibilities and must rank among the highest priorities of all business activities, including banking" and that "environmental risk should be part of the normal checklist of risk assessment and management" (Schmidheiny and Zorraquin, 1996, pp. 177-178). In order to limit their risk, bankers are beginning to consider the concept of eco-efficiency in their lending decisions.

Along with the growing awareness of bankers on the environmental issues, bankers can encourage customers toward eco-efficiency in the industrialized countries, but it is difficult for banks to be cost-effective in encouraging eco-efficiency in small and medium sized enterprises (SMEs). International organizations can provide funds to encourage eco-efficiency in SMEs and transfer of environmental technology to the SMEs. The Global Environmental Facility, a funding source for environmental work, can be used to channel technical and financial resources for sustainable development to developing countries. UNEP, UNDP, and the World Bank are also jointly providing funds for environmental problems with four priorities: biological diversity, global warming, ozone depletion, and the protection of international water (UNIDO, 1991).

Insurance companies have already suffered direct financial damage from environmental problems. United States' insurers faced on estimated US$2 trillion in pollution cleanup and asbestos-related claims (Schmidheiny and Zorraquin, 1996). The direct relations of insurers with environmental problems means insurers have key role in improving environmental sustainability.

As briefly reviewed, there is an emerging trend in financial markets towards eco-efficiency to conserve resources and limit pollution in the industrialized countries. However, the newly emerging financial markets are mainly related to businesses in the developed countries or to global environmental problems. Industrial firms in the developing countries, and

SMEs in general, have poor access to this source of finance. Considering that environmental problems in developing countries become more serious with rapid industrialization, international financial organizations should seriously consider ways to encourage eco-efficiency and sustainable industrialization in developing countries. The role of international organizations, such as UNIDO, should be strengthened to support transfer of environmental technology, development of appropriate environmental technologies in the developing countries, worker training for environmental technology, and coordination and networking for environmental sustainability in the developing countries. Public-private financing partnership and financial sector reform programmes for supporting sustainable industrialization are also important for providing funds for environmental change.

Conclusion

During the last decade environment and development have been seen more and more as inseparable sides of same coin, and ecological sustainability has appeared on research and political agendas. Nonetheless, significant differences exist over views on sustainability, by countries and by level of industrialization. Developing countries are now newly industrializing and priority is given to economic and social sustainability. But because it is expected that serious environmental degradation could accompany rapid industrialization in these countries and the environmental impacts extend from local to global, we need environmental policies for sustainable industrialization in the global society.

This chapter has emphasized the importance of the industrial ecology approach, the role of MNCs, networks for sustainability, and financing environmental change as important issues in policies that seek to enhance sustainable industrialization. These directions should be integrated as a whole to reorganize industrial systems for sustainable industrialization in the global society with an emphasis in ecological sustainability beyond economic and social sustainability. However, individual paths of sustainable industrialization are different by region and by industry. Thus, relative emphasis can be given to the industrial ecology approach in the food industry, while networking is more important for software industry or SMEs. The roles of MNCs and international organizations are more important in developing countries, while market mechanisms of financial markets for financing environmental change will be more important in the industrialized countries.

Innovative partnerships between the private and public sectors and environment are needed for sustainable industrialization. In addition, governments, private firms, NGOs, NPOs, and international organizations should have cooperative networks at local, regional, and global level for innovation of environmentally sound products and processes, diffusion of low waste and emission-free environmental technologies, spread of environmental information, and environmental education and training. The basic principle of the creation of knowledge through the interaction of tacit and codified knowledge can be applied to the creation of environmental knowledge and the innovation of environmentally sound products and processes.

Given that shortages of capital, information, and expertise are hampering the innovation and diffusion of cleaner production in developing countries, new industrial systems should be organized by encompassing the possible directions of sustainable industrialization identified in relation to the needs of developing countries. However, without significant changes in individual attitudes toward the environment and without cooperative networking, sustainable industrialization cannot be easily accomplished. For sustainable industrialization in the knowledge-based economy, information and communication technologies and the trend of globalization should be positively utilized for the creation of knowledge for environmentally sound products and process innovations.

References

Angel, D. P. and Rock, M. T. (eds) (2000), *Asia's Clean Revolution - Industry, Growth and Environment*, Greenleaf Publishing, Sheffield.

Camagni, R. (ed) (1991), *Innovation Networks, Spatial Perspective*, Belhaven Press, London.

Chertow, M. R. (1998), 'Waste, Industrial Ecology, and Sustainability', *Social Research*, vol. 65, pp. 31-53.

Cooke, P. and Morgan, K. (1993), The Network Paradigm: New Departures in Corporate and Regional Development', *Environment and Planning D: Society and Space*, vol. 11, pp. 543-564.

Doryan, E. A. (1993), 'An Institutional Perspective of Competitiveness and Industrial Restructuring Policies in Developing Countries', *Journal of Economic Issues*, vol. 27, pp. 451-458.

ECE (Economic Commission for Europe) (1994), *Low-waste Technologies in Engineering Industries*, United Nations, Geneva.

Erkman, S. (1997), 'Industrial Ecology: A Historical View', *Journal of Cleaner Production*, vol. 5, pp. 1-10.

Graedel, T. E. and Allenby, B. R. (1995), *Industrial Ecology*, Prentice Hall, Englewood Cliffs, N. J.

Hudson, R. (1995), 'Toward Sustainable Industrial Production: But in What Sense Sustainable?' in M. Taylor (ed), *Environmental Change: Industry, Power and Policy*, Avebury, Aldershot, pp. 37-56.

Imaizurni, H., Yabuta, M., and Ida, T. (2000), 'Towards a Construction of Administrative Managed System of Common-pool Resources', a paper presented at the 6th PRSCO Summer Institute June 13-17, 2000, Mexico City, Mexico.

Jaffe, A., Steven, B., Peterson, R., and Portney, P.R. (1995), 'Environmental Regulation and the Competitiveness of U.S. Manufacturing: What Does the Evidence Tell Us?', *Journal of Economic Literature*, vol. 33, pp. 132-163.

Malecki, E. and Oinas, P. (eds) (1999), *Making Connections, Technological Learning and Regional Economic Change*, Ashgate, Aldershot.

Park, S. O. (1996), 'Network and Embeddedness in the Dynamic Types of New Industrial Districts', *Progress in Human Geography*, vol. 20, pp. 476-492.

Park, S. O. (1998), 'Environmental Policy and Responses of Industry in the USA', *American Studies*, vol. 21, pp. 153-72 (in Korean).

Porter, M. K. and van der Linde, C. (1995), 'Green and Competitive: Ending the Stalemate', *Harvard Business Review*, vol. 73, pp. 120-123.

Robinson, K. (1995), 'Industrial Location and Air Pollution Controls: A Review of Evidence from the USA', *Progress in Human Geography*, vol. 19, pp. 22-244.

Schmidheiny, S. and Zorraquin, F. J. L. (1996), *Financing Change: the Financial Community, Eco-efficiency, and Sustainable Development*, MIT Press, Cambridge.

Spofford, W. O. Jr. (1996), 'Environmental Management During a Period of Rapid Industrial Development', *Resources*, vol. 123, pp. 13-16.

Taylor, M. (1995), Linking Economy, Environment and Policy', in M. Taylor (ed), *Environmental Change: Industry, Power and Policy*, Avebury, Aldershot, pp. 1-14.

UNIDO (United Nations Industrial Development Organization) (1991), *Striving for Ecologically Sustainable Industrial Development*, UNIDO, Austria.

UNIDO (1996), *Industrial Development*, Oxford University Press, Oxford.

Wallace, D. (1995), *Environmental Policy and Industrial Innovation*, Earthscan Publications Ltd. London.

Wallace, D. (1996), *Sustainable Industrialization*, Earthscan Publications Ltd., London.

Wallner, H. P. (1999), 'Towards Sustainable Development of Industry: Networking, Complexity and Eco-clusters', *Journal of Cleaner Production*, vol. 7, pp. 49-58.

WCED (World Commission on Environment and Development) (1987), *Our Common Future*, Oxford University Press, Oxford.

Womack, J.P., Jones, D.T., and Ross, D. (1990), *The Machine That Changed the World*, Macmillan, New York.

Zhang, W., Vertinsky, I., Ursacki, T. and P. Nemetz, P. (1999), 'Can China be a Clean Tiger? Growth Strategies and Environmental Realities', *Pacific Affairs*, vol. 72, pp. 23-33.

4 Building up Competence, Institutions and Networks: A Perspective on 'Catch-up' in the Knowledge Economy

Ayda Eraydin

Introduction

Over the last fifteen years, a growing literature has emphasized the importance of knowledge and learning for national and local development (Lundvall, 1993; Audretsch and Feldman, 1996; Amin and Cohendet, 1999). Initially, the discussion focused on the knowledge acquisition and learning process of individual firms. Recently, however, interest has shifted towards understanding collective learning, the reproduction of tacit knowledge (Mascietelli, 1999; Lei, 1997) and the capacity to combine the diverse types of knowledge efficiently (Nelson and Winter, 1982). Moreover, it is widely recognized that learning and innovative activities are profoundly affected by socio-economic context and the nature of personal interaction, that is, space matters (Audretsch and Feldman, 1996; Sweeney, 1996; Kirat and Lung, 1999). Indeed, the leading theoretical formulations of the relationships between knowledge and local development have defined innovative or learning clusters, variously labelled 'learning regions', 'innovative milieu, and 'regional systems of innovation', as the key to competitive advantage (Camagni, 1991; Florida, 1995; Asheim, 1996; Cooke, Uranga and Etxebarria, 1997, 1998; Bell and Albu, 1999).

However, the regions that best deploy the advantages of the knowledge economy control only a small share of production in the world.

Can other regions build the competencies, know-how and networks and 'catch-up' with leading edge learning economies? This chapter addresses this question on the basis of a selected review of the literature. The first part of the paper identifies the causes underlying the spatial structure of learning. Second, it defines the difficulties facing regions wishing to more fully participate in knowledge networks while the last part of the paper presents different policy recommendations and measures that aim to induce regionalized learning economies.

Spatial Aspects of Learning

The theory of knowledge is important in understanding the cumulative character of the learning process. According to Popper (1972, p. 71) knowledge never begins from nothing, but always from some background knowledge. He claims that knowledge arises from clashes between, on the one side, expectations inherent in background knowledge and, on the other side, new findings rooted observations or hypotheses derived partly from these expectations. That is, expectation always precedes observation. "The fundamental role of observations and empirical tests is to show that some of the theories are false, and so stimulate us to produce better ones" (Popper, 1972, p. 258). Popper asserts that knowledge depends upon worlds of logical contents, as understood by individuals. This perspective is supported by cognition theory that argues that people perceive, interpret and evaluate the world according to categories of thought which they have developed in interaction with their physical and social environment (Nooteboom, 1999, p.130). Such categories both enable and constrain cognition. As a result, cognition is to some extent idiosyncratic and path dependent: people evolve separately in different environments, and their cognition varies. People and firms understand only what fits their idiosyncratic, path dependent categories. The knowledge acquired is, therefore, a process of conceptual construction based on elements of experience (Dupuy and Gilly, 1996). Learning is the understanding of existing knowledge or the creation of new knowledge.

One view of the learning process claims that it is sequential; for example, in the process of learning, a firm uses the resources that were developed earlier in the path and combine them to create new ones. According to McKee (1992) learning paths have two dimensions, depth and scope:

learning in-depth deepens existing knowledge, skills and resources, while integrating new types of competencies broadens the scope of learning. Nooteboom's (1999, p. 132) definition of learning as a cyclical process is a related idea.

Learning begins with individuals. Individual learning relies on training through education and research systems as well as acquiring knowledge through learning by operating, using and searching during the involvement of activities (Jin and Stough, 1998). However, individuals interpret their environment in different ways, since they have different interests. This situation necessitates rules and regulations for coordinating individual attempts and leads to creation of collective actors, namely organizations (Harvey and Denton, 1999). Organizational learning is direct and indirect, and may involve unintended aspects. This type of learning depends on some knowledge being shared, typically tacit and embodied in organizational routines and procedures (Lawson and Lorenz, 1999, p. 307). The generation of new knowledge within organizations also means combining diverse knowledge, since innovation depends on searching for new knowledge close to the existing knowledge base of organizations (Pavitt, 1993). While organizations learn by exploiting internal capabilities and competence, there are always interactive relations among organizations that may be direct and planned or random and not deliberate. Transactional, interactive, collective and network learning are some of the concepts which try to define the learning process via interactions between organizations.

Transactional learning is defined as learning that takes place within markets and mainly stems from relations between customers, suppliers and competitors, especially user-producer relations. Learning between the subcontractors in supplier chains (Morgan, 1997) and interactive and complementary relations between firms facilitate interactive learning that may lead to innovative activities (Sweeney, 1996; Sternberg, 1999). This type of learning is supported by vigilant trust, negotiated loyalty and cultural proximity among different actors. Another form of transactional learning is achieved by using new equipment, since new machines embody new knowledge (Jin and Stough, 1998) and help to build new skills and expertise or at least show the different possibilities of improving technology. Even participation in trade fairs contributes to learning processes (Eraydin, 2001; Belussi, 1999).

Moreover, learning that occurs within market or competitively-based transactions can be extended by complementary networks that enable firms to access knowledge and competence beyond their individual capabilities (Jin and Stough, 1998, p. 1268). Labour is also part of interactive learning. Indeed, labour actively committed to the firm is important for problem solving within the organization (Morgan, 1997), and labour mobility contributes to the flow of knowledge among firms in related activities, and networks. For firms within networks, specialized, sometimes highly nuanced learning processes, define "core competencies" (Nelson and Winter, 1982; Cyert and March, 1963; Lawson and Lorenz, 1999) that, in turn, must be renewed, augmented and adapted to create "dynamic capability" over time (Teece et al 1997, p. 18).

Learning processes are further differentiated according to the different types of knowledge. Thus, in several related dichotomies, specifically codified-tacit knowledge (Lawson and Lorenz, 1999; Maskell and Malmberg, 1999b) embodied-disembodied and descriptive-prescriptive knowledge (Massey et al 1992; Saxenian, 1990), and procedural-declarative knowledge (Nooteboom, 1999), a distinction is made between socially embedded knowledge and formal, research-based knowledge. Tacit knowledge, for example, is learned procedurally via imitating observed behaviour or by 'learning by doing' under the guidance of a community, while codified knowledge is encoded in some form. Knowledge is rarely completely tacit or codified, however.

Recent literature emphasizes tacit knowledge because it is vital for complex problem solving and for innovation in distinctive circumstances (Chiaromonte and Dosi, 1993). It is precious, since it is not easily codified and communicated (Nonaka and Takeuchi, 1995). Yet, any transfer of tacit knowledge implies some form of codification. As Nooteboom (1999, p. 141) says "the transfer of tacit knowledge requires richer media (mutual observation of conduct and on-line feedback) and embedding in the community of practice".

Spatial Learning: Proximity Dynamics

The building blocks of learning, namely its cumulative nature, its path dependent characteristic, the importance of knowledge transactions among firms, networks and organizations, and the essential role of tacit knowledge in innovative activities, directly imply the spatial cumulativeness of knowledge and the increasing role of clusters in the learning process.

The cumulativeness of knowledge denotes the importance of proximity in the learning process (Torre and Gilly, 2000; Schmitz, 1999; Kirat and Lung, 1999; Maskell and Malmberg, 1999a). However, the context and the meaning of proximity differs. Thus, Kirat and Lung (1999, p. 29) contemplate four types of proximity, namely geographical, technological, organizational and institutional. It is usually stated that tacit knowledge is more dependent on proximity than codified knowledge in capturing the benefits of knowledge spillovers. Tacit knowledge is the basis of localized capabilities that are sticky and difficult to transfer within and between organizations, producers and customers in different places. Although sharing the same location does not guarantee localized networks of knowledge exchange and collective learning processes, spatial proximity facilitates interactive activities and competitiveness based on learning (Amin and Cohendet, 1999; Maskell and Malmberg, 1999b). This situation explains the strong interest in spatial agglomerations that have the ability to learn, create knowledge and innovate.

Recent attempts to theorize territorial/regional development are based on definitions of the dynamics in innovative milieux, industrial districts, new industrial spaces, innovation systems and learning regions. These models of territorial development-evolution provide slightly different perspective and points of emphasis on the processes of learning, knowledge dissemination and innovation. They reach a consensus on the advantages of geographical proximity, however.

Thus, the industrial district literature emphasizes collective learning based on small firms that are specialized in different steps of production and innovative capacities. Belussi (1999, pp. 734-736), based on the experience of Italian industrial districts, lists the factors that enable collective learning processes and the diffusion of technical change and know-how within local clusters. He emphasizes the sunk nature of knowledge, fluid interactions and many channels where information can quickly circulate among the firms in spatial and social proximity, higher levels of inter-firm cooperation, low transaction costs and stimulating environment for enterprises to adopt innovation process more rapidly. In this approach, learning is a collective process, which can only be understood by historical and socio-economic factors within that industrial cluster. The transmission of tacit knowledge, which is facilitated by trust and reciprocity among firms, is important for small firms, which lack management information systems, resources, and have limited access to codified information sources (Fuellhart, 1999).

The literatures on high technology industrial clusters or new industrial spaces also concentrate on local interdependencies and transfer knowledge among firms, while giving special emphasis to research and development (R&D). In this approach the cluster is a place where knowledge for new products and processes appears and disseminates under the existing social regulation mechanisms of that cluster. According to Scott and Storper (1988, p. 29) social regulations define the new industrial spaces by coordinating inter-firm transactions, organizing local labour markets and supporting community formation and social reproduction.

The theory of regional innovation systems focuses on the institutional basis of learning following the debate on national innovation systems. The argument is that different kinds of R&D institutions complement and compete with one another in support of learning processes and innovative activities (Gregersen and Johnson, 1997). At the regional scale, Cooke, Uranga and Etxebarria (1997) define an innovative industrial cluster as the area likely to have firms with access to others in similar or complementary sectors as customers, suppliers and partners. They also have access to such knowledge infrastructure as universities, research institutes, contact research organizations and technology transfer agencies. The interactive learning process in these areas are promoted by governance structure of business associations, chambers of commerce and public economic development, training and promotion agencies as well as government departments. In innovative milieux, learning and innovation depends on the capacity of firms through relationships with other agents within a 'co-operative atmosphere'. Finally, the learning region model integrates these ideas in order to indicate the conditions of building knowledge-based dynamic competitive capacities. All these new models of territorial innovation dynamics are based upon the fundamental importance of industrial clusters. They have less to say about why regions may become such a cluster. There is also another literature that questions the role of such clusters.

Skepticism on Clustering and Decentralized Networks in Learning

Several studies have argued that the formation of clusters does not necessarily realize high rates of innovation (Malmberg and Maskell, 1997; Lyons, 2000), increased access to information sources (Fuellhart, 1999) and the assimilation of tacit knowledge (Rallet and Torre, 1998). Furthermore, there is increasing recognition of the limits of local tacit learning (Amin and Cohendet, 1999) and the role of trust and cooperation in economic development (Blois, 1999; Moore, 1999). In fact, the tendency to glorify the successful industrial clusters in the past obscured some of the conflicts and tensions evident in these areas (Staber, 1996).

These doubts are expressed according to different lines of argument. First, it is recognized that clusters can experience the lock-in effects of interactive learning as a result of limited inflows of external knowledge, resistance to change and delay in generating response to changing economic conditions (Camagni 1991; Glasmeier, 1994). Second, institutional thickness, which was believed to be a critical factor in interactive learning, has been criticized as some institutional structures, which were crucial in the early phases of growth process, can act as barriers to innovative activities (Amin and Thrift, 1995; Coriat and Bianchi, 1995; Amin, 1999). In addition, 'institutional overload' occurs if many institutions in the same field create various problems of domination (Glasmeier, 1994; Heidenreich and Krauss, 1998).

Third, there are increasing number of examples indicating that locally embedded relations may be less effective in the later stages of growth due to increasing competition (Amin, 1999). This condition raises the question as to whether embeddedness is temporary or insufficient to enable local capacities to cope with changing conditions. The experience shows that the main achievements of many regions are path dependent.

Fourth, there are increasing numbers of studies that point out the importance of non-local networks in learning processes (Harrison, 1994; Bell and Albu, 1999). Exogenous stimuli are widely recognized as important in enhancing creativity. In an era of globalization, it is easier to reach external knowledge networks and transform tacit knowledge into codified. This tendency changes the importance of locally based tacit knowledge and erodes the advantages of spatial cumulativeness of knowledge. Maskell and Malmberg (1999b, p.12) claim that globalization converts localized inputs into ubiquities. However, the process of codification does not mean tacit knowledge is less important. It is impossible to codify all knowledge, the cost of transforming tacit knowledge to codified knowledge is high, there is uncertainty about the value of tacit knowledge and the tacit knowledge is needed to use codified knowledge (Maskell and Malmberg, 1999b). While ubiquitification processes have important consequences for the relative advantages of firms and regions, tacit knowledge can still be effectively utilized, even if this implies shifts from local embeddedness into network embeddedness (Yeung, 2000).

Different Perspectives on the Manipulation of Learning Processes

In practice, the theories of industrial districts, new industrial spaces and innovative milieu, have been strongly affected by changes in the evolutionary path of the 'have regions', that is the ones with appropriate competencies, capacities and institutions for learning-based competitive advantages. This emphasis has overstated endogenous processes in development, and neglected the conditions faced by 'have-nots'.

Difficulties Facing the Manipulation of Learning Processes

It is difficult to shift from descriptive to prescriptive approaches towards learning for local development. First, the cumulative character of knowledge and its path dependent character creates certain barriers, since established holders in given technologies accumulate a differential advantage with respect to potential entrants (Saviotti, 1998). Knowledge spillovers and innovative activities are likely to occur within geographically bounded areas, which means that firms in these areas are in a favourable position for subsequent rounds of innovation compared to the others (Breschi, 2000). Second, although knowledge is increasingly codified, codification is difficult in newer developments. Even if codified knowledge is available, its use can still entail costs. Only the agents who know the code can use knowledge without paying additional costs, whereas others have to pay to learn the code. To reach tacit knowledge is even more difficult, since it refers to actions that are performed, at least in part by internalized skills that are socially constructed (Keane and Allison, 2000).

Third, there are difficulties related to competence, experience and other problems in preparing a transcoding function in order to translate external information into a local/firm language (Glasmeier, 1999). Camagni (1991) calls this process of transcoding within networks of firms as emergence of a new language and culture. Fourth, technological frontiers can be very far away from most firms (Romijn, 1998), despite rapid improvements in telecommunication technology. If the digital communication makes data transmission between places easy and rapid, for learning processes data transmission is not adequate. Fifth, power relations help shape learning processes, and vice versa (Hayter et al 1999). There are asymmetrical power relations among firms and local actors, because firms can have different resources and status, such as competence, information and property rights. The non-egalitarian and asymmetrical power relations obviously mean that the benefits of the interactive process among different organizations may be unequal (Hudson, 1999, p. 68).

Other arguments concentrate on the activities that support learning processes. The regions with limited learning infrastructure have important handicaps to integrate themselves in the knowledge economy. It is widely accepted that the quality of human infrastructure and institutional mechanisms can foster interactive learning (Keane and Allison, 2000) while an actively committed, engaged workforce dedicated to enhancing corporate performance is crucial (Hudson, 1999). Human capital inputs, know-how, skill, competence and expertise are features that distinguish one region to another in competitive conditions. This is also true for complementary activities. Since the core competence of corporations cannot lead alone to competitive advantage, there is also a need for complementary knowledge (Saviotti, 1998), activities (Lawson and Lorenz, 1999) and assets (Teece, 1986).

Policies for a Learning Economy

Arguably, the literature on regional innovation systems, which explains the roles of different institutions and actors in the work that leads to innovation (Cooke et al 1997 and 1998), has direct policy relevance for disadvantaged regions. However, this literature has been erratic in defining its position towards regional policy and intervention. In general, the debates related to different policy options concentrate on overcoming the difficulties faced in the learning process. These options relate to: increasing access to knowledge; improving learning activities by facilitating learning environment and by creating learning infrastructure; and by initiating new local and central government institutions for implementing policies for creative learning and innovation. The role of policy makers in these types of policies can be defined as 'facilitating' or 'enabling'.

Increasing access to knowledge is essential for regions wishing to be a part of the knowledge economy or to sustain their position by importing knowledge, preferably by attracting the source of knowledge. Although local clusters provide advantages in the learning process, their innovation capacity can be limited by lock-in situations. In order to break such vicious circles there is a need for extra-regional information flows. Cassiolato et al (1992), for example, argue that accessing foreign technology on favourable terms is an important way of speeding up the process of technological learning. Another way to increase access to knowledge is to form direct linkages with main sources of knowledge, for example, by creating research units in the cores of specific knowledge accumulation or forming subsidiary firms in knowledge-rich clusters. On the other hand, many firms locate their branch plants in certain areas to become a part of the localized system, thereby reaching locally embedded knowledge (Cohendet et al 1993; Blanc and Sierra, 1999; Yeung, 2000). Closeness to the source of knowledge may not imply innovation, but it may facilitate imitation, assimilation and creative synthesis. The positive atmosphere brought by the leading innovative firms via learning infrastructure is also important in generating learning processes in regions. That is why the attraction of leading firms with innovative capacities in certain fields can be effective in creating images for certain areas. The experience of the Sophia Antipolis region showed that learning infrastructure promoted via different government schemes can be more influential, if there are some leading firms (French Telecom in this region) with power to attract the others (Longhi, 1999).

The second group of policy recommendations is directed to strengthening the learning support mechanisms of public bodies. As the case studies of Minneapolis and Cambridge demonstrate, the facilitation of the learning environment by indirect measures and support mechanisms can help knowledge openness and collective action among the partners of the economy (Lawson and Lorenz, 1999). This help can be very important for recombining existing technologies or developing new ones. Direct measures to create learning infrastructure by public institutions are also important. The most effective role a state can play in promoting learning and learning capability in a region is the provision of learning infrastructure and effective policy assistance to promote learning capability of firms and industries (Jin and Stough, 1998, p. 1272). In addition, local and regional authorities, as well as communities and voluntary associations, have influential roles in developing existing competence and learning capabilities. Thus, these agencies can provide different collective goods that vary from educational and training facilities, technical infrastructure including transportation and communication to technological infrastructure and R&D units (Florida, 1995; Jin and Stough, 1998). In this context, Cooke et al (1997) emphasize the importance of specialized financial formulas and regional budgets devoted to learning activities. Universities (and their science parks and spin-off firms) are increasingly accepted as the key infrastructure in order to create learning regions (Goddard, 1997). The provision of real service facilities and learning sites are also important to supplement the information deficiencies presumed to exist within the local milieu.

As Belussi (1999, p. 743) claims, local policies are vital for the creation of new knowledge. New local policies are increasingly concentrated on promoting cooperative institutional environments for facilitating the exchange of information and knowledge. Likewise, beginning from the 1990s, it is possible to define the new regionalism, mainly in EU countries, which is characterized by strong top-down policy pressures to encourage existing and newly created regional development agencies and institutions to adopt innovation oriented strategies. Governments that wish to improve technological and innovation capacity, try to develop policies in order to support learning processes, such as the European Union programes for supporting regional systems of innovation. Technopolis Programmes, Regional Technology Plans and Regional Innovation Strategies are various forms of policy tools initiated by either national/regional governments or EU in order to create a collective learning process in several parts of Europe (Morgan, 1997). This new regionalism has been influenced by the concepts of network paradigm and learning regions (Thomas, 2000, p. 190), mainly designed to support the innovative basis of clusters.

Concluding Remarks

In this paper I reviewed the theoretical discussions and policy measures related to the possibility of firms and regions 'catching-up' in the knowledge economy. However, it is evident that uneven development within and between regions and their constituent social groups are unavoidable, since both firms and regions learn by producing and protecting their tacit knowledge (Hudson, 1999). The distribution of available learning infrastructures and the new communication facilities seem to support the inequalities in the learning process. There are several theoretical and empirical studies that indicate the problems facing local development from cumulative learning that results in lock-in situations, whereby firms in clusters are unable to unlearn unsuccessful habits (Maskell and Malmberg, 1999a). Meanwhile, the areas with strong competence and learning infrastructure have comparative advantages.

In the paper these factors are summarized under the heading of existing barriers that increase the difficulty of being a part of the knowledge economy. The barriers stem from the characteristics of knowledge and the learning processes, the competence of the actors and the lack or insufficiency of complementary units within the milieu. They mean that units or places with already accumulated knowledge and a strong basis for interactive learning (among the organizations or firms in a spatial cluster) have relative advantages over others, whereas firms and places without a worldwide competitive knowledge base have important difficulties. The firms, networks of firms, and clusters with already accumulated knowledge have advantages over those without such knowledge. Obviously this situation increases the discrepancies among the firms and clusters. That is why the 1990s became the revival period for regional policies on innovation, in order to tackle the problem of being excluded in the knowledge economy. However, whether these policies and measures will be successful, or whether they will 'fetishize' knowledge and learning and lead to a neglect of the institutional factors that underlie regional competitiveness, will be more evident in the near future.

References

Amin, A. (1999), 'The Emilian Model: Institutional Challenges', *European Planning Studies*, vol. 7, pp. 389-405.

Amin, A. and Cohendet, P. (1999), 'Learning and Adaptation in Decentralised Business Networks', *Environment and Planning D: Society and Space*, vol. 17, pp. 87-104.

Amin, A. and Thrift, N. (1995), 'Globalisation, Institutional Thickness and the Local Economy', in P. Healey, S. Cameron, S. Davoudi, S. Graham and A. Madani-pour (ed), *Managing Cities: The New Urban Context*, John Wiley and Sons: London, pp. 91-108.

Asheim, B.T. (1996), 'Industrial Districts as Learning Regions : A Condition for Prosperity', *European Planning Studies*, vol. 4, pp. 379-397.

Audretsch D.B. and Feldman, M.P. (1996), 'Innovative Clusters and the Industry Life Cycle', *Review of Economic Organization*, vol. 11, pp. 253-273.

Bell, M and Albu, M. (1999), 'Knowledge Systems and Technological Dynamism in Industrial Clusters in Developing Countries', *World Development*, vol. 27, pp. 1715-1734.

Belussi, F. (1999), 'Policies for the Development of Knowledge- intensive Local Production Systems', *Cambridge Journal of Economics*, vol. 23, pp. 729-747.

Blanc, H. and Sierra, C. (1999), 'The Geography of Organisation of TNC R&D: Benefiting from External and Internal Proximities', *Cambridge Journal of Economics*, vol. 23, pp. 187-206.

Blois, K.J. (1999), 'Trust in Business to Business relationships: An Evaluation of Its Status', *Journal of Management Studies*, vol. 36, pp. 197-215.

Breschi, S. (2000), 'The Geography of Innovation: A Cross-sector Analysis', *Regional Studies*, vol. 34, pp. 213-229.

Brusco, S., (1986), 'Small Firms and Industrial Districts: The Experience of Italy', in D. Keeble and E. Wever (eds), *New Firms and Regional Development*, Croom Helm, London, pp. 184-202.

Camagni, R. (1991), 'Local Milieu, Uncertainty and Innovation Networks: Towards a New Dynamic Theory of Economic Space', in R. Camagni (ed), *Innovation Networks*, Belhaven, London, pp. 121-144.

Cassiolato, J., Hewitt T. and Schmitz, H. (1992), 'Learning in Industry and Government: Achievements, Failures and Lessons', in H. Schmitz and J. Cassiolato (eds), *Hi-tech for Industrial Development: Lessons from the Brazilian Experience in Electronics and Automation*, Routledge, London, pp. 273-306.

Chiaromonte, C. and Dosi, G. (1993), 'The Micro Foundations of Competitiveness and Their Macroeconomic Implications', in D. Foray, and C. Freeman (eds), *Technology and Wealth of Nations*, Pinter, London, pp. 107-34.

Cohendet, P., Heraud, J.A. and Zuscovitch, E. (1993), 'Technological Learning, Economic Networks and Appropriability', in D. Foray and C. Freeman (eds), *Technology and Wealth of Nations*, Pinter, London, pp. 66-76.

Cooke, P., Uranga, M.G. and Extebarria, G. (1997), 'Regional Innovation Systems: Institutional and Organisational Dimension', *Research Policy*, vol. 26, pp. 475-491.

Cooke, P, Uranga, M. G. and Extebarria, G. (1998), 'Regional Systems of Innovation: An Evolutionary Perspective', *Environment and Planning A*, vol. 30, pp. 1563-1584.

Coriat, B. and Bianchi, R. (1995), 'A European Response to the Japanese Challenge', in L.A. Andreason (ed), *Europe's Next Step*, Frank Cass, Ilford, pp. 59-77.

Cyert, R. M. and March, J. (1963), *A Behaviourial Theory of the Firm*, Prentice Hall, Englewood Cliff, N.Y.

Dupuy, C. and Gilly, J.P. (1996), 'Collective Learning and Territorial Dynamics: A New Approach to the Relations Between Industrial Groups and Territories', *Environment and Planning A*, vol. 28, pp. 1603-16.

Eraydin, A. (2001, in press), 'The Roles of Central Government Policies and the New Forms of Local Governance in the Emergence of Industrial Districts', in M. Taylor and D. Felsenstein (eds), *Promoting Local Growth: Process, Practice and Policy*, Ashgate, Aldershot.

Florida, R. (1995), 'Towards the Learning Region', *Futures*, vol. 27, pp. 527-536.

Fuellhart, K. (1999), 'Localisation and the Use of Information Sources: The Case of Carpet Industry', *European Urban and Regional Studies*, vol. 6, pp. 39-58.

Glasmeier, A.K. (1994), 'Flexible Districts, Flexible Regions? The Institutional and Cultural Limits to Districts in an Era of Globalization and Technological Paradigm Shift', in A. Amin and N. Thrift (eds), *Globalization, Institutions and Regional Development in Europe*, Oxford University Press, Oxford, pp. 73-84.

Glasmeier, A. K. (1999), 'Territory Based Regional Development Policy and Planning in a Learning Economy: The Case of Real Service Centers in Industrial Districts', *European Urban and Regional Studies*, vol. 6, pp. 73-84.

Gregersen, B. and Johnson, B. (1997), 'Learning Economies, Innovation Systems and European Integration', *Regional Studies*, vol. 31, pp. 479-490

Harrison, B. (1994), 'Concentrated Economic Power and Silicon Valley', *Environment and Planning A*, vol. 26, pp. 307-328.

Harvey, C and Denton, D. (1999), 'To come of Age: The Antecedents of Organizational Learning', *Journal of Management Studies*, vol. 36, pp. 897-918.

Hayter, R., Patchell, J and Rees, K. (1999), 'Business Segmentation and Location Revisited: Innovation and The Terra Incognita of Large Firms', *Regional Studies*, vol. 33, pp. 425-442.

Heidenreich, M. and Krauss, G. (1998), 'The Baden-Württemberg Production and Innovation Regime: Past Success and New Challenges' in H.J. Braczy, P. Cooke and M. Heidenreich (eds), *Regional Innovation*

Systems: The Role of Governance in a Globalized World, UCL Press, London, pp. 214-44.

Howells, J. (1996), 'Tacit Knowledge, Innovation and Technology Transfer', *Technology Analysis and Strategy Management*, vol. 2, pp. 91-105.

Hudson, R. (1999), 'The Learning Economy, the Learning Firm and the Learning Region: A Sympathetic Critique of the Limits to Learning', *European Urban and Regional Studies*, vol. 6, pp. 59-72.

Jin, D.J. and Stough, R. (1998), 'Learning and Learning Capability in the Fordist and Post-Fordist Age: An Integrative Framework', *Environment and Planning A*, vol. 30, pp. 1255-78.

Keane, J. and Allison, J. (2000), 'The Intersection of the Learning Region and Local and Regional Economic Development: Analysing the Role of Higher Education', *Regional Studies*, Policy Review Section, vol. 34, pp. 896-901.

Kirat, T. and Lung, Y. (1999), 'Innovation and Proximity: Territories as Loci of Collective Learning Process', *European Urban and Regional Studies*, vol. 6, pp. 27-38.

Lawson, C and Lorenz, E. (1999), 'Collective Learning, Tacit Knowledge and Regional Innovative Capacity', *Regional Studies*, vol. 33, pp. 305-317.

Lei, D.T. (1997, 'Competence Building, Technology Fusion and Competitive Advantage: The Key Roles of Organisational Learning and Strategic Alliances', *International Journal of Technology Management*, vol. 14, pp. 208-237.

Longhi, C. (1999), 'Networks, Collective Learning and Technology Development in Innovative High-tech Regions: The Case of Sophia-Antipolis', *Regional Studies*, vol. 33, pp. 333-42.

Lundvall, B.A. (1993), 'User- producer Relationships, National Systems of Innovation and Internationalisation', in D. Foray and C. Freeman (eds), *Technology and Wealth of Nations*, Pinter, London, pp. 891-908.

Lyons, D. (2000), 'Embeddedness, Milieu, and Innovation Among High Technology Firms: A Richardson, Texas Case Study', *Environment and Planning A*, vol. 32, pp. 891-908.

Malmberg, A. and Maskell, P. (1997), 'Towards an Explanation of Regional Specialisation and Industry Agglomeration', *European Planning Studies*, vol. 5, pp. 1997.

Mascietelli, R. (1999), 'A Framework for Sustainable Advantage in Global Hi-tech Markets', *International Journal of Technology Management*, vol. 17, pp. 240-258.

Maskell, P. and Malmberg, A. (1999a), 'Localised Learning and Industrial Competitiveness', *Cambridge Journal of Economics*, vol. 23, pp. 167-185.

Maskell, P. and Malmberg, A. (1999b), 'The Competitiveness of Firms and Regions: Ubiquitification and the Importance of Localized Learning', *European Urban and Regional Studies*, vol. 1, pp. 9-25.

Massey, D., Quintas, P. and Wield, D. (1992), *High-Tech Fantasies*, Routledge, London.

McKee, D. (1992), 'An Organisation Approach to Product Innovation', *Journal of Product Innovation Management*, vol. 9, pp. 232-245.

Moore, M. (1999), 'Truth, Trust and Market Transactions: What Do We Know?', *The Journal of Development Studies*, vol. 36, pp. 74-88.

Morgan, K. (1997), 'The Learning Region: Institutions, Innovation and Regional Renewal', *Regional Studies*, vol. 31, pp. 491-503.

Nelson, R.R. and Winter, S. (1982), *An Evolutionary Theory of Economic Change*, Harvard University Press, Cambridge.

Nonaka, I and Takeuchi, H. (1995), *The Knowledge Creating Company*, Oxford University Press, Oxford.

Nooteboom, B. (1999), 'Innovation, Learning and Industrial Organisation', *Cambridge Journal of Economics*, vol. 23, pp. 127-150.

Pavitt, K. (1993), 'What Do Firms Learn from Basic Research?', in D. Foray and C. Freeman (eds), *Technology and Wealth of Nations*, London, Pinter, pp. 29-40.

Popper, K.R. (1972), *Objective Knowledge*, Oxford University Press, Oxford.

Rallet, A. and Torre, A. (1998), 'On Geography and Technology: Proximity Relations in Localised Innovations Networks', in M. Steiner (ed), *Clusters and Regional Specialization*, Pion, London, pp. 65-78.

Romijn, H. (1998), *Acquisition of Technological Capabilities in Small Firms in Developing Countries*, Macmillan, Basingstoke.

Saviotti, P.P. (1998), 'On the Dynamics of Appropriability of Tacit and of Codified Knowledge', *Research Policy*, vol. 26, pp. 843-856.

Saxenian, A.L. (1990), 'Regional Networks and Resurgence of Silicon Valley', *California Management Review*, vol. 33, pp. 89-112.

Schmitz, H. (1999), 'Collective Efficiency and Increasing Returns', *Cambridge Journal of Economics*, vol. 23, pp. 465-483.

Staber, U. (1996), 'Accounting for Variations in the Performance of Industrial Districts: The Case of Baden-Württemberg', *International Journal of Urban and Regional Research*, vol. 20, pp. 299-316.

Sternberg, R. (1999), 'Innovative Linkages and Proximity: Empirical Results from Recent Surveys of Small and Medium Sized Firms in German Regions', *Regional Studies*, vol. 33, pp. 529-540.

Storper, M. (1993), 'Regional Worlds of Production: Learning and Innovation in the Technology Districts of France, Italy and USA', *Regional Studies*,vol. 27, pp. 433-455.

Sweeney, G. (1996), 'Learning Efficiency, Technological Change and Economic Progress', *International Journal of Technology Management*, vol. 11, pp. 5-27.

Teece, D.J. (1986), 'Profiting from Technological Innovation', *Research Policy*, vol. 15, pp. 286-305.

Teece D.J., Pisano, G. and Shuen, A. (1997), 'Dynamic Capabilities and Strategic Management', in N. Foss (ed), *Resource, Firms and Strategies*, Oxford University Press, Oxford, pp. 268-288.

Thomas, K. (2000), 'Creating Regional Cultures of Innovation? The Regional Innovation Strategies in England and Scotland', *Regional Studies*, Policy Review Section, vol. 34, pp. 190-198.

Torre, A. and Gilly, J. P. (2000), 'On the Analytical Dimension of Proximity Dynamics', *Regional Studies*, vol. 34, pp. 169-180.

Yeung, H.W.C. (2000), 'Embedding Foreign Affiliates in Transnational Business Networks: The Case of Hong Kong in Southeast Asia', *Environment and Planning A*, vol. 32, pp. 201-222.

5 Global-local Networking of PC Manufacturing in Dongguan, China

Tong Xin and Wang Jici

Introduction

Dongguan, a young export-oriented manufacturing city in the Pearl River Delta, has developed rapidly following the introduction of China's policy of 'openness' in the late 1970s. Like many other developing cities, Dongguan's rapid industrialization has been based on exploiting low costs of land and labour. This chapter focuses on the evolution of the agglomeration of personal computer (PC) manufacturing activities in Dongguan, with particular emphasis on the role of the global-local ties that have shaped the nature of manufacturing networks in the city and its neighbouring areas. A significant issue underlying the formation of these networks is the creation and transfer of various kinds of skills and expertise. This issue is addressed in the latter part of the chapter. In general, the chapter seeks to contribute towards understanding the geography of fast growing industrial districts in developing countries (see chapter 7 by Hwang).

In recent decades, regional development has become increasingly problematic, increasingly challenged by the deepening forces of globalization, themselves rooted in fast changing technologies and institutional innovations, including a strong trend towards deregulation. Rapid change is an especially strong feature of the PC-related industry. In this industry, technological and market dynamics have generated strong competition among firms, further stimulated by tendencies among national and regional governments throughout the world to attempt to stimulate 'high tech' activities, most notably micro-electronics. Moreover, in PC-related industries, competition among firms features the development of global-local networks. They imply various forms of cooperation that in turn have consequences for regional development.

PC Manufacturing in Dongguan

In the 1990s, the PC industry, especially hardware manufacturing, was booming in China under both pressures from the global shift of multinationals (MNCs) and the fast growing domestic market (Wang, 1998). In contrast to most developed countries, PC hardware manufacturing in China has not declined. As the growth of the Internet has helped stimulate the market for low-price PCs, multinational or trans-regional PC manufacturers, especially those from Taiwan, the third largest PC exporter of the world, have moved their production lines to Mainland China to reduce their costs. In addition, the potential of China's vast domestic market attracts both foreign and domestic firms. In China, the rapid growth of the Internet (see chapter 6 by Liu) has occurred in concurrence with the expansion of PC hardware production. Indeed, it is often said that the 'PC industry is right in its prime in China'. Certainly, many PC-related manufacturing agglomerations have developed recently in the coastal areas of Mainland China, including those in the provinces of Jiangsu, Zhejiang, Fujian, Shandong and Guangdong. Competition within and among these regions is furious. How future patterns will unfold, however, remains uncertain.

Dongguan city is part of the PC-related manufacturing agglomeration of the Pearl River Delta. It was among the earliest export-oriented processing areas to boom in China after the introduction of the policy of openness and reformation in the late 1970s. Multinational PC corporations began to set up plants in the region in the late 1980s. At present, there are over 1800 information and communication technology firms within 100 kilometres from each other. These firms supply 95 percent of parts and components in the final assembly of PCs. Virtually all manufacturing operations are assembly only and in the case of monitor production, about 70 percent of parts and materials can be bought locally. Some products account for a huge share in the global market, such as Hard Disk Drive (HDD) heads and cases (40 percent); floppy disk drivers (30 percent); advanced alternating current capacitances and transformers (25 percent); scanners and mini-motors (20 percent); keyboards (16 percent); and motherboards (15 percent).

As a Taiwanese entrepreneur said in an interview, the density of the PC-related plants in the Dongguan area has nearly reached that of the corridor of Hsinchu-Taipei of Taiwan about 10 years ago. Famous foreign and domestic PC companies, including IBM, Compaq, Dell, Acer, Founder and Legend, have established plants or purchased parts in the Dongguan area. But value added production is quite low. Thus, the value of the parts and components bought locally only accounts for 10-15 percent in the typical price of products. The high value products, such as Central Processing Units (CPU) and Integrated Circuits (IC), must be imported from overseas. The high value added activities, such as R&D, marketing, are also offshore. Indeed, the local government has noted its desire to attract R&D institutions and improve the technological level of the products manufactured in the area.

However, since 1998, as the central government has shifted the focus of development from the eastern coastal region to the western inland area, and as China has joined the WTO in late 2001, domestic industry, including in Dongguan, will face a disturbed market environment for at least a few years. While the preferential policy conditions for investment tend to be common throughout the country, the advantages of tax incentives and low costs of labour and land in Dongguan, which contributed to its fast growth in the past decade, will vanish in the near future. Yet, the concentration of industry, once established, tends to be self-sustaining (Krugman, 1991) so that location advantages will not suddenly vanish and local production networks within Dongguan will have the capacity to facilitate restructuring.

The most important organizational feature of indigenous firms within Asian capitalist economies is that they are organized through networks of firms (Hamilton, 1991). The fundamental source of strategic effectiveness in overseas Chinese firms is the strength of family-based network ties, which was characterized by Redding (1991) as "the weak organization and strong linkages". Chinese business networks, based largely on personal trust and organized primarily through kinship circles and ties of common origins, have helped firms to transcend their inherent limitations and attain both economies of scale and scope (Hamilton, 1991). These features underpin the competitive advantages of Chinese overseas businesses in traditional industries such as textile, clothes, shoes and furniture (Chen, 1994). What is interesting is that such networks are also evident in the rapid growth of PC manufacturing in China.

Since IBM opened the structure of its personal computer to facilitate involvement by other firms, while agreeing to common technical standards, flexible specialization has become an important feature of the PC industry

(Piore and Sabel, 1984). Time control is the central concern in the whole production chain (Curry and Kenny, 1999). The transformation from traditional sectors to PC manufacturing was doubtless facilitated by the ability to exploit the advantages of flexible specialization. However, PC manufacture is more sophisticated than former, traditional labour-intensive manufactured products. The networking patterns differ in that the linkages of firms tend to be more formalized in the PC industry. The government's high tech policies have also shaped the production system. Furthermore, the spontaneous integration of the electronic industry within the 'China Circle' (Naughton, 1997) has been affected by the complicated political relationships between China and Taiwan.

In this study of the PC industry in Dongguan, we selected three towns: Qingxi, Shijie and Changian. Among the 33 towns of Dongguan, these three are the most important PC manufacturing 'centres', or areas, since manufacturing is relatively dispersed throughout their jurisdictions. These towns are 'typical' sites for 'overseas' investment, including that from Hong Kong, Taiwan and Macao, the main source of such investments. For this study, we trace the trajectory of the networking process created by these investments in the three towns. The institutional bottlenecks created by reliance on trade based on export processing activities are highlighted. This chapter now turns to the characteristics and structural evolution of this local production system.

The Trajectory of the Global-local Networking of PC Manufacturing in Dongguan

The period of the development of PC manufacturing in Dongguan can be divided into two stages. First, in the late 1980s and early 1990s, the initial wave of PC manufacturers created an export-based boom and development model based on low cost labour and land, proximity to Hong Kong, and the various kinds of financial support provided by governments, especially to overseas investment. This initial wave of investment established (or rented) assembly plants, and had strong linkages with overseas markets. Linkages with local plants were weak, however. Second, after the middle 1990s, the PC-related industry in Taiwan kept moving offshore, and Dongguan became the gathering place for Taiwanese PC makers, in fact, one of the densest concentrations of Taiwanese PC makers in Mainland China. In this stage, the local PC production network became stronger. Furthermore, this agglomeration attracted more firms, foreign and domestic, large and small. Simultaneously, the interface between the global and the local in these evolving production networks became more complicated.

The Global Network of Export-based PC Manufacturing

Dongguan is located between Guangzhou, the capital of Guangdong province, and Shenzhen, one of the five Special Economic Zones (SEZs) in China. Transportation accessibility is good. It is close to Hong Kong and there are strong kinship ties and much shared family backgrounds among the people of these two cities. Following China's 'open door' policies, these connections gave Dongguan some advantages in inducing foreign direct investment (FDI) especially as Hong Kong accounted for 64.3 percent of the total amount of the FDI pledged during 1979 to 1992 (Hayter and Han, 1997; Qu and Green, 1997). In 1997, foreign firms accounted for 70 percent in the gross industrial output value in Dongguan, and over 90 percent of these firms are invested from Hong Kong; over half of them deal with export processing activities (Statistical Yearbook of Dongguan, 1998). The industrialization of Dongguan stemmed from the export processing activities stimulated by overseas investment.

In the late 1980s and the early 1990s, the demands of PC products in the global market were growing fast, while average profits were decreasing. Mature technological products in the PC industry kept moving to low cost regions, notably Mainland China. In Dongguan, prior to the establishment of PC manufacture, the consumer electronics industry from Hong Kong had relocated and thrived since the early 1980s. This development provided some initial facilities that subsequently benefitted the PC industry. Indeed, some consumer electronic factories just transferred to the PC-related industry as the entry-barriers between electronic toys and such PC components as mice or keyboards are low. These firms inherited the export processing character naturally.

Most MNCs that invested in the PC industry in Dongguan have built plants for export in cooperation with local Chinese partners. In this context four points need to be noted. First, the government offered inducements to MNCs to export. Second, Dongguan's main initial advantages were low costs in labour and land while most input materials had to be imported. Third, many MNCs were not confident about the political stability of China and this attitude limited local linkage development. Fourth, in a relatively short time local linkages have been generated. As one of the Taiwanese entrepreneurs recalled, when his company first set up in Dongguan in 1990, 'even screws had to be brought from Taiwan.' Now the company sources virtually all of the necessary raw materials and parts on the mainland and relies on Taiwan for very little. Thus, the network pattern that was initially established in Dongguan was typical of those in other satellite platforms in many fast growing industrial districts in developing countries (Park, 1995). The local linkages between the plants were weak while most producers kept strong linkages with the global market through Hong Kong (Figure 5.1). This has now changed considerably.

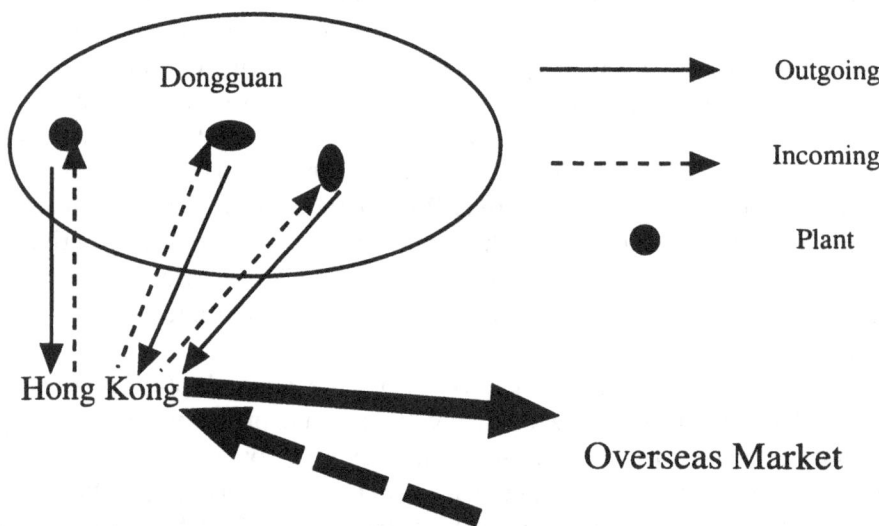

Figure 5.1 First stage export processing activities in Dongguan: Plants keep global linkages through Hong Kong while local linkages are weak

The Local PC Manufacturing Network Constitution

There are two stimuli to the rapid constitution of the local PC manufacturing network since the middle 1990s. One comes from the relocation of PC companies in Taiwan to low cost regions such as Dongguan. The other is the booming PC market of Mainland China.

PC manufacturing network extended from Taiwan According to the Market Intelligence Centre (MIC) of Taiwan, the ratio of offshore to domestic manufacturing for Taiwan's computer hardware industry has grown from 28 percent in 1995 to just over 47 percent in 1999. Mainland China is the largest destination. In 1999, the Mainland China had a 33 percent share of all offshore manufacturing of Taiwan, followed by Thailand with only a little over five percent.

Taiwan is recognized as the world's third largest supplier of PC products, after the United States and Japan, and is the world leader in the production of motherboards, notebooks, monitors, scanners, CD-ROMs and desktop PCs. As more and more labour-intensive industries, such as cases and power supplies, move away from Taiwan, the government has placed more emphasis on the high-tech sector, such as ICs and networking parts.

Dongguan holds the highest concentration of Taiwan companies anywhere in Mainland China (Tong and Wang, 1999). According to the Taiwan Business Association of Dongguan (TBAD), there are now 30,000 Taiwanese companies on the Mainland. Guangdong province accounts for about one-third of the total, of which 3,200 are in Dongguan. There are two main reasons why Dongguan became such a magnet for Taiwan firms.

First, Hong Kong became the bridge for Taiwanese businessman travelling between the Mainland China and Taiwan Island. Dongguan is not close to Taiwan in physical distance and the commonality of society and culture is also less if compared with Fujian province, which directly faces Taiwan Island, separated only by the Taiwan Strait. For historical reasons, however, there is no direct transportation and communication between Taiwan and Mainland China. There are still many restrictions, even after 'Regulations on Transportation across the Taiwan Straits' were promulgated in 1996. Taiwanese business people have to travel to Mainland China indirectly through Hong Kong or Macao. Some investment that in name originates from Hong Kong actually comes from Taiwan as government regulations in Taiwan prevent direct investment on the Mainland. As mentioned above, the propinquity to Hong Kong gave Dongguan priority in the location decisions of Taiwanese firms.

Second, the industrial upgrading of Shenzhen forced manufacturing to migrate northward. Shenzhen is located across the border from Hong Kong and as the first Special Economic Zone in South China it experienced fast industrialization and urbanization after the early 1980s, developing into a modernized city. The labour cost and land availability advantages of Shenzhen soon declined and it lost its competitiveness for labour-intensive manufacturing. This pressure forced Shenzhen to shift to more high tech development and high value added production activities. Consequently, peripheral areas to the north of Shenzhen, including Dongguan and Huizhou, became targeted manufacturing centres.

Investment from Taiwan has increased quickly in Dongguan since the government of Taiwan rescinded some restrictions on investment to Mainland China in 1987. Although some political affairs, such as the Tiananmen incident in 1989, the presidential election in Taiwan in 1996, and most recently, the tension across the Taiwan Straits in 1999, affect the investment decisions of Taiwanese entrepreneurs, total investment has increased steadily (Table 5.1). Thus, with respect to the number of PC-related firms classified by origin of investment, about 80 percent of Dongguan's 1,800 computer-related companies are from Taiwan, Hong Kong and Macao. In terms of production value, they account for at least half of this Ren Min Bi (RMB) 50 billion industry.

H.D.Yeh, the former president of the TBAD and the spiritual leader of the Taiwan businesspeople in Dongguan, runs his own electronics firm (Primax) in Shijie town of Dongguan. The company rented a factory in Dongguan in 1989 and initially produced labour-intensive power surge protectors. Three years later it began to invest massively on the mainland to establish mouse production lines. Thereafter, Primax launched new products every year: cellular phone parts, overhead projectors, paper shredders, and scanners. Constantly expanding its product lines and production capacity, Primax has grown from a single plant with 25 workers in 1989 to its current nine factories and workforce of over 4,000. Its outstanding performance contributed much to its mother company in Taiwan in going public seven years ago. As H.D.Yeh noted, the competition in the 21st century is all about speed. When customers in America make purchase orders, the whole process, including drawing up the contract and manufacturing the goods, requires four months. But Taiwanese firms can deliver in 15 days. "There are two keys," he said, "One is that all the upstream and downstream factories are here. You can assemble a computer completely from parts made within 50 kilometres". The second key is that when dealing with Taiwanese companies, there is no need to endure a long contract process. Everything starts with a single telephone call. Taiwanese business people are always playing golf, singing karaoke or chewing betel nut together. As Yeh said, "Everyone is always asking how every one else is doing – it's a way of networking and building up good relations".

Table 5.1 Origins of PC-related firms invested in Dongguan

Origin of Investment	Number of Firms	Share of PC-Related Firms in Total
Hong Kong	898	50.3
Taiwan	497	27.9
Japan	44	2.5
The United States	20	1.1
Singapore	7	0.4
Korea	7	0.4

Sources: Statistical Yearbook of Dongguan (1999), Chinese Statistical Press.

The PC industry is highly vertically disintegrated. The plants depend on external resources to achieve economies of scale and scope. Taiwanese business networks have been created step by step in Dongguan, closely connected to Hong Kong and Taiwan. The Dongguan area has already developed an economic circle in which Taiwan firms can conduct most of their business among themselves. A frequently cited example is Delta, a corporation controlled from Taiwan, that produces electrical power for PCs and has established several plants in Shijie town in Dongguan since 1992. Many suppliers of Delta in Taiwan have also moved to Dongguan for the convenience of delivery. The once rural area, where Delta plants locate now, has developed into Delta Town, with thousands of workers living and working there.

Qingxi, located in the southwest of Dongguan, was a comparatively disadvantageous area. It was mainly rural until 1994, when the PC industry was established by a few big PC companies from Taiwan creating booming conditions. It even won the title 'Computer Town' in the global market of PC products (Table 5.2).

Figure 5.2 reveals the local PC-related production chain in Dongguan. By the late 1990s a local production network had emerged supplying many of the components used in final assembly from within Dongguan. As Figure 5.2 shows, hard disks are not produced in Dongguan. However, in Shenzhen, Seagate and Great Wall are two plants that manufacture this product, and could be said to belong to local production capabilities. As already noted, technology and capital-intensive parts, such as integrated circuits and memories, are imported from the United States, Taiwan, Japan or Korea, as is processing equipment, some key components and raw materials.

The PC-related products in the three towns in our study are similar. Their dominating outputs are keyboards, mice, cases, and monitors that are mature commodity lines usually generating little profit. As demand increases, prices continue to fall. To quench market thirst, manufacturers not only have to increase volumes but also find lucrative niches. Now they recognize that flexibility is the key to larger profit margins. Confronting fierce competition, manufacturers are forced to cut costs without discounting the quality of their products.

The booming PC domestic market in Mainland China The growing PC domestic market in Mainland China is another force stimulating the PC industry in Dongguan. This trend will also affect the network pattern of the local PC production system. At present, the PC industry in Dongguan is still mainly export-oriented. Most informants declared that over 70-80 percent of their products were shipped to the global market. The Mainland market occupies only a small share in their output now. However, all of our interviewees are interested in the huge potential of the Chinese PC market. Indeed, China is already the fifth-largest global PC market, following the USA, Japan, Germany, and the UK. The shipments of PCs in Mainland China have increased at an average speed of over 40 percent per year from 1991 to 1997. Although this speed decreased to 13.6 percent in 1998 due to the Asia Financial Crisis, it recovered to 16.7 percent in 1999. It seems that the stable increasing trend will continue after 2000 (Wang, 2000).

Table 5.2 Top ten largest PC-related firms in Qingxi, Dongguan

Name	Investment	Products	Export in 1999 (million US$)
Zhixin (Dongguan) Ltd. Company	US$10.0 mil.	Monitors	25.59
Dongguan Zhili Computer Ltd. Company	HK$151.80 mil.	Computer Components (Cases, Motherboards)	35.64
Shengrong Electronic (Dongguan) Ltd. Company	HK$67.00 mil.	Powers	57.53
NPF Monitor (Dongguan) Ltd.Company	US$18.0 mil.	Monitors	81.06
Huangjia Electronic (Dongguan) Ltd. Company	US$101.18 mil.	Monitors	43.92
Qunguang Electronic (Dongguan) Ltd. Company	US$10.00 mil.	Keyboards	31.96
Dongguan Zhanyu Hardware Ltd. Company	HK$37.86 mil.	Keyboards	28.54
Dongguan Qingxi Guangping Electronic Plant	HK$13.01 mil.	Monitors	2.18*
Dongguan Qingxi Hanyang Computer Plant	HK$77.87 mil.	CD-ROM	1.81*
Dongguan Changjian Computer Ltd. Company	HK$14.60 mil.	Cases	14.35

Note: *Manufacturing fee paid by the export processing plant.

Source: The government of Qingxi Town (1999): The Condition and Stategy of the computer industry in Qingxi Town.

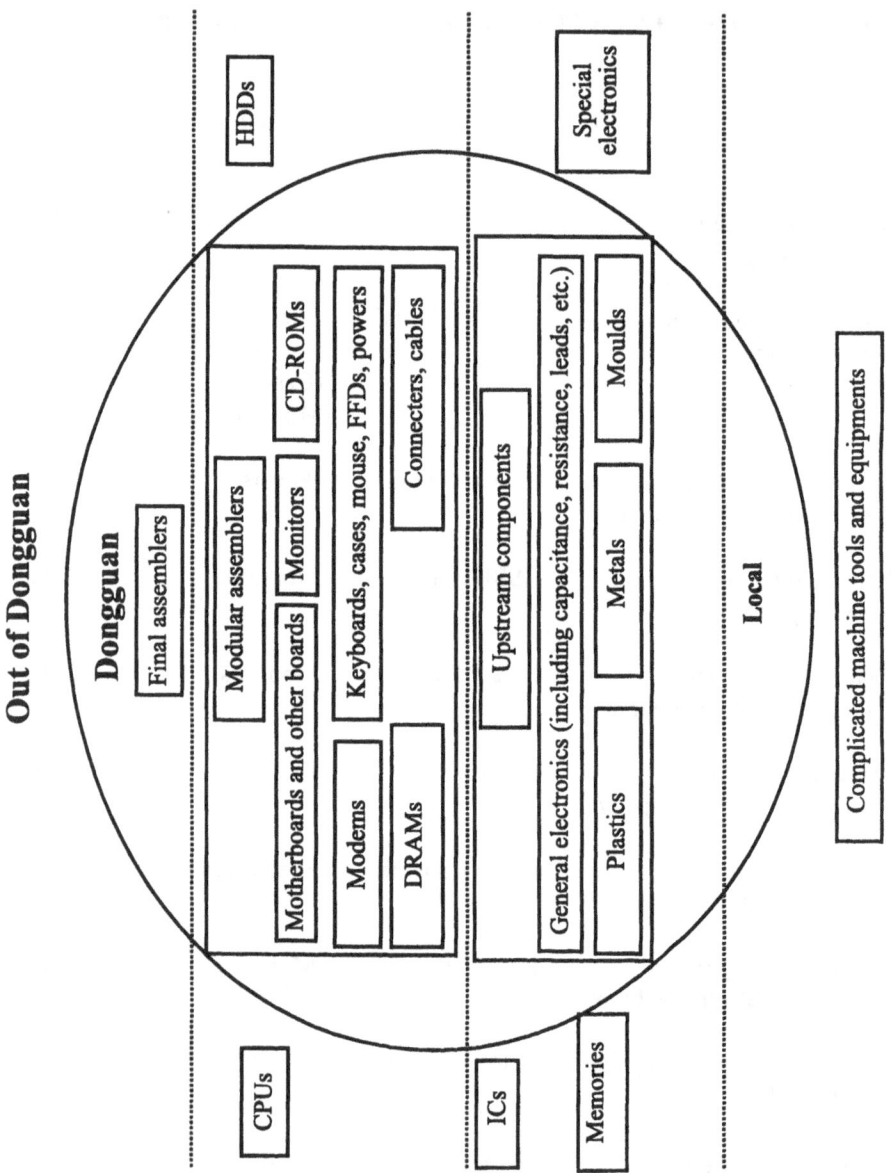

Figure 5.2 Local PC-related hardware processing activities and products in Dongguan

The PC markets in Shenzhen and Guangzhou are of great influence in Mainland China. In this regard, Dongguan, which is located between these two cities, has advantages in using its manufacturing agglomeration based on the export-oriented processing activities to meet interregional competition in the Chinese PC market. Relatively advanced manufacturing technology and management methods exist in Dongguan. Subcontracting networks, just-in-time production and inventory management are also well developed among all kinds of firms, both foreign controlled and domestic PC corporations. The intensive, intermediate production chains have also attracted PC assemblers who aimed at domestic market.

Knowledge Transfers in Dongguan's Networks

Although there are still tariff barriers and import quotas to protect the domestic PC market, the PC industry faces stiff challenges from global competition. From the perspective of local development, a vital concern is the nature of knowledge transfers within the production networks that have been established in Dongguan.

Learning Through Global Networks

Taiwanese PC firms are fierce global competitors, even for small local markets. They tapped into the opportunities that IBM created by opening up its PC structures and they engaged in the original equipment manufacturing (OEM) business for world leading PC companies. Through close linkages with global markets, Taiwanese PC companies kept up with the changes of technology in this industry and sharpened their competitive advantages in manufacturing processes. Taiwanese firms extended this kind of global linkage development to Dongguan.

The companies in Mainland China have confronted increasing competition from global giants in domestic markets following the dramatic lowering of import taxes and the removal of import quotas in 1993, although the list of products lowering taxes in 1993 did not include most components and raw materials of PC-related products. Many firms failed, but some accumulated capital by engaging in distribution for global brand PCs in China, while establishing self-owned brands and production lines at the same time.

This fierce competition has forced manufacturers to learn from the best practice firms in the market. In Dongguan, the industrial agglomeration has helped the imitation and transformation of local producers.

Interactions Between Local Players

In the Dongguan PC industry subcontracting is extensive. For example, in the assembly plant of Founder, the second PC producer in Mainland China, we found that the managers shared common knowledge with those producers from Taiwan in relation to subcontracting and it sought to further engage in cooperation with overseas-invested companies. This behaviour is quite different from the state-owned companies in the hinterland. Many local suppliers felt pressure to improve the quality and speed of reaction while cooperating with big buyers. In this regard, the quality management departments of buyers play an important role in the technological communication between suppliers and customers. This interaction could probably lead to further integration of the local production network.

The Chinese PC market is different from that in North America or Western Europe. Many foreign PC companies have found it necessary to cooperate with Chinese partners. As the PC market expanded in China, many domestic consumer electronic corporations, such as Changhong, Haier, TCL, Hisense, Pander, Chunlan, Peony, Konka, and so on, planned to enter the PC market in 1998. These corporations have large distribution and service channels throughout Mainland China. These channels are what the Taiwanese PC manufacturers lack, especially those centred in original equipment manufacturing (OEM) businesses. An example of the cooperation between these two kinds of companies occurred between TCL, a famous consumer electronics brands in Mainland China, and GVC, the third largest PC manufacturer in Taiwan that specializes in OEM businesses. GVC established many plants in Dongguan, especially in Qingxi town and Shijie town, which integrated almost all the modular processing in desktop PC except for hard disks and CPUs. GVC promotes self-owned brand components, such as modems and motherboards in Mainland China's domestic markets, while cooperating with TCL to produce TCL computers for family use. However, this cooperation ended in 1999 when the TCL group purchased 35 percent of the stock of TCL-Zhifu Computer Co. Ltd. from GVC, in order to exploit Chinese Information Technology (IT) markets by itself.

Finally, the interactions between local producers and local governments have played an important role in the development of a local PC manufacturing cluster. In particular, local governments have strongly supported overseas investors, which account for over 70 percent of local industrial output and make important contributions to local economic development. There is also considerable competition among different towns in Dongguan in inducing overseas investment. Because the PC industry contains important high-tech activities, it has received policy priority. Guarantees from local governments have helped overseas investors, especially small and medium investors, reduce the risk of investing in Dongguan. Occasionally, personal relations between entrepreneurs and local officials play important roles in the decision-making process of the investors. For example, many interviewees complained that the unstable import and export regulations of Mainland China created uncertainties additional to those associated with technology and marketing. Local governments often helped them access information on regulation adjustments quickly. Sometimes, local officials can help interpret the many regulations promulgated by the central government in ways that are helpful to local producers.

Export Processing and Institutional Bottlenecks

The local production network in Dongguan has emerged rapidly. This highly integrated market for intermediate products is closely integrated with developments throughout the Pearl River Delta, Hong Kong and Taiwan. The specialization and industrial restructuring of these regions continues. The local PC manufacturing network is also part of a global production network. Rigid export and import regulations designed solely to support export trade, however, hold back this trend.

Figure 5.3 demonstrates the different ways through which the local PC-related firms participate in material flow transactions. In Dongguan, it is difficult to distinguish the place of original manufacturing of a product. One product imported from Hong Kong by a plant might have been just manufactured by an adjacent plant the day before. Because export is the premise for preferential tax treatment, many manufacturers export their products to Hong Kong, and sell them back to customers in China. Apart from ethical questions, such movements waste time and render just-in-time production difficult as the total amount of the trade based on export processing activities increases quickly.

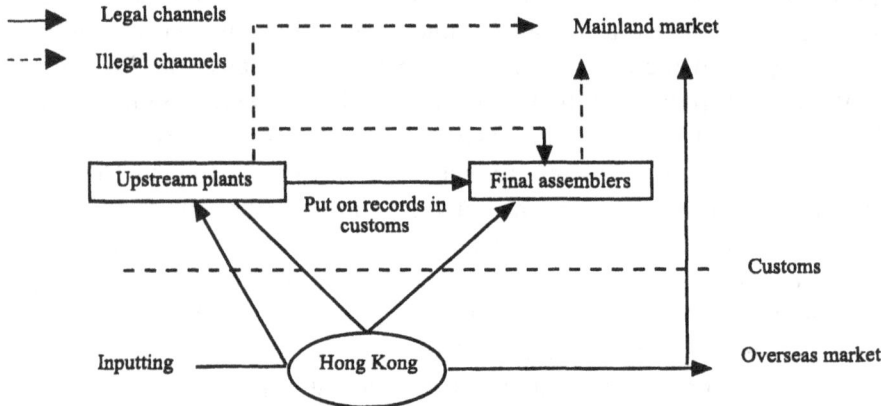

Figure 5.3 Dongguan: Interplant linkage and export institutions

Transfer regulations established in 1990 permitted suppliers to delivery duty-free products directly to the local customers. But the customs authorities must record the contracts, and the material in the final exports must match the recorded imported material. The process is quite complicated. If there is any mistake in the process, the offender faces the great risk of being treated as smuggler. The customs faced the dilemma between preventing smuggling and simplifying the process of products passing customs or contracts inspection. Meanwhile, producers found it difficult to respond rapidly to market changes without offending the regulations. The tariff barriers increase prices significantly on imported computers which in turn encourages smuggling by many firms. For example, a smuggled Notebook in Beijing is about $100 cheaper than a legal product. Many smuggled goods come from coastal provinces and businessmen are also willing to bribe custom officials. As an interviewee said, 'First, it just cost RMB100 or less for a cargo of goods if you want to avoid troubles, but the cost increased quickly as the flows of goods grew fast'. This institutional bottleneck affects the flexibility and fluency of local PC production.

The government has tried to adjust trade regulations, but its actions have often increased the uncertainty of the production environment. All of our interviewees are concerned about the instability of import and export regulations. For example, in 1996, the Chinese customs stipulated that all foreign investors must pay a 30 percent deposit on all materials purchased abroad in order to avoid smuggling. This deposit is not returned until the investors have exported the same amount that was imported. Many small companies found this policy imposed too much of a burden. Later, Chinese customs classified firms into four classes, but while class A firms were exempted from this deposit, over 80 percent of the firms are in class B or lower. In other words, this particular policy adjustment did not help.

The WTO and China

In late 2001, China gained membership in the World Trade Organization, following 13 years of efforts to do so. The entrance of China to the WTO, however, will create new uncertainties for the network patterns of the PC Industry in Dongguan. Although the duty and taxation regulations will supposedly become more predictable, in practice this may not be true. Indeed, PC makers in Donguan are concerned about whether exports will enjoy the same tax treatment as those imported, if sold on the domestic market, and whether the process of passing customs can be really simplified.

Moreover, as the entrance barriers of the domestic market decreases, many PC companies that intend to enter the market of Mainland China are planning to transfer investment to such cities as Suzhou or Shanghai, which are in close proximity to the majority of domestic consumers. The industrial background of these areas is also better than that found in the Pearl River Delta.

However, the global PC market is mature and the advantage of export-oriented manufacturing in Dongguan will last for years. Import tariffs have been decreasing for several years in China and entrance into the WTO will further open markets. As China's competitive advantage in the labour-intensive industries is so overwhelming at present, some researchers are concerned that entrance into the WTO might have negative effects on the technological upgrading of China's domestic industries. Based on this research, we argue that it is institutional bottlenecks that distort market relationships within China. As such, we argue that China's entrance into the WTO will be helpful for creating a more equitable and transparent local industrial environment in Dongguan.

Conclusion

This paper has traced the trajectory of PC manufacturing networking in Dongguan, China. This foreign-investment driven district with export-oriented manufacture experienced a fast growing industrialization process. Its experience is similar to other satellite industrial platforms, with the characteristics of inter-district mobility of labour, non-local linkages and non-local embeddedness (Park, 1995). However, there are reasons to be optimistic that the PC-based agglomeration in Dongguan has created opportunities for local linkages to develop. Networks of firms manufacturing PC products now exist, and the industrial district may become a 'sticky' one (Markusen, 1996).

As a border area, Dongguan has grasped the developing opportunity and has realized fast industrialization in the past decade. But in an increasingly 'borderless' world, such localities should adapt to changing circumstances by making themselves more competitive, fending off emerging competitors and identifying the economic opportunities offered by new global market relations. The technological interdependencies of the local production network have been criticized for being too weak. We feel, however, that problems within the PC industry in Dongguan are more related to institutional bottlenecks that are holding back the growth of spontaneous interaction among local firms. It is important and pressing for institutional reforms to create a business environment that is more equitable and transparent.

Acknowledgements

We are grateful for the financial support of the Chinese National Science Foundation (grant number ƒ®4977102ƒ©) and for the cartographic help of Li Tianhong.

References

Chen, Chieh-Hsuan (1994), *Subcontracting Networks and Social Life*, Lien-Jin Press, Taipei (in Chinese).

Curry, J. and Kenney, M. (1999), 'Beating the Clock: Corporate Responses to Rapid Change in the PC Industry', *California Management Review*, vol. 42, pp. 8-37.

Hamilton, G. (ed) (1991), *Business Networks and Economic Development in East and Southeast Asia*, Center of Asia Studies, Hong Kong.

Hayter, R. and Han S. (1997), 'Reflections on China's Open Policy Towards Foreign Direct Investment', *Regional Studies*, vol. 32, pp. 1-16.

Krugman P. (1991), *Geography and Trade*, MIT Press, Palatino.

Markusen, A. (1996), 'Sticky Places in Slippery Spaces: A Typology of Industrial Districts', *Economic Geography*, vol. 72, pp. 293-31.

Naughton, B. (1997), *The China Circle: Economics and Electronics in the PRC, Taiwan and Hong Kong*, Brooking Institution Press, Washington.

Park, S.O. and Markusen, A. (1995), 'Generalizing New Industrial Districts: A Theoretical Agenda and An Application from a non-Western Economy', *Environment and Planning A*, vol. 27, pp. 81-104.

Piore, M. and Sabel, C. (1984), *The Second Industrial Divide*, Basic Books, New York.

Qu, T. and Green, M.B. (1997), *Chinese Foreign Direct Investment: A Subnational Perspective on Location*, Ashgate, Aldershot.

Redding, S.G. (1991), 'Weak Organizations and Strong Linkages, Managerial Ideology and Chinese Family Business Networks', in G. Hamilton (ed), *Business Networks and Economic Development in East and Southeast Asia*, Center of Asia Studies, Hong Kong, pp. 30-48.

Research group on migrant workers in Guangdong province (1995), 'The Study on Migrant Workers in Guangdong Province', *Strategy and Management*, May, pp. 112-120 (in Chinese).

Statistical Yearbook of Dongguan (1998), *Chinese Statistical Press*, (in Chinese).

Statistical Yearbook of Dongguan (1999), *Chinese Statistical Press*, (in Chinese).

Tong, X. and Wang, J. (1999), 'An Analysis on the Global Production Network of the Information Technological Industry', *Science and Technology Review*, Sept, pp. 14-16, (in Chinese).

Wang, J. (1998), *The Trajectory of the PC-related Industry on the Mainland China: Enlightenment for SME Development*, Working paper.

Wang, L. (2000), 'Four Issue on Chinese Computer Market', *China Information Industry Policy and Decision-making*, April, pp.17-29, (in Chinese).

6 Development of the Internet in China and Its Spatial Characteristics

Weidong Liu

Introduction

The Internet is a major symbol of the contemporary world that is predominated by two interwoven tendencies, namely globalization and the rapid growth of information. It is now widely accepted that the world is moving fast towards a digital and information age and the Internet is a major channel in this historical transformation. Arguably, no technological progress in the last several decades can match the advent of the Internet in terms of its profound and extensive influences. Since the early 1990s, when most users were a few government sponsored or university sponsored researchers (Hamill, 1997), the spread of the Internet has been incredibly fast. From 1990 to 1998, the Internet hosts (domains) increased by more than 100 times and in 1999 there were over 290 million Internet users in the world. The increasingly wide use of the Internet is causing fundamental changes in the daily lives of people, and hence, probably in the entire socio-economic system of the world. In particular, the Internet makes the propaganda of information and knowledge much easier and faster than ever. The explosion of electronic business/ commerce (e-com) on the Internet implies vast, rapidly growing information and knowledge flows. In this context, the development of the Internet as an industry is not only the core of the so-called 'new economy' but also an accelerator of the development of the entire economic system.

According to research jointly conducted by the Centre for E-Com Studies of the University of Texas and Cisco Systems Inc., the output value of the Internet economy in the USA, which consists of the Internet infrastructure, Internet application facility, net agent and e-com sectors, amounted to US$523.9 billion in 1999, a 62 percent increase over the previous year (Sohu, 2000a). It indicates that the Internet economy is a major contributor to recent economic growth in the USA. In Asia, the recent economic recovery from the 1997-1998 financial crisis has also been attributed partially to the rapid growth of business-to-business (B2B) e-com (Rykken, 1999). Besides, the Internet/e-com is building a new industrial order, as new kinds of organizations are being spawned by the rapid reshaping of global business by information technology, a trend that Fortune names 'E-corporation' (Hamel and Sampler, 1998).

The implications of rapid Internet developments for industrial spatial changes are becoming important and interesting research topics for economic and industrial geographers. This chapter investigates the development of the Internet in China with special attention to its geographical features and then analyses the spatial implications of such development. It hopes to offer a general background, and a stepping-stone, for future more detailed studies on the issue.

An Internet Era: The Death of Geography?

A brief review of the scope and origin of the Internet and academic concerns about its widespread use is presented in this section to provide context for the case study of China. Facing the overwhelming uses of the words 'Internet' or 'e-something' in the media, few people would deny that an Internet era is coming. What the word 'Internet' means exactly, however, is ambiguous. Hamill (1997) gives a simple definition of the Internet as: "a network of interlinked computers throughout the world operating on a standard protocol which allows data to be transferred between otherwise incompatible machines". Recently, Hamill's definition has become too narrow as Internet technologies have developed very fast. At present, people can explore the Internet via their Wireless Access Protocol (WAP) mobile phones, and they are able to also do so by using their home televisions via set-top-box (STB). That is, computers are no longer the only machines/devices connected to the

Internet. Thus a key factor to understanding the scope of the Internet is the worldwide connectivity of machines/devices for the purpose of information and knowledge exchanges.

The prototype of the Internet is the Advanced Research Projects Agency net (ARPAnet) established by the USA defense Department in the 1970s. A major achievement in the project was the development of a standard communication protocol, known as TCP/IP (Transmission Control Protocol/ Internet Protocol). In the late 1980s, institutions in other countries started to register domain names to the network information centre under the ARPA net, which initiated the formation of the worldwide Internet. It was in the early 1990s, however, that the Internet was opened to groups other than academic institutions, including individuals and companies. In 1991, the Commercial Internet Exchange Association was established to encourage greater business participation in the development of the Internet (Hamill, 1997). Since then, the nature of the Internet has gradually shifted from the original free exchange of information to more and more commercial exploration.

The increasingly commercial use of the Internet is creating new possibilities for business, a trend that has profound implications for industrial spatial organization. Ellsworth and Ellsworth (1996) list a number of the main business uses of the Internet, namely communications (via e-mail), corporate logistics control, achieving globalization, achieving comparative advantage, cost savings, on-line support of inter-firm collaboration, information search, marketing and sales promotion, and transmission of data (see Hamill, 1997). Moreover, the use of the Internet has become an effective method of improving corporate images (Sterne, 1995). Indeed, few large companies have not been involved in the expansion of the Internet, and the Internet/ e-com will leave no industry untouched. Hutchinson's model shows that e-com is emerging in four distinct but not mutually exclusive areas (Hutchinson, 1997), including global geodesic commerce, members-only sub-nets, electronic middlemen and new consumer marketing channels. These four areas are briefly summarized.

In the 'geodesic commerce' model, the traditional middleman is removed and goods and services are purchased over the public Internet, directly from the source. This behaviour facilitates market dynamics, and markets become truly global.

The 'members-only sub-nets' model refers to B2B e-com occurring within private, industrial-strength sub-networks that guarantee high levels of service and security. A good example of this model is the automotive parts/components sourcing network jointly established by General Motors, Ford and Daimler-Chrysler. Here, the power of large corporations is extended by e-com systems.

In the 'electronic middlemen' model, suppliers make their products or services available through independent third-party electronic distribution channels on the Internet, while buyers purchase through these organized electronic 'stores'. Companies are able to reduce inventories substantially by just-in-time (JIT) and build-to-order (BTO) manufacturing. The middlemen have to maintain tight electronic links to both manufacturers and distributors that manage delivery logistics.

In the 'new consumer marketing channels' scenario, the traditional marketing media collapses into a unified consumer-centric e-com medium on the Internet. All Internet-connectable machines/ devices will drive affordable e-com access into homes.

No matter which model an industry is likely to take, there is little doubt that industries, particularly their organization, are being reshaped by e-com. Since the major procedures of e-com are finished on the Internet where information, data and knowledge cross space 'freely', the significance of geography has been questioned. Some observers argue that the Internet is causing the death or end of geography (e.g. Bates, 1996; Hamel and Sampler, 1998), as companies can easily break their geographic moorings and 'fly' over the digital globe by using the new information technology (IT).

Does geography really not matter to industrial developments in the Internet era? The answer seems to be positive – geography still works, if we agree that a digital visual world cannot substitute for a real material world. At least, a complete and successful e-com cannot be achieved without successful material flows and distribution, except for those industries dealing with soft knowledge-products, for example, software, digital books, etc. Logistics systems, however, have self evident geographic structures. As argued by Bradley (2000), the key role of logistics in e-com is often forgotten when people applaud the convenience of the new business pattern. Although the Internet is changing the ways business works, the physical fabric of production and distribution is still fixed; transportation systems and machines cannot run on the Internet. Actually, the business-to-consumer (B2C) e-com is raising new geographic issues, as 'fulfilling all those small orders that must be delivered to individual households is a far different proposition from delivering truckload or less-than-truckload shipments to stores' (Jedd, 2000). Besides, the electronic sales and delivery of hard goods are still subject to the 'old' international laws and tariffs. Thus, 'many businesses limit e-sales to a more confined geography', although e-com is labelled as space-free (C.H.B., 1998).

Moreover, the Internet/e-com is fostering new production patterns while it highlights the significance of JIT in manufacturing. The BTO operation is one of the potential new production patterns. According to Dr. David M. Anderson, a noted management consultant, the reality of spontaneous BTO is "a supply chain where the inflow of raw materials and parts matches the outflow of finished products" (Cottrill, 2000). That is, ordering, manufacturing and delivery are part of the same continuum. The operation of Dell Computers represents the BTO model to a certain degree. Such a new production pattern will consequently cause changes in industrial organization, and in the spatial relations of companies as well. Those vertically integrated companies may take advantages of their close internal division of labour, and spatial proximity between assemblers and suppliers may be a necessity of BTO operation, as in the JIT model. All these potential changes imply new factors in the location choices of companies. Therefore, the internet/ e-com adds complexity to geography, though it does make some 'soft' industries footloose. That is, the Internet is not obliterating Geography, but challenging geographers to rethink their notions of Geography.

Development of the Internet in China and its Spatial Features

China was a late comer in the development of the Internet. Before the 1990s, there were only several experimental e-mail communications between Chinese scholars and their collaborative foreign partners. In 1990, China formally registered a national domain name 'CN' to the network information centre under ARPA net. In 1993, the Institute of High-energy Physics of the Chinese Academy of Sciences (CAS) was allowed by the USA to connect to the Stanford Liner Accelerator Center. In the next year, the National Sciences Foundation (NSF) of the USA approved CAS's application for Internet connection, and CAS leased an international connection line with a bandwidth of 64K from Sprint and connected to NSFNET. Later that year, China Telecom signed an agreement of Internet connections with the USA Trade Department, initiating the development of the Chinese public network, CHINANET. In 1995, CHINANET and two other networks, CERNET and ASTNet that were established by the Ministry of Education and CAS respectively, were opened for public Internet connection services. In 1996, China Golden Bridge Network (CHINAGBN), a public network for economic information exchange, began its Internet services. By that year, therefore, China had four independent national networks that connected to the Internet and were able to offer Internet connection services.

Recent Internet Boom: Empirical Evidence[1]

In early 1997, the Chinese central government decided to build the Internet as one of the key national information infrastructures, which greatly encouraged and stimulated public participation, particularly business participation, in the Internet development. Since then, China has witnessed an Internet boom. In November 1997, there were only 620,000 Internet users, defined as those having or sharing an Internet connected computer or Internet account, while the figure jumped to 16.9 million by mid-2000 (Figure 6.1). In particular, from mid-1999 to mid-2000, Internet users in China increased two fold. Of the contemporary Internet users, nearly 70 percent are dial-up users, 15 percent use leased lines, and another 15 percent use both dial-up and leased lines. The group using other equipment (e.g. WAP mobile phone, STB) is still quite small, consisting of only 590,000 users. The Internet in China is still the world of the youth, as users aged 35 and below accounted for 88 percent of the total and the percentage changed very slowly in the last three years.

In the same period, the number of computers connected to the Internet rose from 299,000 to 6.5 million. Among the computer hosts in mid-2000, 1.0 million were connected to the Internet through leased lines and the other 5.5 million were through dial-up connections. The increase of domains shows a similar pattern (Figure 6.2). In mid-2000, the total domain names registered under 'CN' reached 99,743. Among them, 78,878 were under '.com' and 10,719 were under '.net'. The number of academic and educational domains was only 624 and 812 respectively, and that of government domains was 3,665. Such a structure indicates that most domains registered were for commercial purposes, following the same tendency as the world. The worldwide web-sites in China also developed very fast. In late 1997, there were only 1,500 web-sites, but the number increased to 27,289 by mid-2000. According to a study by China Labs.com, the web-sites with market value more than Ren Min Bi (RMB) 1.0 billion will grow to 40 in 2000 (*Beijing*

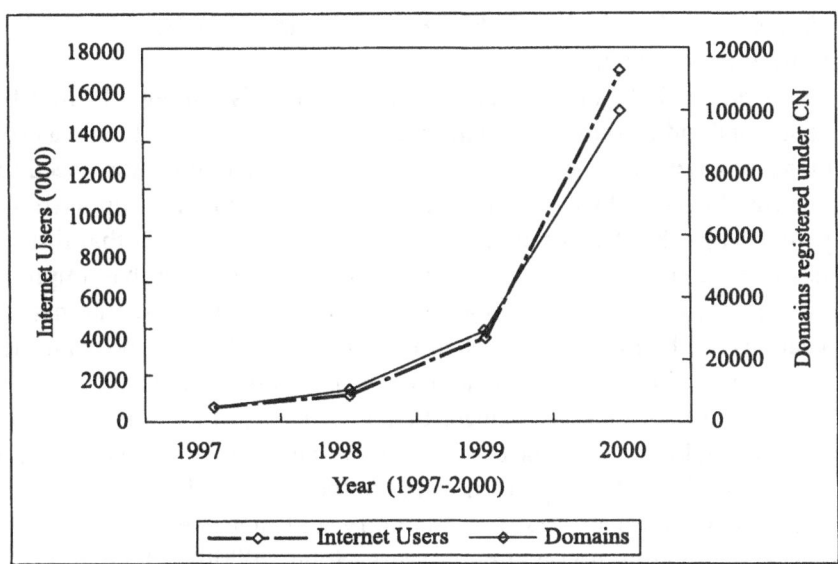

Figure 6.1 Number of Internet users and domains under 'CN' (1997-2000)

Source: CNNIC's Survey Report on Internet Development in China.

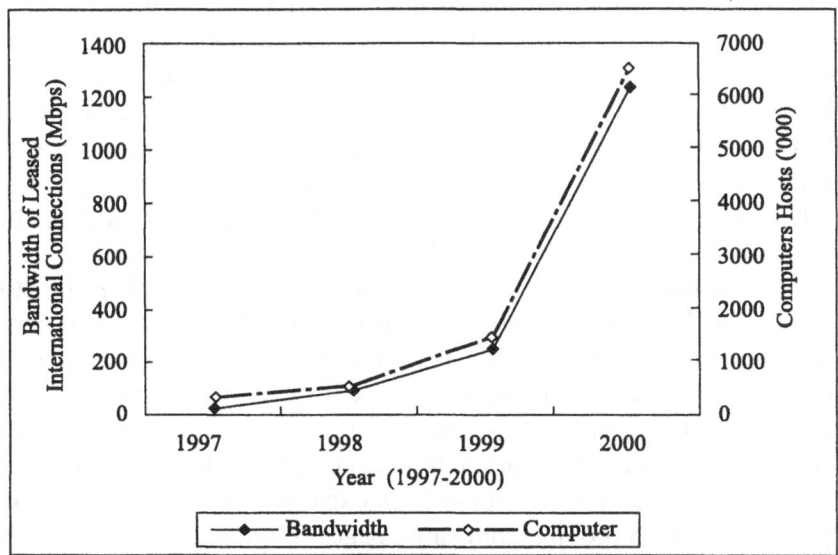

Figure 6.2 Number of computer hosts and bandwidth of leased international connections in China (1997-2000)

Source: CNNIC's Survey Report on Internet Development in China.

Youth Daily, 2000a), including SINA, SOHU, and CHINANET that have been listed in NASDQ.

Another indicator of Internet development is bandwidth. In 1997, the total bandwidth of international Internet connections of China was only 18.64 Mbps (megabytes per second). The then fastest connection speed that the Internet Services Provider (ISP) could offer to public dial-up users was only 14.4 Kbps (kilobytes per second). It was quite normal at that time for users to suffer from slow data down loading. The situation has improved quickly in the last three years. In mid-2000, the total bandwidth of international connections reached 1234 Mbps, 66 times of that in 1997. Now the connection bandwidth that ISP offers to dial-up users is normally 56 Kbps.

These data clearly indicate that Internet use in China has grown incredibly quickly over the last several years. The growth rate is among the fastest in Asia. The Ministry of Information Industry of China estimates that the number of Internet users in the country will increase to 60 million in three years, including non-computer users (Sohu, 2000b), which will make China the second largest Internet country in the world.

The rapid development of the Internet has been paralleled by the fast growth of e-com. In early 1999, there were less than 100 e-com companies (defined as Hutchinson's e-middlemen model) in China, but the figure rose to 600 by mid-2000. Like the early days of e-com in the USA, the current e-stores in China are mainly e-book stores. By the end of June 2000, there had been 300 e-book stores in China, an increase of 258 percent since the previous year. Of these e-book stores, 80 percent take the B2C model, four percent take the B2B model and 16 percent take *both (Beijing Youth Daily*, 2000b). In 1999, the total e-sales in China reached RMB 100 million (*Beijing Youth Daily*, 2000c). Although the figure is tiny, it represents an increase of 150 percent from 1998. By early 2000, the monthly sales of the biggest e-store, '8848', had reached RMB10 million. The logistics system of this e-tailer has covered more than 450 cities in China. Besides, more and more industrial companies are establishing their own e-com networks (defined as both Hutchinson's members-only subnets and geodesic commerce models). Like their foreign competitors, Chinese companies have learnt that the involvement in e-business is not only an expansion of business but also an important way of improving company images (*Beijing Youth Daily*, 2000d; Sohu, 2000c).

Spatial Features of the Internet Development

The Internet development in China in the last several years shows very clear features of spatial concentration in major cities and developed regions, including Beijing, Shanghai, Guangdong, Zhejiang, Jiangsu and Shandong. In mid-2001, domains registered under 'CN' were mainly concentrated in Beijing, Guangdong and Shanghai. These three provincial units account for 60.6 percent of the total domains under 'CN', and in particular, Beijing accounts for 38.2 percent of the total. This domain distribution structure has not changed markedly in the last year. In mid-1999, Beijing, Guangdong and Shanghai accounted for 36.7 percent, 14.9 percent and 7.7 percent, respectively, of the total domains under 'CN'.

The distribution of Internet users is a little more scattered than that of domains though they are also concentrated in the Beijing, Guangdong and Shanghai areas, respectively accounting for 18.7 percent, 12.8 percent and 10.8 percent of the total users. Other regions with a big Internet population include Shandong (8.5 percent of the total), Jiangsu (6.8 percent), Zhejiang (4.9 percent) and Hubei (4.5 percent). These seven regions/cities in total account for 67 percent of the total Internet users in China. There has been a prominent change in user spatial distribution in the last three years. In 1997 when public use of the Internet had just started to spread, users were mainly concentrated in Beijing. The city itself accounted for 36 percent of the total users in the country. Guangdong and Shanghai registered the fastest growth of Internet users in 1997-2000. Their share of the total users increased by 4.5 and 2.8 percent respectively. It could be expected that the spatial distribution of users would get more and more scattered along with the development of the Internet in China.

The geography of international connection bandwidth of the Internet in China is listed in Table 6.1. In mid-2000, six nation-wide networks in China have established international connections for Internet services. More than half of the bandwidth of international connections is located in Shanghai while Beijing is the centre of Internet development in China as shown in the above discussion. Guangzhou ranks the second with a total capacity of 355 Mbps. In 1997, the Internet in China was connected to the USA, Germany, France and Hong Kong only, but the connections have been extended to other countries like Canada, Australia, Britain, Japan and South Korea in 2000.

Table 6.1 Geography of international connection band width of the Internet in China

Networks	Beijing	Shanghai	Guangzhou	Shenzhen
CSTNet	10 Mbps	NA	NA	NA
CHINANET	170 Mbps	214 Mbps	327 Mbps	NA
CERNet	12 Mbps	NA	NA	NA
CHINAGBN	49 Mbps	NA	12 Mbps	8 Mbps
UNINET	NA	47 Mbps	8 Mbps	NA
CNCNet	4 Mbps	365 Mbps	8 Mbps	NA
Total	245 Mbps	626 Mbps	355 Mbps	8 Mbps

Source: CNNIC, 2000, Survey Report on Internet Development in China. NA: not applicable.

Another interesting spatial phenomena of Internet development in China is that local people tend to explore local web-sites. According to a survey on the use of the Internet in Beijing, Shanghai and Guangzhou, users in Beijing and Guangzhou seldom visited 'Shanghai-on-Line', which is a popular local web-site in Shanghai, while 75 percent of local Internet users in Shanghai were fond of it. This happened to the websites of Capital-on-Line (Beijing) and 163.net (Guangzhou) also, which reflects somewhat the regional features of local web-sites and perhaps the cultural differences between the three cities. Those web-sites recognized as not locality-specific (e.g. SINA and SOHU), however, got quite similar visit rate between regions.

In order to have a better understanding of the spatial development of the Internet in China, I have employed a multiple regression model to identify the factors behind the spatial distribution of Internet users (Table 6.2).

Table 6.2 Results of the multiple regression: model summary

Model	R	R Square	Adjusted R Square	Std. Error of the Estimate
1	.942[a]	.888	.884	1.3792
2	.951[b]	.904	.897	1.3007

[a] Predictors: (Constant), UNIVS. See Appendix for more details of the analysis

[b] Predictors: (Constant), UNIVS, CH

The percentage of Internet users in total population (USER) is the dependent variable. Through correlation analysis, GDP per capita, telecommunication infrastructure (TI), the percentage of university students in the total population (UNIVS) and the share of college degree or higher degree holders in total population (CH) are chosen as independent variables among a handful of factors. Both dependent and independent variables are standardized by dividing by the national average. The spatial unit is a province or province-level city, as data are available only at this level. The regression calculations are attached as an Appendix. It shows that UNIVS and CH, but not GDP and TI, are the major factors affecting the spatial development of the Internet in China. A final regression equation is as follows:

USER = (-2.2614) + 2.183*UNIVS + 1.307*CH (un-standardized coefficient)
USER = 0.655*UNIVS + 0.314*CH (standardized coefficient)

Thus, university students are clearly important in deciding the provincial distribution of the Internet population in China, while the influence of population quality in terms of degree levels is also significant. These results suggest that university students are the most active absorbers of new knowledge and new things, and they are also the propagators of new ideas.

The results further indicate that Internet development in China is still in an initial stage in which the Internet is new to the majority of the population.

Spatial Implications of Increasingly Wide Use of the Internet in China

To some extent, it is premature to summarize the spatial results of the increasingly wide use of the Internet in China, and perhaps in the world, since the profound transformation is ongoing. At this stage, I am able to only briefly introduce spatial tendencies.

An Industrial Application of the Internet in Wenzhou City

As mentioned previously, the explosion of e-com highlights the significance of JIT and promotes the formation of BTO manufacturing since the nature of e-com is speedier response and better service. Such a tendency may cause spatial concentration of industries from the side of production efficiency. However, the logistics system has driving forces from the other side, that is, transportation costs may drive more scattered production so as to make it easy and economic to fulfil delivery. In China, industrial companies also face these two driving forces in making spatial choices.

In Wenzhou City of Zhejiang province, which is one of the most dynamic regions in China, I interviewed a manufacturing company, the Renben Industrial Group, which produces bearings and now extends its business to B2B e-com. In collaboration with the former Ministry of Machinery Industry (the ministry has now become a bureau as a result of the administrative reform), the company is establishing a nation-wide logistics system of machinery and electrical components/products by employing the Internet technologies. The system aims at providing JIT delivery of components/ products to manufactures/retailers so that the latter will have 'zero' inventories. It consists of an e-trading sub-system, a delivery sub-system, and several big exhibition and sales centres in cities. To ensure safe delivery, the company will have to set up hundreds of small warehouses across the country. It could be reasonably expected that in the next stage, other companies would have to take into consideration the spatial structure of this logistics system when making decisions of new production sites. That is, the regional branch plant model may attract attention again. Other companies like Haier and Meide are working towards similar spatial networks of e-com.

A Cyber Manufacturing Society in Dongguan City

The Internet can also enable small and medium-sized enterprises (SMEs) to gain access to resources that were once available only to large firms, and thus offers SMEs new opportunities to develop. Lack of internal technological capability and marketing and information channels has been major obstacles to the survival and expansion of SMEs. The Internet technology, however, is taking away such obstacles by allowing SMEs to share on-line resources of technology, market, talents, and equipment, that is, to achieve cyber-integration on the Internet. The Internet application in Dongguan City, Guangdong province provides a good case for observation on the issue.

Dongguan is a fast growing city in the Pearl River Delta and a well-known export-oriented computer-manufacturing base in China and in the world as well (chapter 5). At present, electronics and IT industry account for 41 percent of the total industrial output value in the city. The intensive involvement of its economy in the IT industry lets the city have an early awareness of the significance of the Internet in future development. In mid-2000, subsidized by the municipal government, the China Manufacturing Cooperation Network (CMC) was established in Dongguan. The CMC network aims to provide an Internet-based cooperation environment for SMEs to seek dynamic alliances, technological cooperation/assistance and commercial opportunities and cooperation and to get help in corporate resources management.

A Survey of the Growing Internet Generation in Beijing

The spatial impact of the use of the Internet is of course not limited to industrial organizations. It may reach every aspect of the socio-economic system. In order to have some understanding of the spatial tendency of the growing Internet generation, I conducted a questionnaire survey of high school students in Beijing. I chose to do such a survey because these young students will be an energetic new force in both business and consumption in several years. Five hundred questionnaires on the use of, and attitude towards, the Internet were delivered to Grade 10 students from ten randomly selected high schools in the Haidian District of Beijing. Four hundred and ninety-two questionnaires were returned and four hundred and eighty-seven of them are valid. According to this survey, 60 percent of high school students (that is, 293 respondents) in the district have used the Internet. They use the Internet mainly to communicate (via e-mail), chat, read news, look for reading materials, and some of them (52 respondents) attend cyber-schools.

Several statistical results from this investigation are worthy of note:

- 117 respondents now go to the post office less frequently than they did before using e-mail, and 159 respondents tend to consider that e-mail will eventually substitute letters;
- 156 respondents approve of cyber-schools though two-thirds of them also consider that face-to-face communication is important;
- 246 respondents consider that the use of the Internet enhances their knowledge of other regions, and 227 respondents think that the Internet is more attractive than textbooks in learning geographic knowledge;
- 232 respondents approve of working at home via the Internet in the future;
- 155 respondents think that they would feel inconvenienced without the Internet;
- 243 respondents consider that they are more likely to visit a place of interest after having seen a very detailed visual introduction of it on the Internet;
- 175 respondents consider the Internet is more attractive than TV;
- 194 respondents approve of e-com and 185 respondents consider that e-com will become their major purchasing method in 5-10 years, even though only 54 respondents have had purchasing experiences via the Internet; and
- 221 respondents will encourage others to use the Internet.

These brief data indicate the behavioural tendencies among the growing Internet generation, which will have implications for businesses and planning in China. For present purposes, several points are worthy of attention. First, the new generation is getting into the habit of using the Internet, as more than half of the students who have the experience of using the Internet feel inconvenience without access to it. Second, most of these young Internet users tend to accept new education, working, communication and marketing patterns based on the Internet. Third, B2C e-com may soon be a strong competitor of traditional commercial patterns, as two-thirds of the young Internet users approve of the new commerce pattern. Fourth, the Internet is becoming one of the major channels of learning and receiving information. How to meet these potential changes in spatial behaviour is a critical factor for successful business and planning in the near future.

Conclusion: New Challenges for Geography

The growth of the Internet has been greatly accelerated in the world since public and business participation started to increase in the early 1990s. In China, the Internet tide came quite late. It is since 1997 that the public and businesses have been widely involved in the development of the Internet. The recent growth, however, has been incredibly fast. In three years, the number of Internet users in China has increased to nearly 17 million from several hundreds of thousand. The huge Internet population means numerous commercial opportunities, which has resulted in a rapid growth of e-com in China and in the world. The four e-com scenarios of Hutchinson have occurred in China, and Chinese companies have also realized that the involvement in Internet development is a way of improving corporate images. The development of the Internet is not even, however. Major Internet facilities and users are concentrated in Beijing, Shanghai and Guangdong. The empirical study finds that the university students and population quality are the major factors behind the contemporary spatial distribution of the Internet users, indicating that the Internet is still fresh to the majority of population in China.

The spread of the Internet has profound spatial implications, but systematic data are still not available for a detailed study in China from this perspective. In the business field, the Internet has been integrated into manufacturing, marketing and logistics. Thus, theoretically, at least two forces produced by the application of Internet technologies will affect the spatial organization of industries. They are the need for speedier responses to customer preferences in the manufacturing side and the requirement of better and quicker fulfilment of delivery from the logistics side. The interactions between these forces and their dynamics are far from clear, and lack empirical studies. Besides, the questionnaire survey of the growing Internet generation raises a handful of issues that are closely related to potential change in spatial behaviour of people. Therefore, the fast development of the Internet is raising academic questions to geographers and challenging the significance of geography.

From the viewpoint of industrial organization, three issues are critical for the geographic understanding of the Internet era. These issues are the geographic laws governing the development of the Internet economy, the changes in location factors of the 'traditional' industries transformed by the Internet technologies, and the spatial patterns and organizations of the logistics system that can adapt to the Internet era. In short, to avoid being obliterated by the Internet, Geography as a subject must catch up with the rapid pace of the Internet and reveal the geographic dynamics behind the new developments.

Acknowledgements

The author acknowledges the financial support of the Chinese Academy of Sciences (Research Grant No. KZCX2-307) for the research conducted. He also thanks colleagues attending the 2000 Annual Residential Conference of the IGU Commission on the Organization of Industrial Space for their comments and suggestions on an early version of this paper.

Notes

[1] Statistical data in this section come mainly from the Survey Reports of Internet Development in China conducted by CNNIC (http://www.cnnic.net.cn). There was no official statistics of Internet development in China until late 1997 when China Internet Network Information Center (CNNIC) carried out the first survey on the issue. From then on, CNNIC conducted the same kind of survey every half-year, and the latest one (the seventh) was finished in July 2000.

References

Bates, S. (1996), 'The Death of Geography, the Rise of Anonymity, and the Internet', *American Enterprises*, vol. 7, 2, pp. 50-52.

Beijing Youth Daily (2000a), 23, 3, p. 19

Beijing Youth Daily (2000b), 26, 7, p. 42

Beijing Youth Daily (2000c), 3, 4, p. 15

Beijing Youth Daily (2000d), 19, 5, p. 55

Bradley, P. (2000), 'Logistics and E-commerce', *Logistics Management and Distribution Report*, vol. 39, 2, downloaded from the digital library of the University of Hong Kong.

C.H.B. (1998), 'Making Peace with Global E-commerce', *Network Computing*, vol. 9, 23, p. 50, downloaded from the digital library of the University of Hong Kong.

CNNIC (China Internet Network Information Center), (1997), *Survey Report of Internet Development in China*, vol. 11, http://www.cnnic.net.cn/develst/.

CNNIC (1998), *Survey Report of Internet Development in China*, vols. 1 and 7, http://www.cnnic.net.cn/develst/.

CNNIC, (1999), *Survey Report of Internet Development in China*, vols. 1 and 7, http://www.cnnic.net.cn/develst/.

CNNIC (2000), *Survey Report of Internet Development in China*, vols.1 and 7, http://www.cnnic.net.cn.

Cottrill, K. (2000), 'Speedier Supply Chain', *Traffic World*, vol. 26, pp.16-17.

Dyson, E. (1997), 'A Map of the Network Society', *New Perspectives Quarterly*, vol.14, 2, pp. 25-27.

Ellsworth, J.H. and Ellsworth, M.V. (1996*), Marketing on the Internet: Multimedia Strategies for the WWW*, John Wiley, New York.

Hamel, G. and Sampler, J. (1998), 'The E-Corporation', *Fortune*, vol. 138, 11, pp. 80-87.

Hamill, J. (1997), 'The Internet and International Marketing', *International Marketing Review*, vol. 14, 5, downloaded from the digital library of the University of Hong Kong.

Hutchinson, A. (1997), 'E-commerce: Building a model', *Communications Week*, Issue 654, 3/17, pp. 57-58.

Jedd, M. (2000), 'Sizing up Home Delivery', *Logistics Management and Distribution Report*, vol. 9, 2, downloaded from the digital library of the University of Hong Kong.

Preston, H. H. (2000), 'E-Commerce Explosion', *Computer-World*, vol. 34, 10, pp. 58-59.

Rykken, R. (1999), 'The World's Most Web Centric Cities', *Export Today*, vol. 15, 8, pp. 32-37.

Sims, M. (2000), 'Tomorrow's Warehouse Today', *Global Cosmetic Industry*, vol. 166, 2, pp. 38-42.

Sohu (2000a), http://news.sohu.com (08/06).

Sohu (2000b), http://news.sohu.com (23/2).

Sohu (2000c), http://news.sohu.com (10/03).

Sterne, J. (1995), *World Wide Web Marketing: Integrating the Internet into Your Marketing Strategy*, John Wiley, New York.

Appendix 6.1 Statistical basis of model summarized in Table 6.2

ANNOVA[c]

	Model	Sum of Squares	df	Mean Square	F	Sig.
1	Regression	436.325	1	436.325	229.395	.000[a]
	Residual	55.160	29	1.902		
	Total	491.485	30			
2	Regression	444.113	2	222.056	131.250	.000[b]
	Residual	47.372	28	1.692		
	Total	491.485	30			

Coefficients

Model		Unstandardized Coefficients		Standardized Coefficients		
		B	Std. Error	Beta	t	Sig.
1	(Constant)	-2.286	.360		-6.355	.000
	UNIVS	3.141	.207	.942	15.146	.000
2	(Constant)	-2.614	.372		-7.024	.000
	UNIVS	2.183	.488	.655	4.474	.000
	CH	1.307	.609	.314	2.145	.041

Excluded Variables

Model		Beta In	t	Sig.	Partial Correlation	Collinearity Statistics Tolerance
1	GDP	-.089[a]	-.832	.412	-.155	.343
	CH	.314[a]	2.145	.041	.376	.161
	TI	-.010[a]	-.078	.938	-.015	.226
2	GDP	-.090[b]	-.900	.376	-.171	.343
	TI	-.057[b]	-.447	.659	-.086	.219
2	GDP	-.090[b]	-.900	.376	-.171	.343
	TI	-.057[b]	-.447	.659	-.086	.219

[a] Predictors: (Constant), UNIVS
[b] Predictors: (Constant), UNIVS, CH.
[c] Dependent Variable: Users

7 Characteristics and Development of Industrial Districts: The Case of Software Clusters in Seoul, South Korea

Joo-Seong Hwang

Introduction

Since the 1980s, discussions among economic geographers have generated a range of theories and concepts about the origins and dynamics of 'New Industrial Districts (NIDs)'. Despite this theoretical debate, however, efforts to reproduce industrial districts of the same calibre as Silicon Valley in California in other nations and regions have had little success. The work of economic geographers does not seem to have met the needs of policy makers who have tried to imitate the success of Silicon Valley.

This chapter argues that part of the failure of theory lies in the paucity of studies that synthesize related theories. Orelemans et al (1999) contend that the work of theoreticians in Economic Geography and related fields is not sufficiently cumulative and often subject to fads. If studies of industrial districts are to have social value, they need to provide implications for policies that seek to stimulate industrial districts.

In South Korea, much attention was paid to 'information technology' (IT) industrial districts' around 1996. Many local governments made plans to build industrial complexes to foster their local IT industry. The central government remained indifferent to these local government initiatives. Instead, the federal government announced its own plans to develop the 'Kangnam-

Seocho area', an existing area of software activity, as a substantial software industrial district. Questions arise as to the nature of these industrial districts and their similarities and differences.

This chapter has two specific purposes. First, drawing from the work of Capello (1999) and other studies, it outlines a conceptual model of the evolution of industrial districts based on five 'core' dimensions, namely those of localization, networking, embeddedness, collective learning and innovative synergy (see also chapter 4 by Eyradin). Second, this theoretical framework is explored empirically through a comparison of the 'Kangnam-Seocho area' with other clusters in Seoul, namely 'Jung-Chongro', 'Mapo-Yongsan', and the 'Youngdeungpo' areas. The evidence indicates that the 'Kangnam-Seocho area' is different from the others.

The first part of the chapter identifies five core dimensions of industrial districts and specifies variables for empirical analyses that draw from questionnaire and interview surveys, conducted in 1999, of intra-/inter-firm and regional level features. A total of 869 companies located in the four software industrial clusters (Kangnam-Seocho, Jung-Chongro, Mapo-Yongsan, Youngdeungpo) were approached. Twenty-seven percent (238 companies) responded. The companies were chosen from the roster of Korea Software Industry Association (KOSA) and surveyed in November and December, 1999. In terms of the analysis, the characteristics of 'Kangnam-Seocho area' as an industrial district are first assessed with reference to the five core dimensions and this assessment is used as a basis for comparing the other industrial districts.

The Characteristics of Industrial Districts: A Conceptual Framework

Theories about 'New Industrial Districts (NIDs)' can be categorized into five groups: new industrial spaces, district theory, cluster theory, milieu innovateur, regional innovation theory (Oerlemans et al 1999; Lagendijk, 1997). They differ with each other in the mechanism of agglomeration, the role of firms, and evolutionary tendencies. Each theory has its own strength and weakness. However, not much effort has been made to synthesize these theories, possibly because each theory has focused on borrowing new concepts from other academic fields instead of complementing existing ones.

Indeed, Lagendijk (1997) concludes that "the literature seems to be far removed from developing a truly evolutionary model of spatial-industrial development which can provide a more systematic basis for the study of spatial agglomeration and clustering."

In this chapter, the five theories are considered as parts of an interconnected and complementary system. While each of them has implicit relations and context in common, any one theory alone cannot explain complicated phenomenon. I argue that the above-mentioned five theories and their derived concepts have been developed along a consistent trajectory to provide an explanation of 'the basis for industrial districts' (Figure 7.1).

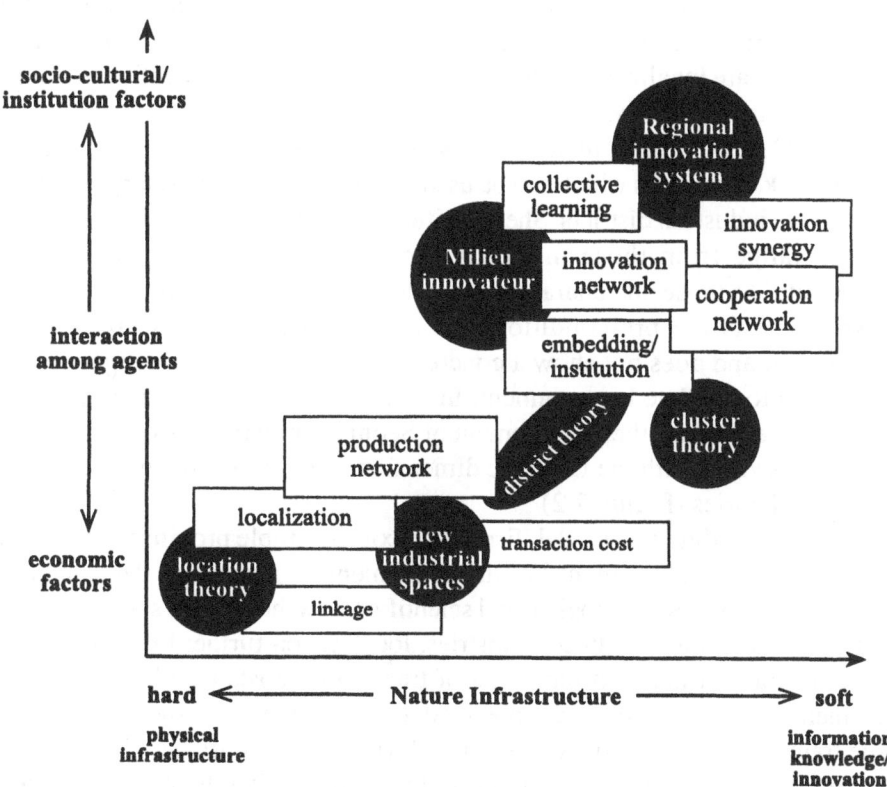

Figure 7.1 Basis for industrial districts

'The basis of industrial districts' is defined by two axes or dimensions: the nature of infrastructure and the nature of interaction among agents. In early stages of development, physical infrastructure, such as buildings and roads, is of importance. Later, intangible infrastructure including information, knowledge, and innovation, becomes more important, and these features have been stressed in recent studies. 'The interaction among agents' is the dimension of factors that motivate inter-firm relations. Interactions have been considered to be mainly motivated by economic factors such as reduction of transaction costs and operational costs. Recent studies, however, put the focus on non-economic factors such as trust, social proximity, and roles played by governments, universities and non-governmental organizations (NGOs). Figure 7.1 shows the relative position of existing NID theories and their related concepts on these two axes.

The synthesis presented in this figure draws five core dimensions for the formation and development of industrial districts. These five dimensions are localization, networking, embeddedness, collective learning, and innovative synergy.

Capello (1999), in her recent study on collective learning, suggests a framework of analysis which can be used to explain the evolutionary processes that create industrial districts. She identifies different types of spatial clustering which range from those in which firms exist in 'simple geographical proximity' to those she characterizes in terms of a 'milieu innovateur'. She also suggests the pre-conditions for each stage. But her framework is theoretical and does not show a concrete methodology of investigation that is needed to validate her argument. In order to complement her framework, this paper examines the development of Seoul's software clusters in terms of a model revised with the five core dimensions derived from the synthesis of existing theories (Figure 7.2).

According this model, firms that exist in simple proximity can create a 'specialized area' when the localization dimension guarantees the continuity over time of local technological and scientific know-how in the specific sector. Given a clustering of related industries, localization further implies a stable local labour market, and reduction of transaction costs (see Figure 7.1). If elements of networking and embeddedness are added to a specialized area, the framework for an industrial district develops, or is enriched. Networking means that customer-supplier relationships are sustainable while increasing the prospects of local labour market stability. Embeddedness gives a different meaning to inter-firm relationships. In particular, the quality of the relationships among firms improves through greater levels of trust and social interaction, and by use of support systems such as universities, chambers of commerce, associations and local governments. These institutional bodies set in motion informal and tacit transfers of information, know-how and other untraded assets among local firms.

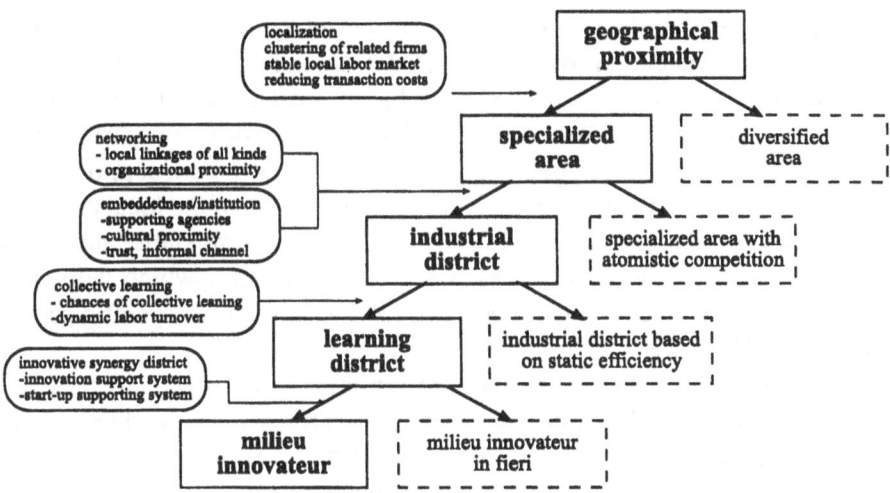

Figure 7.2   Five core dimensions of industrial districts in the evolutionary process

Source: Modified from Capello (1999, p. 358).

When cooperation and tacit transfer of knowledge are transformed into collective learning capacity rather than simply social solidarity and interaction, an industrial district becomes a milieu. The milieu is characterized by a collective learning dimension by which local firms feed each other with their knowledge in constant, mutually supportive exchanges. For this reason, the milieu can also be called a learning district. The interest and capacity of local actors to grasp collective learning may explain the difference between a learning district and a milieu innovateur. Innovative synergy is a capacity that turns socialized knowledge by collective learning into actual profits. Incubator and venture capital are important aspects of this dimension.

Attention is now turned to the operational definition of the core dimensions of industrial districts, namely localization, networking, embeddedness, collective learning and innovative synergy. After a brief definition of each dimension suitable indicators are outlined.

Localization

Localization refers to 'external economies caused by spatial agglomeration of the same or related kinds of firms'. This definition implies common access to production factors and decrease of production costs. The former includes specialized functions, skilled labour, and specialized machinery, the latter the decrease of transaction costs caused by easy access to both suppliers and customers. Localization is also one of the pre-requisites of the local division of functions and labour. Moreover, localization often forms stable inter-firm linkages and local labour markets, and it can be a factor which helps the region accumulate technological edge and know-how in the specific sector (Capello, 1999, 357).

In order to measure the localization level of software clusters, the rate of product specialization and the ratio of professional workers in the total workforce provide useful indexes at the firm level. These variables are derived under the following assumptions; first, the more specialized the firms of a cluster, the higher the localization level of that cluster, and second, the firms in a cluster that have a more technical workforce will exhibit higher degrees of localization. At the inter-firm level, the extent of externalization of production process is measured under the assumption that the more externalized firms are, the higher is the localization level. At the regional level, the relative location advantage of a cluster is evaluated in terms of three variables: the information and access to high quality labour force, the relationships with local firms in the software industry, and the cooperation with related local IT firms.

Networking

Networking refers to sustaining cooperative relations among firms that are initiated by the inter-firm division of labour. It embraces not only backward and forward linkages but also all kinds of relationships with other firms. Networking is not only defined by material (production) linkages but also by linkages with companies that provide producer services like legal, financial, and consulting and accounting services.

In order to measure networking within the software clusters, the quantity of production linkages is investigated at the intra-firm level. At the inter-firm level, supplier linkages, customer linkages and service linkages are studied. Finally, at the regional level, the relative locational advantage of an area is evaluated in terms of three variables: the stability of existing supplier relationships, the stability of existing customer relationships, and the ease of access to specialized services such as legal, financial, and consulting services.

Embeddedness

While embeddedness and institutional thickness have different meanings and origins, both of these concepts seem to depict a highly correlated phenomenon from different points of view. Oinas and Malecki (1999, p. 16) regard the notion of institutional thickness as referring to specific characteristics of the social relations (in total) in which actors are embedded, while embeddedness refers to the nature of the relation to the total set of relations from individual actors' points of view.

Embeddedness means that inter-firm relationships are adhered to social networks. In this context, inter-firm relations can build 'trust' through this social network. Trust is assumed to contribute to economic development by changing frequency and intensity of inter-firm knowledge transfers. Institutional thickness is a comprehensive characteristic of social relations, and is defined in terms of social custom, government structure, common perspective and operation of various related organizations: corporations, chambers of commerce, educational institutions, trade unions, local governments, and development agencies.

Three levels of data are used to understand embeddedness and institutional thickness. At the intra-firm level, the use of informal recruiting channels is surveyed. The degree and effects of cooperation between a company and other agencies (e.g. associations, societies, local governments, universities and research institutes) are considered at the inter-firm level. Finally, a region's relative advantages are surveyed in three ways; informal interaction and cooperation among firms in the same sector, inter-firm trust formation by associations or societies, and institutional environment such as incubator and venture capital presence.

Collective Learning

Collective learning is defined as the presence of common knowledge that goes beyond the boundaries of the firm. This common knowledge remains within the spatial boundaries of the learning district, giving rise to a process of cumulative local know-how (Capello, 1999, p. 354). If an industrial district is an innovation promoter, the results of the innovation process rapidly diffuse among all firms within the region, as no one in the district is denied access. Since knowledge is assembled and accumulated in a socialized way through inter-firm linkages and labour force mobility, this process tends to be facilitated by common cultural and organizational rules and codes.

At the individual firm level, labour force turnover rates and internal systems of learning and training are analyzed to ascertain spatial differences in collective learning. At the inter-firm level, external learning opportunities, and interactive learning with suppliers, customers and rival firms are probed. External learning outside a firm includes such learning opportunities as seminars, conferences, exhibitions and learning from education institutions. At the regional level, the location merits of areas are evaluated through accessibility to exhibitions and conferences, external learning from universities and institutions and interactive learning with suppliers, customers, and rival firms.

Innovative Synergy

Innovative synergy refers to the overall condition and environment for transforming creative ideas into real profits in highly innovative districts. Most of all, the internal ability of an individual firm, which can transfer its learning to profit-making activities, is important in this context. Besides essential cultural and institutional elements, such as 'venture spirit', a willingness to take risks and uncertainty is important. Open access to socialized knowledge plays a role in innovative synergy.

At the firm level, measures of innovative synergy are types of start-up, achievements of innovation and internal innovation support systems. At the inter-firm level, relationships by spin-off companies with their host company is an important consideration, addressed by the survey. Information was also obtained in terms of access to new business opportunities, stimulus to technological innovation, and stimulus to management innovation.

The Characteristics of Kangnam-Seocho Area as an Industrial District

The four software clusters surveyed account for 80 percent of the total software companies of Seoul and 56 percent for the South Korea. Kangnam-Seocho dominates, however, with half, of Seoul's and 35 percent of the nation's software firms. The survey responses were largely confined to the Kangnam-Seocho and Youngdeungpo areas of Seoul. The low response rate from the Jung-Chongro and Mapo-Yongson clusters is dealt with by grouping their responses into a category called 'Other'.

Firm and Inter-firm Characteristics

In terms of product specialization, the average ratio of software revenue against all revenue in the companies is approximately 30 percent, with little variation between areas, suggesting a low degree of software specialization at the firm level. The technical workforce was on average about 70 percent of the total workforce, with again only minor inter-area variation. However, Kangnam-Seocho firms outsourced to a noticeably higher extent (54 percent) compared to the other areas (less than 30 percent).

The networking data further confirmed the distinctive aspects of Kaugnam-Seocho area's development. While inter-firm links for production were low (per project in 1998, each of the area's firms made about 10 contracts with suppliers), financial service linkages were high uniformally and very local (as much as 90 percent locally). Kangnam-Seocho featured much greater levels of legal and consulting and accounting services. The local market was equally important in each of the four areas, accounting for approximately 50 percent of sales.

In terms of the embeddedness (and institutional thickness) dimension the evidence is similar for the survey areas. The use of informal recruiting channels, which rely for example on private human networks, for labour is a key indicator for measuring local embeddedness. Informal recruiting channels are asserted to be based on trust and may be a form of local convention (Storper, 1997). The convention could be interpreted as a common framework of action with other actors engaged in that activity. According to the survey, half of the firms recruit new employees through formal channels and about 35 percent through informal channels, but there is no significant difference among the areas. The software industry association is one of the key industry organizations. However, 85 percent of firms are registered to more than one related industry association. The helpfulness of business associations differed between areas (Kangnam-Seocho area 61 percent, Youngdeungpo area 40 percent, Other 44 percent), reflecting the fact that most of the associations in the software industry are located in Kangnam-Seocho area. Despite local government playing a visible role in promotion of industry, few connections were found. Universities and research institutes contribute to the innovation process of an industrial district by several means. They are the focal agencies of knowledge transfer, suppliers of technical labour and are often involved in joint R&D with local companies. In the survey, 46 percent of software companies in Kangnam-Seocho have a continuous relationship with universities or research institutes, while this ratio slightly decreases to 32 percent in the Youngdeungpo and 39 percent in the Other grouping.

We can assume that if a higher rate of labour turnover exists in the professional work field, the greater the chance of collective learning. Tacit knowledge is especially important in a field which changes rapidly like the software industry, and this knowledge can be best exchanged by transfer of workforce. To measure the mobility of the workforce, the turnover rate of each firm during 1998 is calculated. Overall, the turnover rate is 16 percent, with 17 percent in Kangnam-Seocho, 13 percent in Youngdeungpo, and 16 percent in the Other area.

Regarding internal learning, dependence on 'on-the-job training' is far higher than 'in-house training programmes'. Firms in the Other area show a relatively higher tendency to resort to 'on-the-job training', but these differences are not statistically significant. In the meantime, software companies tend to prefer such external learning opportunities as workshops and exhibitions as opposed to regular training programmes. This seems to reflect the rapid technological changes in this sector, which is mainly led by the firms rather than universities and research institutes. However, there is no significant difference among the areas in this type of learning. In the interactive learning with related companies, Kangnam-Seocho shows the highest rate (72 percent) ahead of that in the Other area (58 percent).

Interactive learning is classified into three categories: learning with customers, learning with suppliers, and learning with rival companies. According to the analysis, more than 70 percent of firms in Kangnam-Seocho area and Youngdeungpo consider learning with customers to be important while Other firms show a far lower rate (43 percent). This pattern is echoed in the importance of learning with suppliers. These results suggest that software companies in both Kangnam-Seocho and Youngdeungpo consider inter-firm learning to be important; they especially stress interactions with customers and suppliers.

The findings relating to innovative synergy point to low levels of development. Spin offs amount to just 10 percent of all start-ups, with minimal difference between the areas. Nearly two-thirds of all companies had a patent or software copyright. Innovation support was slightly higher in Youngdeungpo, because of the presence of foreign companies.

Regional Perspective

For the analysis of characteristics at the regional level, the location merits of each area for software companies were evaluated. Fifteen questions regarding the location advantages were asked.

According to the analysis, we found that firms in Kangnam-Seocho area generally consider their location to be more advantageous compared to firms in other areas. Kangnam-Seocho when viewed across the range of indicators, appears to be perceived as more advantageous in terms of the collective learning, embeddedness and institution thickness dimension. Meanwhile, firms in Youngdeungpo consider their area more efficient in 'information and access to high quality labour force', 'stability of existing customer relations', and 'stimulus to management innovation' (Table 7.1).

A more detailed comparison was undertaken by standardizing the data. A simple illustration of such a step is possible by calculating the relative values for the Kangnam-Seocho and Youngdeungpo areas supposing that the original data of the Other area is standardized as one. (Specifically with respect to the last three columns of Table 7.1, divide each of the percentages in the first two by the percentages in the last column). According to this analysis, Kangnam-Seocho is highly differentiated from the Other area. In this regard, the following factors are important: 'institutional environment such as incubator, venture capital', 'accessibility to exhibitions and conferences', 'interactive learning with suppliers, customers, and rival firms', 'informal interaction and cooperation among firms in the software industry', and 'cooperation with related local IT firms'.

Table 7.1 Perceived locational advantages by areas and dimensions of industrial districts

Industrial District Dimension	Locational Advantage	% of Firms Considering Regions Advantageous		
		K	Y	O
Localization	• information and access to high quality labour force	52.3	60.5	34.3
	• relationships with local firms in the software industry	57.5	55.3	31.4
	• cooperation with related local IT firms	54.6	31.6	23.5
Networking	• stability of existing supplier relationships	55.6	44.7	34.3
	• stability of existing customer relationships	50.3	60.5	40.3
	• ease of access to specialized services	44.7	39.5	25.7
Embeddedness/ Institution	• informal interaction and cooperation among firms in the software industry	50.0	31.6	20
	• inter-firm trust formation by associations	39.1	31.6	20
	• institutional environment such as incubator, venture capital	35.5	21.1	5.7
Collective Learning	• accessibility to exhibitions and conferences	53.3	21.1	8.6
	• external learning firm universities and institutions	14.7	13.2	11.4
	• interactive learning with suppliers, customers, and rival firms	35.1	21.6	11.4
Innovation Synergy	• access to business opportunities	39.7	29.7	17.1
	• stimulus/technological innovation	45.7	21.6	28.6
	• stimulus/management innovation	27.2	29.7	25.7

Notes: K: Kangnam-Seacho; Y: Youngdeungo; O: Other areas

Summary and Conclusion

The analysis of the empirical data is summarized in Table 7.2. The development level of the Kangnam-Seocho area as an industrial district with respect to the five core dimensions, compared to the other two industrial areas examined in the study, is categorized into four levels; significantly developed, developed, at the same level, and less developed.

First, the dimension of 'localization' can be said to be more or less developed. Product specialization of individual firms and their technical workforces in the software industry are not significantly different from other areas. But the degree of externalization and perceived location merits turn out to be at a higher level. This suggests that the agglomeration of software firms in Kangnam-Seocho has a substantial effect on the localization dimension.

The networking dimension shows similar results to localization. The locality of supplier linkages and service linkages in Kangnam-Seocho area is much higher than those in the Other area. But the Kangnam-Seocho area is at the same level in production networking and perceived location merits, and is less developed in the locality of customer linkages. Thus, we can infer that the level of the networking dimension of this area cannot yet be considered as at a self-perpetuating stage. The interviews with associations and companies confirmed that inter-firm networking within the area is not higher than the other areas.

The Kangnam-Seocho area is particularly strong on the embeddedness and institutional thickness dimension. Cooperation with universities and research institutes is especially prominent. Also, the location merits of Kangnam-Seocho area at the regional level exceed those of other areas.

Collective learning is more developed in Kangnam-Seocho area as well. Generally speaking, interactive learning with customers and suppliers is preferred in both the Kangnam-Seocho and Youngdeungpo areas. Meanwhile, Youngdeungpo demonstrates a higher rate of adoption in internal learning systems. This seems to be due to the 'global networking' associated with the Youngdeungpo area. Foreign companies from developed countries are a good channel for 'external relation', through which more diverse technologies and innovations are imported.

Finally, the Kangnam-Seocho area shows a lower level in the 'Innovative synergy' dimension than the other regions. In particular, spin-

Table 7.2 Comparative characteristics of the Kangnam-Seocho industrial district

Dimension	Intra-firm characteristics	Inter-firm characteristics	Regional characteristics
Localization	• product specialization[3] • technical workforce[3]	• externalization of production process[2]	• perceived location merits:[2] • labour force • relations (software firms) • relations (IT firms)
Networking	• production networking[3]	• locality of: • service linkages[1] • supplier linkages[2] • customer linkages[4]	• perceived location merits:[3] • supplier relations • customer relations • producer services
Embeddedness/ Institution	• informal recruiting channel[3]	• cooperation with: • association (society)[2] • local government[2] • university & research institute[1]	• perceived location merits:[1] • informal cooperation • trust building • institutional environment

Table 7.2 continued

Dimension	Intra-firm characteristics	Inter-firm characteristics	Regional characteristics

Collective Learning	• labour force turnover rate[3]
	• internal systems of learning & training[3]
		• external learning opportunities[3]
		• interactive learning with related companies[1]
			• perceived location merits:[1]
			  • exhibition, conferences
			  • universities, institutions
			  • interactive learning

Innovative Synergy	• types of start-up[4]
	• achievements of innovation[3]
	• internal innovation support systems[4]
		• relations with the original company[4]	• perceived location merits:[3]
			  • business opportunities
			  • technological innovation
			  • management innovations

Notes: Each characteristic is compared with the other industrial districts in terms of:
[1] significantly developed; [2] developed; [3] at the same level; [4] less developed

offs and innovation support systems at the firm level are lower than those of the Youngdeungpo area. The same can also be said of relations with the original company at the inter-firm level.

In summarizing information from all five dimensions, it can be said that the Kangnam-Seocho area is above average as an industrial district in four out of five dimensions. Namely, firms in Kangnam-Seocho area are enjoying the significant effects of embeddedness, institutional thickness and collective learning. In the localization and networking dimensions, this area scores in between 'developed' and 'at the same level'. However, it turns out to be less developed in innovative synergy.

It is important to recognize that the variables at the intra-firm level are not differentiated among the areas generally. It is another research issue whether dimensions of industrial districts are developing from firm-level to regional-level or not, and if this diffusion direction varies from dimension to dimension.

Despite the limitations of the research, it may be concluded that the Kangnam-Seocho area stands at the level in between 'learning district' and 'innovative milieu' of the evolutionary model in Figure 7.1. However, caveats to this conclusion are that localization and networking dimensions are not much different between the Kangnam-Seocho area and the other two software areas of Seoul. Another fact that we should keep in mind in accepting this conclusion is that it is based on the relative superiority of Kangnam-Seocho area to the other areas in Seoul. To overcome these limitations of a single metropolitan study, there needs to be an international comparative study on industrial districts with a uniform framework of analysis.

References

Amin, A. and Thrift, N. (1995), 'Globalisation, Institutional "Thickness" and the Local Economy', in P. Healey, S. Cameron, S. Davoudi, S. Graham and A. Madanipour (eds), *Managing Cities: The New Urban Context*, John Wiley and Sons, Chichester, pp. 91-108.
Antonelli, C. (1999), 'The Evolution of the Industrial Organisation of the Production of Knowledge', *Cambridge Journal of Economics*, vol. 23, pp. 243-260.

Asheim. B.T. and Cooke, P. (1999), 'Local Learning and Interactive Innovation Networks in a Global Economy', in P. Oinas and E.J. Malecki (eds) *Making Connections*, Ashgate, Aldershot, pp. 145-178.

Capello, R. (1999), 'Spatial Transfer of Knowledge in High Technology Milieux: Learning Versus Collective Learning Processes', *Regional Studies*, vol. 33, pp. 353-366.

Cooke, P. (1998), 'Introduction: Origins of the Concept', in H-J. Braczyk, M. Heidenreich, and P. Cooke (eds), *Regional Innovation Systems*, UCL Press, London, pp. 2-27.

Dankbarr, B., Hassink, F., and Covers, F. (1995), 'Technology Networking in Border Regions', *European Planning Studies*, vol. 3, pp. 63-84.

Hassink, R. (1999), 'What Does the Learning Region Mean for Economic Geography?', *The Korean Journal of Regional Science*, vol.15.

Jin, D.J. and Stough, R.R. (1998), 'Learning and Learning Capability in the Fordist and Post-Fordist Age: An Integrative Framework', *Environment and Planning A*, vol. 30, pp. 1255-1278.

Keeble, D., Lawson, C., Moore, B., and Wilkinson, F. (1999), 'Collective Learning Processes, Networking and Institutional Thickness in the Cambridge Region', *Regional Studies*, vol. 33, pp. 319-332.

Lagendijk, A. (1997), 'From New Industrial Spaces to Regional Innovation Systems and Beyond: How and From Whom Should Industrial Geography Learn?', *EUNIT Discussion Paper* 10, CURDS, University of Newcastle.

Maskell, P. and Malmberg, A. (1999), 'Localized Learning and Industrial Competitiveness', *Cambridge Journal of Economics*, vol. 23, pp.167-185.

Oerlemans, L.A.G., Meeus, M.T.H., and Boekema, F.W.M. (1999), 'Innovation and Space: Theoretical Perspectives', *Eindhoven Centre for Innovation Studies Working Paper* 99.3.

Oinas, P. and Malecki, E.J. (1999), 'Spatial innovation systems', in P. Oinas and E.J. Malecki (eds), *Making Connections*, Ashgate, Aldershot, pp. 7-34.

Park, S.O. (1996), 'Network and Embeddedness in the Dynamic Types of New Industrial Districts', *Progress in Human Geography*, vol. 20, pp. 476-493.

Park, S.O. (1999), *Contemporary Economic Geography*, Seoul: Arche (in Korean).

Park, S.O. and Markusen, A. (1995), 'Generalizing New Industrial Districts: A Theoretical Agenda and an Application from a Non-western Economy', *Environment and Planning A*, vol. 27, pp. 81-104.

Scott A.J. and Storper, M. (1992), 'Regional Development Reconsidered', in H. Ernst and V. Meier (eds), *Regional Development and Contemporary Industrial Response: Extending Flexible Specialization*, Belhaven, London, pp. 3-24.

Storper, M. (1993), 'Regional 'Worlds' of Production: Learning and Innovation in the Technology Districts of France, Italy and the USA', *Regional Studies*, vol.27, pp.433-455.

Storper, M. (1997), *The Regional World*, The Guilford Press, New York.

STEP Group, (1994), *What Comprises a Regional Innovation System? An Empirical Study*, Science, Technology and Economic Policy Group (STEP), Oslo

8 Qualified Labour Migration and Regional Knowledge Economies

Martina Fromhold-Eisebith

Introduction

Extensive research efforts have been made to identify trajectories of regional learning with respect to clusters of innovative, technology-intensive industries, especially in developed countries, and recently for developing countries. However, there is one factor of knowledge creation and dissemination that explicitly links developed and developing countries that so far has hardly received appropriate scientific attention. This research lacuna relates to the mobility of academically qualified people who are crucial carriers of know-how and learning capabilities and are associated with distinct characteristics of social behaviour. This chapter seeks to contribute towards closing this gap.

A central assumption of the chapter is that to a considerable degree the migratory moves of highly educated labour, broadly defined as those carrying a college or university degree, determine the kind and amount of know-how that is transferred from place to place (see Salt, 1997; Gaillard and Gaillard, 1998; and Iredale, 1999 for definitions). This migration is often referred to as the 'brain drain' or mutual 'brain flows'. In principle, the brain drain occurs at a variety of spatial scales, including several 'blank spots' on the research map of labour mobility about which very little is known. In the present discussion, emphasis is placed on aspects of international movements, specifically between Asian Newly Industrialized Countries (NICs) and Western developed ones, in connection with related sub-national dynamics, that are propelled by a significant spatial selectivity of skilled migration. This perspective promises to be highly relevant with respect to the topical question as to what actually triggers global-local dialectics in technology industries. An analysis of major brain flows, with respect to their geographic direction and spatial clustering tendencies, potentially offers important clues for a better understanding of (inter) regional learning processes based on interacting international and local information exchanges, and of the emergence of localized knowledge economies in globalized industrial space. Such an understanding is vital for developed and developing countries that are striving to support economic progress and innovativeness.

This chapter seeks to contribute towards such an understanding by composing a theoretical approach that, in a preliminary way, interprets qualified migrants as agents of a continuous international connection of specifically structured, internally networked regions. This approach draws on insights from recent studies of international migration and brain drain issues in Asia and beyond, as well as from the author's own investigations of networking features of the Indian technology cluster of Bangalore (Fromhold-Eisebith, 1999, 2001). The first part of the chapter briefly refers to important literature on qualified labour mobility, demonstrating a theoretical gap with respect to the central idea put forward here. The second part provides statistical information on the size and significance of brain flows, not only for developing countries but also for highly industrialized nations and regions, specifically with reference to the prominent example of California's Silicon Valley. The third part highlights the systemic factors that are currently discussed in migration research and introduces the concepts of interregional migration systems and transnational communities. They provide a crucial point of departure for theorizing on the interdependencies of international brain flows between less and highly developed countries and local knowledge accumulation. In the fourth part of the chapter, these assumptions are elaborated and cast into a model. This model depicts mechanisms by which the relevant migration processes nurture regional clusters of know-how-intensive industries in those areas that are internationally connected in a 'people-intensive' way. In the conclusion further research issues are suggested.

Qualified Labour Migration: An Overview of Topics and Tasks

Existing empirical and conceptual work on the issue of international movements of qualified labour can roughly be divided into four partly overlapping themes. In this section the widely varying objectives and findings of this research are selectively introduced as a basis for a theoretical approach towards the international migration of qualified labour and its implications for the dynamics of regional knowledge economies.

A first theme of research, which has dominated this literature over several decades, centres on the problem of the brain drain comprising the mass emigration of academics and professionals from many less developed countries. These studies try to reveal the main reasons behind the phenomenon, emphasize crucial impacts on the nations of origin, and search for adequate development policies (see literature overviews by Martin et al 1995; Salt, 1997; Gaillard and Gaillard, 1998). A second theme, closely linked to the first, is characterized by the theory-oriented view of (classical) economics on the overall costs and benefits of internationally shifting allocations of human capital (e.g. Karnath 1998; Stark et al 1998; Khadria, 1999).

Thanks to these two themes of investigation the (predominantly negative) socio-economic effects of skilled emigration on developing source countries are well explored, though merely on national levels of analysis. In this context, the greatest concerns relate to the loss of a potentially highly productive workforce and the associated implications for downward economic multiplier effects on employment and investments in the education system. A nation's industrial future is specifically jeopardized by the outflow of personalized know-how and of agents who could have promoted technology development and diffusion at home, as is stressed by Ong et al (1992) and Wong and Chong (1999). The common conclusion is that the international migration of skilled labour strongly hampers the formation of know-how-driven industry clusters in the developing world.

However, recent research has revealed positive 'second-generation effects' of the brain drain on source countries (Khadria, 1999). These effects not only occur in the form of financial remittances, but are also linked to the impacts of increasing educational attainment and knowledge upgrading. Thus, the success stories of highly skilled emigrants serve as exemplars and incentives for those left behind to raise their academic and related qualifications (Stark et al 1998; Chau and Stark, 1999). Further, educational policies in source countries that seek to sustain a sufficient output of graduates in order to keep the national economy and administration running, despite the loss of skilled people, can contribute to generally higher educational levels (Mountford, 1997). In particular, return migration from earlier brain drains provides important benefits for source countries. In many affected areas the unidirectional 'drain' evolves into bilateral flows of people back and forth over time, supporting a recognition that professional migration to advanced countries involves both 'brain losses' and 'brain gains' to source countries (Cheng and Yang, 1998, p. 626). These arguments support the idea that qualified labour mobility induces mutual knowledge exchanges between less and highly developed nations. These arguments are further elaborated in the theoretical discussion below.

A third theme of research examines international migration processes in a wider sense and from a sociological angle. The main objective of these studies is to explore and explain the economic, psychological, and social forces underlying people's movements in space (Stahl, 1995; Pries, 1997; Jones and Findlay, 1998; Vertovec and Cohen, 1999). Although this literature rarely explicitly refers to the specific subgroup of skilled migrants, it implicitly offers useful findings with respect to the main thrust of this chapter. In particular, important systemic factors are discussed which link certain places, motives and implications of migration, in association with the notions of migration systems and transnational communities. These studies support the assumption that international movements of (qualified) people are connected with a sub-national spatial selectivity, since migration flows tend to gravitate towards distinct regions in source and host countries. Because of the implications of these concepts for elucidating global-local interdependencies of labour and knowledge exchanges, they are developed in more detail in the third part of this chapter.

The fourth category of literature, which has emerged recently, integrates aspects of labour migration into studies with a more general focus on questions of global-local patterns in economic development. The underlying ideas come close to the core argument put forward here: that the spatial mobility of professionals to certain places is an indicator, manifestation, and interconnection of global economic hierarchies (e.g. Koser and Salt, 1997). Such mobility represents an important dynamic within the nexus of industrial globalization and regional development in knowledge-intensive industries, partly connected with the activities of internationally operating companies (Saxenian, 1999; Beaverstock and Boardwell, 2000; Hsu and Saxenian, 2000). Hsu and Saxenian's detailed analysis of people-driven connections within and between California's Silicon Valley and the Hsinchu-Taipei area in Taiwan provides a particularly valuable example of how some specific regional developments and learning processes are adequately captured and explained only by looking at international flows of an highly educated workforce. Indeed, this study hopefully represents a starting point for further investigations along similar lines and incorporating other places.

These four themes of research provide a basis for a more comprehensive, synthetic theoretical framework linking the role of the international-interregional migration with the accumulation of economic knowledge. Such an approach should combine insights into the systemic reasons for, and impacts of, qualified labour mobility, its spatial selectivity regarding national and sub-national levels, and its function in economic globalization. In this approach, a central concern is the elucidation of the relationships between the behaviour of knowledge-intensive multinational companies in both highly and less developed countries with those factors promoting technology-driven regional development, particularly with respect to functional and social networks of collaboration and information exchange.

Qualified Labour Migration: Selected National and Regional Statistics

In spite of the importance of 'brain flow' issues for national and international development, reliable numbers of the size of this migration are not readily available, neither at national nor especially at regional scales. One obstacle to obtaining accurate statistics relates to general problems of definition (Salt, 1997; Iredale, 1999). For example, how should highly qualified people be distinguished from other types of migrants – according to their academic degree, their professional status, or their true work characteristics? It is even a matter of debate as to how to exactly define a foreign migrant. For example, should birth place or ethnic background be the criterion? Incomplete and internationally incompatible statistical accounting is another major problem, as each nation and international organization uses different modes of data collection, classification, and display. Moreover, the changing nature of migration behaviour impedes attempts to record information. In the course of the growing mobility of highly skilled people, flows to 'permanent' destinations are increasingly replaced by volatile (back and forth) movements and by periodical, circulatory and transient residencies. This is due to fragmented career paths and to the increasing influence of the internal labour markets of multinational firms (Salt, 1997; Iredale, 1999; OECD, 1999; Beaverstock and Boardwell, 2000).

At first sight the overall number of international 'brain' migrants from developing to developed countries does not look impressive. According to UN estimates for the first half of the 1990s, such migrants numbered less than seven million world-wide, a small fraction of overall labour migrants of approximately 80 million. In general, however, people carrying academic degrees – an especially mobile stratum of society – are significantly over-represented among migrants, from both source and host countries (Carrington and Detragiache, 1998; OECD, 1999). Moreover, there are some countries and regions that are over-proportionally affected by an out- or in-flow of 'brains' in relation to their overall educated population. In addition, the potential industrial significance of highly qualified migrants gives them an economic weight much greater than quantitative measures of mobility suggest.

Some rough comparative figures for national levels, reduced to the latest 'common denominator' year of 1990, indicate a movement of numerous 'brains' predominantly from certain Asian source countries to highly industrialized OECD states, above all the USA (Table 8.1). Until then, Taiwan, Korea and the Philippines had been most severely affected in relative terms. Anecdotal evidence from country studies or press reports shows that since 1990, emigration of highly skilled labour from China and India has grown considerably (e.g. Khadria, 1999). In general, the most welcoming nations are in the 'anglophile' world, notably Canada and Australia as well as the USA, and these countries continue to import a substantial influx of human capital from the Asian Newly Industrialized Countries (NICs). Internationally, migratory connections are spatially selective.

Regarding the sub-national distribution patterns of international qualified migration, only scant documentation is provided by case studies. Khadria (1999), for instance, maps the regional immigration patterns, particularly of Indians into the USA, and indicates areas of gravitation for source and host countries. The case of Indian Bangalore, where software firms and universities continue to experience dramatic attrition rates of highly qualified people (Fromhold-Eisebith, 2001), strongly suggests the spatially structured nature of losses of skilled workforce to employers abroad.

As a major point of attraction for Asian-born university graduates and professionals in information technology and computer sciences, the example of Silicon Valley stands out. This area concentrates the 'brain gain' from abroad to an impressive degree, and its continuing success as a global innovation centre depends considerably on attracting creative impulses from different cultural backgrounds and the resulting international connections

Table 8.1 Statistics of the 'brain drain': Highly educated immigrants (tertiary level) from Asian less developed countries to the USA and all OECD states: Numbers and shares in 1990

Country of Origin:	Residence in the USA:		In all OECD states:	
	Absolute numbers*	Share of all highly educated in home country**	Rough absolute numbers*	Share of all highly educated in home country**
Philippines	493,074	7.1%	688,650	9.9%
India	228,270	1.1%	517,620	2.7%
Korea	201,460	6.1%	559,610	17.6%
Taiwan	118,017	9.2%	118,050	9.2%
China (mainland)	165,599	1.4%	321,550	3.1%
Pakistan	36,097	2.5%	102,550	7.4%
Indonesia	23,152	1.4%	25,590	1.6%
All Asia-Pacific	1,462,177		5,216,560	
Central America/ Caribbean	647,244			
S. America	284,904			
Africa	95,159			

*   For the USA only people over 25 years of age are included and graduate students are excluded
**   Including migrants

Source: Migrant data of the US-Census, the OECD Continuous Reporting System and other sources, according to Carrington and Detragiache 1998: 15-23.

(Saxenian, 1999; Hsu and Saxenian, 2000). Already in 1990, a third of all scientists and engineers working in the regional technology industries of Silicon Valley were of foreign, predominantly Asian origin – mostly Chinese/ Taiwanese or Indian. From all technology firms that started-up in Silicon Valley in the period 1995-1998, 20 percent are managed by Chinese and nine percent by Indian Chief Executive Officers (CEOs). In terms of the whole set of companies established since 1980, 17 percent (about 2,001 firms) are run by ethnic Chinese and an additional seven percent (774) by Indian chief executives, altogether generating a regional employment of almost 58,300 (Saxenian, 1999). As the latest statistics show, with respect to the population of Santa Clara Country in 2000, 47 percent are foreign-born and 23 percent are Pacific-Asians (Breslau, 2000).

The backflows of once emigrated 'brains' mainly from the USA to Asian NICs, which are particularly important in the framework of global-local interdependencies of technology-driven development, are increasingly recognized in the research literature. Tentative, recent estimates calculated for a small selection of countries and regions show that the amount of return migration still remains far below the opposite brain drain and might account for a stock of less than 10,000 even in big nations as China or India (Ong et al 1992; Khadria 1999). However, particularly in some of the new technology 'hot spots' of the Asian NICs, for instance Indian Bangalore, return flows of qualified migrants, predominantly from high tech agglomerations in the USA, are in the hundreds (Khadria 1999; Fromhold-Eisebith, 2001). Indeed, in the case of the Hsinchu-Taipei area in Taiwan, return migrants amounted to almost 2,860 by 2000 (Hsu and Saxenian, 2000). Currently, the return migration of highly qualified people to these and other similarly endowed regions reveals a remarkable upswing, a trend that strongly influences the industrial and entrepreneurial progress of those locations.

Generally, there is growing statistical evidence and awareness that clustering is a significant phenomena both with respect to source and destination areas of (skilled) labour migration (OECD, 1999). In particular, it seems that both the outflow of 'brains' from developing countries or NICs and the inflow into developed ones occur to and from urban agglomerations that have high level educational institutions, industries that provide (relatively) sophisticated work experiences, and international contacts. In these respects, these locales offer favourable conditions for knowledge-driven regional economic development. However, wider empirical proof for this hypothesis of spatial clustering still has to be produced. The chapter now turns to qualitative arguments that help explain the dynamics of qualified labour movements between certain places, and the global-local exchanges of knowledge associated with them.

Explaining the Patterns of Qualified Labour Mobility

A closer look at factors that are currently discussed in sociological or anthropological research on migration from less developed countries provides instructive information on the reasons and implications of uneven spatial patterns of mobility caused by cumulative clustering. These investigations, associated with the notions of migration systems, transnational communities, or multi-location transnational social spaces (Pries, 1997; Portes 1997; Vertovec and Cohen, 1999), show that social forces critically influence the amount and direction of (qualified) migration. These studies add to the often cited classical economic push- and pull-factors shaping migration, such as international differentials of wages, costs, or job opportunities (Karnath, 1998).

The systems view on (skilled) labour mobility combines ideas of social and spatial coherence, directed information flows, and dynamic path dependencies (Nogle, 1994; Stahl, 1995; Poot, 1996; Iredale, 1999). This analysis reveals a significant nexus of interregional and regional relationships within a framework of international exchange, as is captured in the term transnational community. Thus, labour migration integrates a set of places in different countries that are linked by flows and counterflows of people and the information they carry. Increasingly volatile migration behaviour further sustains these exchanges. Indeed, the international connections may be characterized as self-reinforcing, evolutionary interdependencies of source and host places. Once migration flows commence, they seem to have their own internal dynamics that operate to sustain and usually increase the flow between a particular source and a particular destination. As Stahl (1995, p. 219) states: "The direction, type and magnitude of a specific migratory flow depends on contemporary and historical relationships – economic, political and cultural – between sending and receiving countries. Within these broader macro-structures, individual migrants make decisions based on personal networks, practices and beliefs."

At the same time, the systems view highlights facets of regional concentration and cohesion of qualified migrants in source and host areas, based on the central idea that networks or 'chains' of people migrate, rather than individuals. The emigrants' pursuit to become embedded in a familiar, secure environment in the foreign country, which also reduces costs (Poot, 1996), propels the phenomena of spatial clustering. Socio-cultural values and personal networks from the source place become transposed and replicated at the destination of migration. Consequently, mutual, multiple linkages are created between source and host areas, which are each characterized by socially coherent migrant communities.

The systems view combines several aspects and perspectives that are worth addressing separately, particularly since some of them explicitly relate to the distribution of knowledge and information. Although the insights are mostly derived from general migration theory, they are valid for the subgroup of highly educated people as well. Four themes can be highlighted:

a) Qualified migration as an evolutionary, path-dependent and cumulative process regarding the mobility of people and knowledge: Once a few 'pioneers' have moved to another place and started to develop a local migrants' network, an increasing influx is induced along the same 'well trodden' path. It is predominantly triggered by continuous information feedbacks to the folks at home, easily transmitted by good telecommunication and transport links (Pries, 1997). Mental forces support this process, such as a reduced 'emotional distance' (Poot, 1996) or a widened 'awareness space' (Stahl, 1995), which eventually might lead to a specific inter-place migration culture. With Poot (1996, p. 70) we can conclude that "the inherent non-linearity of the dynamics in a migration system explains why migrant communities are geographically concentrated in particular regions [...] and why emigration propensities are so much higher in some countries than in others".

b) Qualified migration as a kind of interaction which is functionally and complementary linked to other (knowledge-intensive) exchanges: Skilled migration highly correlates – as a cause or consequence – to other linkages of the respective home and host areas, such as trade, direct investment, political or scientific cooperation (Poot, 1996; Cheng and Yang, 1998). Commonalties in culture (e.g. language), administrative or educational system, for instance originating from a colonial past, might be influential too. This systematically combines different spheres of know-how exchange and enforces concentrated information flows between certain regions. In particular, the location behaviour of multinational firms and their strategies in using internal or external labour markets shape the spatial structure of brain flows, because they link professional careers with the obligation to be mobile between major industrial agglomerations (Beaverstock and Boardwell, 2000). However, the impact of internationalizing companies on the brain drain seems to have contradictory effects. On the one hand, the outward mobility of skilled labour from developing countries partly substitutes for inward foreign direct investment there. On the other hand, inward investment and the outflow of qualified people from an area are positively correlated, because the new branch plants induce 'drains' by feeding the locally available human capital into the border-crossing internal labour markets of their corporations, or by influencing local mind sets towards working abroad. In contrast, a growing number of well-paying investors and employers in a region also substantially reduces qualified emigration. These dynamic interdependencies indicate interesting topics of investigation that cannot be discussed here.

c) Impact of educational systems on qualified migration, its spatial selectivity and effects: This aspect highlights the behavioural characteristics of a specifically mobile and significant group of skilled people and is directly associated with questions of learning and knowledge cross-fertilization. The mass migration of students to universities overseas often serves as the initial stage of a subsequent work settlement there (Cheng and Yang, 1998). Additionally, the international diffusion of educational and science systems to certain areas – a "global articulation of higher education" (Ong et al 1992) – stimulates the migration of elites from certain less, to highly developed, nations. The standardization of university teaching in many developing countries according to 'Western' models and academic partnerships eases the way, because they create "a pool of substitutable labour, which shares common skills, a lingua franca of English and technical terms, and internationalized values as well as professional networks that cut across national borders" (Ong et al 1992, p. 548). These developments help to create distinctive spatial patterns of knowledge exchange and accumulation.

d) Influence of family systems on migration decisions: Often the move of qualified people between particular source and destination areas is less driven by individual will, than by direct pressures from the whole family that tries to use one person as a 'door opener' for acquiring broader access to the advantages of a location abroad. This enforces interdependencies of international and regional cohesion and information exchange.

In summary, there are several major advantages of a systems view on qualified migration. Thus, attention is focused on processes both in home and host regions with respect to their interconnectedness. The mobility of people is regarded in correlation with other kinds of interaction like exchanges of goods, services, money, ideas, and information. Migration is comprehended as an evolutionary and ongoing process.

Theorizing the Role of Qualified Labour Migration for Global-Local Dynamics of Knowledge Economies

Given the arguments presented, a conceptual model is outlined which provides a generalized view of the relevant mechanisms by which qualified migration flows, particularly between Asian first or second tier NICs – especially ethnic Chinese countries, India or South Korea – and highly industrialized nations shape the development of regional knowledge economies in both areas (Figure 8.1). It is supported by the sub-national spatial selectivity of skilled migration and its systemic implications. This model emphasizes the relatedness of processes in certain areas of developing and developed countries, on the one hand, and dynamic aspects of phase developments, on the other hand. In this sense, it interprets qualified migrants as agents of a continuous international connection of internally networked regions.

As is sketched in the model, in the first phase, skilled migration is characterized by a massive outflow of 'brains' almost purely in one direction, from less to highly developed nations (Figure 8.1a). Structural and systemic considerations suggest that the migrants cluster unevenly in space, both in home and host countries. Thus, the source regions are mainly urban agglomerations that contain reputable universities and/or internationally integrated firms, and that experience high shares of qualified emigration. These places also potentially represent locations with comparatively good conditions for a local take-off in technology sectors. On the destination side, the immigrants gravitate toward industrial growth centres that are particularly attractive regarding jobs and show a high demand for skills. Beyond the well-documented case of Silicon Valley (Saxenian, 1999), other outstanding

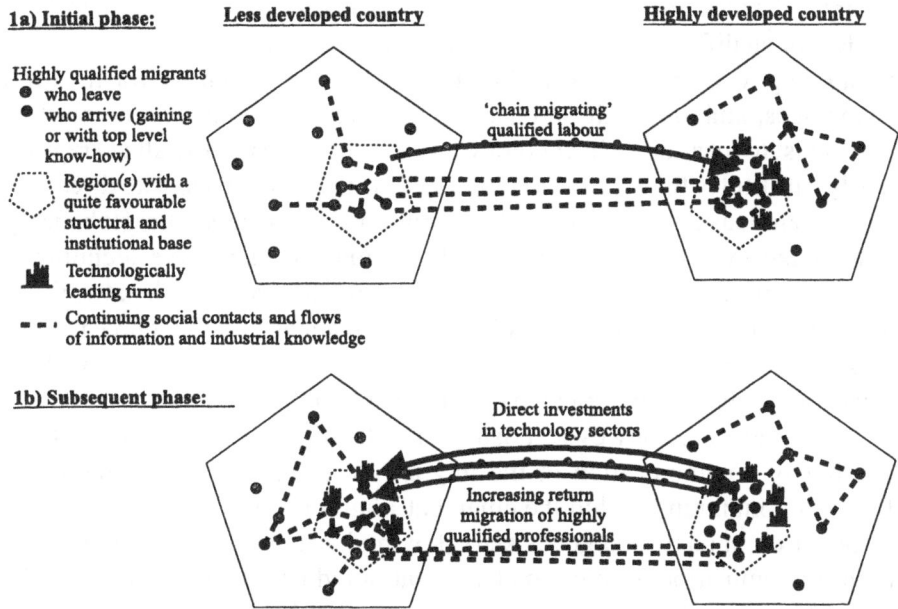

Figure 8.1   Qualified labour migration as a supporting factor of regional knowledge economies – a conceptual model

concentrations of high-wage knowledge-intensive industries of the developed world are likely to profit from such flows. In these places, the resident firms – often leading multinationals – need a constant influx of fresh human capital in order to stay competitive and innovative. Therefore, the regional knowledge economies constituted by them heavily rely on that resource of educational excellence. To facilitate migration, employers often offer excellent opportunities for state-of-the-art learning processes to their foreign 'brain' workers, which are not yet possible in the migrants' home countries.

In this phase, the benefits of the migratory pattern regarding the advancement of industrial know-how predominantly occur in a highly developed region. It can gain substantial creative impulses from attracting highly skilled people with different, but complementary knowledge backgrounds – and their locally emerging ethnic networks, as is illustrated by experience in Silicon Valley dynamics (Saxenian, 1999). 'Injections' from different cultures might enrich the entire regional business atmosphere. In contrast, the less developed area is mostly negatively affected. It suffers from the loss of qualified human resources and know-how, which reduces prospects to build up its own sophisticated industries, despite regional infrastructural advantages, ambitious educational policies and achievements. The outflow of 'brains' to locations of technology multinationals in industrialized regions partly substitutes for the flow of direct investments in the opposite direction, which dampens the integration of the migrants' home area into globalizing knowledge-intensive production systems and the respective technology transfers. In particular, the breaking up of local information-rich networks of acquaintances from study or work forms a disadvantage for the source regions of qualified emigration, because it destroys important foundations on which a successful innovation-oriented development could be built.

There are, however, positive implications. One example is the continuing communication of the migrants with the home community or frequent commuting within the emergent migration system, through which information, money, and goods start flowing back. These flows might lead to the dissemination of some small amount of industrial know-how from the highly to the less developed region. More importantly, the transnational social linkages within the community of skilled migrants pave the way for other economic interactions between the source and destination areas in the future. This is the crucial base for mechanisms of international-interregional transfers of high-value industrial knowledge to migrants' home regions to gradually gain momentum.

In the second phase of the model, migration and knowledge flows increasingly combine backflows of information, capital, and industrially experienced people from the highly to the less developed region (Figure 8.1b). In this context, reverse direct investments and brain re-gains are crucial processes in emerging knowledge economies. The earlier employment of foreign academics from less developed home regions has inspired quite a few technology multinationals of industrial states to directly invest in those areas, where skilled human capital can be sourced at even lower costs, a behaviour which particularly marks Indian experiences (Fromhold-Eisebith, 2001). To facilitate such developments, MNCs can selectively transfer back skilled migrants who are both familiar with the source region and the company's corporate culture. Such migrants provide ideal agents to 'send home' to set up and manage new branch plants there. Subsequently, human and investment capital flows back, which integrates the region of origin into globalized structures of innovation-driven production and provides access to competitive industrial know-how that is transferred as an input to local production or services.

Similar processes are set in motion by emigrated professionals returning independently from internationalizing Western firms. These professionals show a growing inclination to become entrepreneurs 'back home' and to apply the top level knowledge and experiences gained abroad, as the example of technology clusters in Taiwan indicates (Hsu and Saxenian, 2000). The favourable infrastructural and institutional endowment of the main source areas of brain drain that are mentioned above makes them particularly attractive for return investments in knowledge-intensive industries.

The combination of direct investments and reverse qualified migration to a promising NIC region provides specific developmental advantages. It forms the base for a social, functional and institutional embedding of the respective firms into the locality, via their indigenous CEOs or senior staff (on networking in Bangalore, see Fromhold-Eisebith, 1999). Often the linkages of skilled emigrants to professional friends at home had never loosened while working abroad, because of the cohesive forces of transnational communities and their institutional manifestations. In many cases the shared work and residence experiences of skilled migrants of a common origin at a particular destination overseas creates social relationships and a distinct spirit which prevails after their return in support of regional information-rich networks. The specific pride and collaborative coherence of Taiwanese or Indians who had carried out jobs in the Silicon Valley before engaging in emerging technology clusters back home is an example (Saxenian, 1999; Fromhold-Eisebith, 2001).

From a theoretical perspective, the regional socio-functional embedding that results from this behaviour is considered crucial for sustaining a firm's competitiveness, locally and globally. In addition, it accelerates the regional dissemination of state-of-the-art industrial know-how, which can follow lines of trustful social relationships established long ago, including acquaintances now working in foreign or indigenous companies, academic or scientific institutions. The dynamics of migration systems and self-reinforcing processes of regional development in the less developed country also induce wage rises, a growing number of technologically more challenging jobs, and other incentives (as documented for Bangalore by Fromhold-Eisebith, 2001). Consequently, the number of qualified return migrants expands over time, which further promotes entrepreneurship, local learning, and technological 'catching-up' of industries in the home area, depending on adequate policies. Therefore, those NIC regions that had initially suffered most from substantial 'brain losses' might later reap economic and technological benefits that could even outweigh their former detriments. In turn, this trend provides important positive feedbacks for the highly developed technology region as a "two-way process of reciprocal industrial upgrading" (Saxenian, 1999) which promotes its sustainable competitiveness in a globalizing economic environment.

Conclusions and Suggestions

Despite illustrative case studies and a vital political interest in brain drain issues, the role of qualified labour migration between less and highly developed countries for global-local interdependencies of the dynamics of knowledge economies has not received the scientific attention it deserves. This chapter tries to lay the general foundations for understanding this phenomenon by integrating, within a social systems view of migration theory, insights from investigations on brain flows between Asian NICs and Western countries or high-tech centres. With these insights, in association with known common factors of success for innovative regions, a conceptual model is outlined which interprets qualified migrants as agents of a continuous international connection of certain, internally networked and knowledge-accumulating regions. Clusters of technology-driven, learning firms tend to thrive best where the international and local dissemination of industrial know-how interacts in an evolutionary and efficient way. As this chapter emphasizes, mobile professionals serve as crucial, sociable intermediaries. The 'geographical circulation of intellectual elites' induces an important 'de- or multi-nationalization of knowledge' (Ash and Sollner, 1996, p. 6), while the sub-national spatial selectivity of qualified migration supports complementary mechanisms of regional embedding, collaborative application and dissemination of that know-how.

The model presented here should only be seen as a point of departure. Necessarily, wider empirical evidence must be collected, even regarding the basic assumption of a selective flow of 'brains' between certain (Asian) home and (Western) host areas, and beyond the Silicon Valley case. The systemic linkages of skilled (re-)migration and the transfer of industrial knowledge have to be further investigated as well, in order to acquire a more detailed view of the actual learning impacts on the interconnected regions which are induced by mobile professionals and their respective networks. Such studies should help to establish a conviction that qualified labour migration significantly shapes global-local patterns in the development of knowledge economies – with consequences also for respective policies.

Acknowledgement

The author is grateful to the German Research Council (Deutsche Forschungsgemeinschaft) for travel support to attend the Dongguan meetings of the IGU where this paper was presented.

References

Ash, M.G. and Sollner, A. (eds) (1996), *Forced Migration and Scientific Change*, German Historical Institute, Washington DC.

Beaverstock, J.V. and Boardwell, J.T. (2000), 'Negotiating Globalization, Transnational Corporations and Global City Financial Centres in Transient Migration Studies', *Applied Geography*, vol. 20, 3, pp. 227-304.

Breslau, K. (2000), 'Tomorrowland, Today. The Valley's Influx of Hyperachieving Techno-migrants, Along with a Swelling Hispanic Population, Creates a Microcosmos of the Future', *Newsweek*, September 18, p. 40.

Carrington, W.J. and Detragiache, E. (1998), 'How Big Is the Brain Drain?', *IMF Research Department*, Working Paper WP/98/102, Washington DC.

Chau, N.H. and Stark, O. (1999), 'Human-Capital Formation, Asymmetric Information, and the Dynamics of International Migration', in A. Razin and E. Sadka (eds), *The Economics of Globalization. Policy Perspectives from Public Economics*, Cambridge University Press, Cambridge, pp. 333-370.

Cheng, L. and Yang, P.Q. (1998), 'Global Interaction, Global Inequality, and Migration of the Highly Trained to the United States', *International Migration Review*, vol. 32, 3, pp. 626-653.

Fromhold-Eisebith, M. (1999), 'Bangalore – A Network Model of Innovation-Oriented Regional Development in NICs?', in E.J. Malecki and P. Oinas (eds), *Making Connections: Technological Learning and Regional Economic Change*, Aldershot, Ashgate, pp. 231-260.

Fromhold-Eisebith, M. (2001), *Technologieregionen in Asiens 'Newly Industrialized Countries'. Strukturen und Beziehungssysteme am Beispiel Bangalore, Indien, und Bandung, Indonesien*, Series Wirtschaftsgeographie, vol. 18, LIT Verlag: Münster and Hamburg.

Gaillard, A.M. and Gaillard, J. (1998), *International Migration of the Highly Qualified: A Bibliographic and Conceptual Itinerary*, Center for Migration Studies, Staten Island, New York.

Hsu, J.-Y. and Saxenian, A. (2000), 'The Limits of Guanxi Capitalism: Transnational Collaboration between Taiwan and the US', *Environment and Planning A*, vol. 32, pp. 1991-2005.

Iredale, R. (1999), 'The Need to Import Skilled Personnel: Factors Favouring and Hindering its International Mobility', *International Migration*, vol. 37, 1, pp. 89-123.

Jones, H. and Findlay, A.M. (1998), 'Regional Economic Integration and the Emergence of the East Asian International Migration System', *Geoforum*, vol. 29, pp. 87-104.

Karnath, S.J. (1998), '"Brain Drains" and "Brain Gains": A Critical Look at the Literature on the Economics of the International Migration of Human Capital', *The Indian Journal of Economics*, vol. 78, pp. 371-407.

Khadria, B. (1999), *The Migration of Knowledge Workers. Second-Generation Effects of India's Brain Drain*, Sage Publications, New Delhi.

Koser, K. and Salt, J. (1997), 'The Geography of Highly-Skilled International Migration', *International Journal of Population Geography*, vol. 3, pp. 285-303.

Marti, P.L., Mason, A. and Tsay, C.-L. (1995), 'Overview', *ASEAN Economic Bulletin*, vol. 12, no. 2 (special focus issue Labour Migration in Asia), pp. 117-124.

Mountford, A. (1997), 'Can a Brain Drain be Good for Growth in the Source Economy?', *Journal of Development Economics*, vol. 53, pp. 287-303.

Nogle, J. M. (1994), 'The Systems Approach to International Migration: An Application of Network Analysis Methods', *International Migration*, vol. 32, 2, pp. 329-342.

OECD (ed) (1999), *Trends in International Migration. Continuous Reporting System on Migration*, Annual Report, OECD, Paris.

Ong, P.M., Cheng, L. and Evans, L. (1992), 'Migration of Highly Educated Asians and Global Dynamics', *Asian and Pacific Migration Journal*, vol. 1, 3/4, pp. 543-567.

Poot, J. (1996), 'Information, Communication and Networks in International Migration Systems', *The Annals of Regional Science*, vol. 30, pp. 55-73.

Portes, A. (1997), *Globalization from Below: The Rise of Transnational Communities*, Working Paper, Transnational Communities, 98-01. (http://www.transcomm.ox.ac.uk/).

Pries, L. (ed.) (1997), *Transnationale Migration*, Nomos, Baden-Baden.

Salt, J. (1997), *International Movements of the Highly Skilled*, OECD Working Papers vol. V, International. Migration Unit, Occasional Papers no. 3. Paris.

Saxenian, A. (1999), *Silicon Valley's New Immigrant Entrepreneurs*, Public Policy Institute of California, San Francisco. (http://www.ppic.org/publications/PPIC120/index.html).

Stahl, C.W. (1995), 'Theories of International Labor Migration: An Overview', *Asian and Pacific Migration Journal*, vol. 4, 2/3, pp. 211-232.

Stark, O., Helmenstein, C. and Prskawetz, A. (1998), 'Human Capital Depletion, Human Capital Formation, and Migration: A Blessing or a Curse?', *Economics Letters*, vol. 60, 3, pp. 363-367.

Vertovec, S. and Cohen, R. (1999), 'Migration, Diasporas and Transnationalism', *The International Library of Studies on Migration*, vol. 9. Edward Elgar, London.

Wong, K.Y. and Chong, K.Y. (1999), 'Education, Economic Growth, and Brain Drain', *Journal of Economic Dynamics and Control*, vol. 23, 5/6, pp. 699-726.

9 Improving Embedded Knowledge in Old Industrial Districts: Case Studies, from Valencia, Spain

Javier Alfonso Gil, Antonia Sáez Cala, Antonio Vázquez-Barquero, and Ana Isabel Viñas Apaolaza

Introduction

The loss of competitive advantage is one of the risks facing local productive systems that are specialized in mature sectors. Case studies seem to indicate that these systems sometimes respond negatively to technological and commercial change because they are incapable of designing adequate adaptive strategies. We hypothesize that the principal cause of this situation is linked to the relative lack of quality in productive inputs used by local systems. While capital goods show a tendency toward modernization due to investment by firms in new machinery and designs acquired and adapted from abroad, human resources and, in general, the level of knowledge existing in the local productive system, do not conform to a similar pattern. Indeed, there is a potential gap between the technological and training capabilities of managers and workers and the potential levels at the frontier of knowledge. As a consequence, accumulated human knowledge may become stagnant and firms within the system may show considerable inertia.

One possible solution to the problem resides in the transformation of the relatively poor labour inputs through training that is designed to increase knowledge in order to generate greater value added and, ultimately, the competitiveness of local firms. In the Spanish case, Technological Institutes offer training and technical services and hope to act as agents for such a

145

transformation. These institutes also seek to encourage the production of knowledge embedded in machines, notably by creating awareness and providing support for the establishment of firms producing capital goods.

In this chapter, we examine the technological strategies of small and medium size firms (SMEs) in four old industrial districts in Valencia, Spain, notably in relation to four local productive systems, respectively specialized in footwear, ceramics, textiles and toys. In particular, this chapter seeks to assess the extent to which these firms are using technology as a source of competitive advantage. In addition, the role of four Technological Institutes in promoting technological capability among these firms is examined. The fieldwork for this study was carried out from July to December, 1998. Information was compiled by means of personal interviews of owners, managers or executives of the firms and Technological Institutes selected. The four Technological Institutes and a total of 100 SMEs were interviewed (30 percent in textiles, 25 percent in toys, 24 percent in footwear and 21 percent in the ceramics sector). The fieldwork sought to obtain information relevant for an analysis of the current demand for innovation support services by SMEs in the Region of Valencia and to assess the services offered by the Technological Institutes in the footwear (INESCOP), toys (AIJU), textiles (AITEX) and ceramics (AICE) sectors.

This chapter is organized as follows. The first section reveals the sources of knowledge available to firms of mature sectors and how firms protect their knowledge base. The second section focuses on the types of innovations that are introduced by SMEs and the creation and modification of technical knowledge within old industrial districts. The third section studies the significant role of Technological Institutes since their creation in the 1980s. Finally, the paper also proposes changes that should now occur within the Institutes if they are to contribute to competitive advantage and, ultimately, to the performance of local systems.

Reflections on Technology as an Endogenous Factor

The literature of endogenous development has clarified and emphasized the role of knowledge and its accumulation in the economic progress of nations. Traditionally, in models of national economic growth, technical change is considered an exogenous factor. Generally, the origins of technology in these models was unknown, or not considered. The major contribution of new models of economic growth, however, has been to endogenize the search by economic agents for new ideas to obtain profits. This simultaneously accounts for the origin of technical progress, thus defining and substantiating the motor of economic growth in countries (Romer, 1986, 1990; Lucas, 1988).

The contribution of endogenous growth needs to be assessed in relation to the institutions of particular territories. These institutions relate to the rules formulated by every society for governing the economy and are of prime importance for understanding the motivation of individuals within a process of adaptive stability in the face of technological change (Alfonso, 1999). Without clear property rights, adequate judicial neutrality and commercial freedom, economic agents will find it difficult to develop their potential for discovery and invention in a capitalist world. Institutions are significant in that they design the possible paths that economic and social agents may pursue (North, 1990). It is for this reason that the institutional framework is considered the prerequisite or ultimate cause of all economic growth.

But institutional, and of course technological and entrepreneurial, frameworks have to be anchored in particular territories. In fact, the actions of pioneers in a given sector tend to generate spillovers within their milieux, resulting in the specialization of given geographic areas in specific industrial activities (Marshall, 1964). Moreover, the 'bourgeois city' has traditionally been the cradle of enterprising individuals, the place where ideas are created, transmitted and, finally, put into practice. Within a common physical and institutional territory, incentives arise that lead the agents to endogenize their actions.

Nevertheless, the quest for ideas to be developed as products sold in markets creates problems for firms. In particular the question is raised, how can firms retain exclusive rights to ideas put into practice? That is to say, how do firms avoid embedded knowledge being copied or used by rivals? The need for protection, understood as a secret to others, is common to all kinds of firms and activities, although it differs in form. If a relatively small firm performs in a sector where there are no entrance barriers, such as mature sectors with low technology, protection against rivals is personalized in that it depends on what entrepreneurs do with their assets within its territorial environment.

In contrast, in high tech activities, entry barriers are considerable, and firms need to generate processes and products differentiated from those of competitors. These efforts to generate high added value outputs, typically require specific technological knowledge situated at or near the technological frontier, thus creating substantial development costs that need to be covered. Such firms seek to protect their profits with patents that provide a quasi-monopoly advantage. This situation of high technology development costs is the justification of quasi-monopoly rents that firms enjoy, at least for a time.

We have described two extreme situations. Between them are those stable firms, with sufficient capital accumulation, that decide to create technological assets in order to approach the most dynamic firms in the sector. Moreover, mature sectors may comprise a range of high- and low-tech possibilities. The remainder of this chapter examines the technological strategies of firms in the study clusters of the Valencia Region of Spain (Vázquez-Barquero et al 1999).

Technology Strategies Among SMEs in Valencia

There is no general tendency among the surveyed firms to develop research and development (R&D) activities and patent the results. However, a sizeable group of leading, innovative firms is appearing in these clusters and these firms are capable of carrying out R&D activities and protecting the results obtained. Specifically, nearly 25 percent of the surveyed firms protect their innovations, particularly product innovations, with patents. These firms are generally leaders and are medium-size firms who export more than 30 percent of their sales.

The most numerous group of SMEs do not play a relevant role in these innovative processes, because they are dependent on, and often stimulated by, leading local firms in the sector. Small firms usually adapt the technologies of the leading firms suggesting that knowledge is diffused rapidly and consistently throughout the local production system by means of the complex network of productive, social and cultural relations existing within these clusters.

Moreover, significant incremental improvements were made in the districts studied. From 1995 to 1998, 84 percent of firms made some sort of innovative improvement or development. New products (65.5 percent of firms), development of new processes (53 percent) and, to a lesser extent, improvements in existing products (49.5 percent) and processes (44.5 percent) stand out. To a greater or lesser degree, all of these changes caused modifications in the organization of firms. These incremental changes potentially generated increase returns and at least are suggestive regarding improvements to embedded knowledge in the production process.

Table 9.1 Research and development (R&D) indicators for EU, Spain and Valencia, 1994-1997

	1994	1995	1996	1997
R&D expenditure as a % of GDP – EU 15	1.94	1.93	1.91	1.90
R&D expenditure as a % of GDP – Spain	0.85	0.84	0.87	0.86
Total internal R&D expenditures (millions pesetas) – Spain	548,154	590,688	641,024	672,017
% of total internal expenditure in Valencia over Spanish total	6.3	5.9	6.3	6.5
R&D personnel – Spain	80,399	79,987	87,264	87,150
% of personnel employed in R&D in Valencia over Spanish total	6.4	6.7	6.5	6.9

Source: Elaborated from data extracted from *EUROSTAT* and *Instituto Nacional de Estadistica.*

Therefore, innovations developed in the region are primarily incremental and introduced through the improvement of products and processes already on the market. Normally, these kinds of technical changes are difficult to observe; they are typically very small steps that are closely related to the established technological trajectory.

In the Valencian Region, internal investments in R&D have shown a positive trend with regard to the national total since 1995 (Table 9.1). At present, this volume is between 0.5 percent and one percent of the Spanish GDP, similar to the national share of production (at 0.9 percent of the GDP). Although these levels are below those recorded in more dynamic areas, such as Madrid and Catalonia, the trajectory is positive. Universities make the greatest effort in R&D spending in the region (around 60 percent), followed by firms with almost a third of total expenditure. The relative share of the personnel involved in R&D in the Valencian Region with regard to the national total has also increased since 1994, and by 1997 accounted for 6.9 percent of the 87,150 total personnel employed in Spain (Table 9.1). These personnel represent less than 0.6 percent of the active work force in Valencia and most of them are not employed by SMEs.

Table 9.2 National patents applications in Spain by some regions of origin, 1994-1998

Region	1994	1995	1996	1997	1998
Catalonia	539	530	586	504	600
Madrid	509	491	523	567	511
Pais Vasco	176	145	165	163	155
Region of Valencia	271	150	307	253	273
Spain	2136	2047	2274	2236	2270

Source: OEPM.

In Valencia, patent applications are also on the rise; in 1998, the region represented 12 percent of the total in Spain, after Madrid and Catalonia (Table 9.2).

Technological Gaps

The surveyed firms either lack the endogenous incentive to innovate and generate technology or, in the improbable case that they do generate technology, they do not find that obtaining a patent is profitable. The game is to survive without resorting to technological change, or in other words, to move along the assigned production function without decreasing returns. It is a problem of dynamic, but non-endogenous, transition (Jones, 1998). There are two ways, neither mutually exclusive, that firms in mature production sectors can choose to adopt to survive while avoiding endogenous technical change. In particular, firms can invest in capital goods and/or product design:

Capital goods In the first place, SMEs can substitute old capital goods with a new generation comprising greater embedded knowledge. Such is the case among the surveyed firms where process innovations are developed through the installation of new equipment and computer systems for data processing and, to a lesser extent, computerized mechanization and control of testing. Innovative firms consider that these adaptations are new in the local market (36 percent), in the regional market (28 percent), nationally (26 percent), in the EU (54 percent) and internationally (31 percent). Twenty-six percent of innovative firms claim to have developed process innovations rated as totally new in these markets although they still constitute incremental innovations. Ninety-five per cent of the process innovations involve adaptations in processes already in use through the new investments, thus progressively improving processes and, in turn, products. In this way, the productive system adapts exogenous technology. But, in inserting it into the production process, productivity levels can increase and match the levels of those firms that have already incorporated the same generation of machinery.

Although all adaptation is innovation, the level of knowledge necessary to adapt a machine to the productive process is, of course, inferior to that necessary to design and build it. A problem arises when there are no manufacturers of capital goods in the territory where the mature sector firm is located. This can be observed not only in Valencia but across Spain where, in general, advanced equipment is acquired in foreign markets. At present,

66 percent of the surveyed firms in the four districts purchase machinery in the EU. Moreover, 17.5 percent of all imports to the region are capital goods, coming mainly from Germany and Italy.

This trade is clearly an asynchronic situation. Firms and the territory are able to evade investment in direct advancement of knowledge and technology, but in paying for imported equipment, an indirect price they pay is in the form of limited technological possibilities and the relegation to the role of a follower behind leading countries and territories. Inasmuch as the decision not to create machines can be taken as a proxy for the decision not to increase knowledge, the possibility of catching up to leaders is severely limited. The gap between the technological competence of the firms within the region and the frontier is equivalent to the quantity of pure knowledge necessary for the construction of a new generation machine.

In spite of dependence on external technology, we recognize considerable efforts by the firms in achieving high mechanization rates by incorporating and applying international technologies. Along with other factors, this behaviour has allowed them to maintain a significant presence in international markets and sources of know-how. The informational sources on capital goods with advanced technology are international. In this regard, national fairs, exposition events, clients, foreign suppliers of machinery, and as we will see the Technological Institutes of the Valencian Region, are important sources of information.

During the 1990s, firms located in the local productive systems of footwear, ceramics, textiles and toys enjoyed increasing production and sales. Much output is exported, as much as 43 percent in the case of exporters of final consumer goods. In all, the four sectors analyzed accounted for 36 percent of regional exports in the first quarter of 1997. Ceramic tiles and footwear, which each account for 12 percent of total regional exports, deserve special mention (Vázquez-Barquero et al 1999).

Besides the increase in exports, the destination of industrial products from the region is significant. The principal destination is the European Union which imports 54 percent. These firms have not only maintained traditional markets but have also entered new markets, preferably in the geographic area of the European Union and internationally (the USA, Eastern European and North African countries, China and Latin America). Thus, the surveyed firms have not only maintained their domestic market but have also been able to significantly increase their presence in foreign markets. Only the toy sector and a part of the textile sector – rugs – showed a decrease in their share of the Spanish market.

Design Conceptually, design has many of the characteristics of the machine. Both are the result of the accumulation of knowledge in the sector and the trajectory of product design tends to run parallel to that of embedded knowledge. In fact, territories that are situated at the technological frontier are usually also those that maintain superiority in the production of design. If the result of all goods on the market is determined by the acceptance of consumer demand, the aspect of product design is a decisive factor in accounting for success, particularly in mature sectors where the concept of fashion design is crucial.

Firms in the districts analysed develop innovative products based on new designs, and are characterized by greater quality, better adaptation to demand, exclusiveness and standard technological level. As new products are introduced, firms make changes in their product lines. Sometimes these are small changes in design in one or more products made by the firm to adapt to new demands or to give the firm access to new markets.

However, as in the case of capital goods, it is difficult for firms in mature sectors to create designs endogenously. The same arguments used to describe the lack of technological development on the part of firms are also valid in accounting for the lack of creative design. The solution is similar since it is the entrepreneur, within his/her milieu, who forms the fashion tendencies for the next season, usually through fairs in the sector. In this way, they avoid the risk of facing the market alone. It should be emphasized that the technological yields will be much briefer in the case of design. This accounts for the high price of first-rate design (as in fashion). The firm must quickly recover the fixed and variable costs. A problem arises, however, as patents are not particularly successful form of protection and a way of obtaining monopolistic rents. The need for protection based on industrial secrets is evident in such cases. Speed of product design is also a vital variable and can mean the success or failure of the season; for many SMEs, it can even mean survival. Herein lies the importance of design for the firm.

By using imported capital equipment in association with product design, the survey firms are able to compete in the international arena. Indeed, many of the surveyed firms compete, in design and quality, with Italians firms, their direct competitors and, in some sectors, worldwide leaders. Participation in the EU meant new opportunities by providing a destination for their goods in a continental market of equal opportunity and, perhaps more significantly, a stimulus and a technological source of processes and products. Thus, in a little over 10 years, product image, uniformity and competitiveness in price and quality have positioned the region's products at respected international levels. The proverbial Valencian entrepreneurs have managed to consolidate themselves in the European panorama. Nevertheless, the significant difference is that the Italian producers have generally endogenized their production processes, both in the building of the machine and in design, while the technological capacity of the Spanish territory lags behind.

There is a paradoxical gap in products pertaining to the cluster study: their content of embedded endogenous knowledge (machine and design) is inferior to what their market value would suggest. This gap should be closed over the long term. If not, there is a certain risk that other territories and firms will catch up with them in knowledge endogenization, thus displacing their products from the international market. It is even possible, given today's globalization processes, that they could be displaced from the national market as well.

Education and Training Gaps

For firms in mature sectors, the existing level of education of the labour force is usually inferior to the formation embedded in the machine. Consequently, a gap exists on the factory floor between human capacity and the machine's possibilities. The problem is further complicated if observed from a dynamic perspective. When workers in firms have finally mastered the machine, there is a risk that a new machine will be introduced into the firm's stock of capital goods, once again causing a gap between capital and labour factors. This eternal starting-over-again occurs in all production areas. Thus, the adaptation of the newly purchased machinery requires firms to provide additional training for specific groups on their staff; in particular machine operatives, technicians and managers. This training is usually received on-the-job although Technological Institutes also play a significant role in this regard.

The gap will be greater, the larger the difference between worker capacity and the position of the machine at its technological frontier. If the gap is wide, this means that the firm will find it impossible to access the technology represented by the machine. In the sectors and territories under consideration, the gap does not generally reach alarming levels. That is, although the educational levels of the factory populations is less than the potential offered by machines they use, the differential gap in capacities will likely disappear over time. This is due, in part, to the existing education levels of the populations; in 1996 only 3.2 percent of the populations over 16 years old in the regions studied was illiterate, compared to 3.9 percent registered for Spain as a whole. In addition, the capital goods designed for technologically mature sectors, such as those examined in this chapter, are not technologically sophisticated. However, the incorporation of exogenous machinery into the local productive system and the need to adapt it requires improvements of human capital. Such adaptation poses problems, for the firms in the industrial districts studied. In particular, bearing in mind the relative lack of skilled human capital, firms need to upgrade worker skills in order to confront future technological change. For SMEs, however, such training can be expensive. In this context, Technological Institutes exist, in part, to coordinate the entrepreneurial system by providing training for workers throughout the industry. In particular, they encourage and promote relations between firms and the public sector and, more importantly, they provide training services adapted to the real necessities of firms.

In the Valencian Region, a positive trend in the qualification of human resources can be observed. In 1996, 6.7 percent of the population over 16 years old in the region had university studies. At present, there are five public universities in the region. During the academic year of 1998/1999, these institutions employed 8,155 teachers (8.9 percent of the national total) and admitted 145,510 students (9 percent of the national total), putting the student/teacher ratio slightly higher than the Spanish average (17.8 percent as opposed to 17.3 percent).

However, the principal shortcomings of university training can be found in the coursework offered. Of a total of 145,510 students, approximately 30 percent have chosen a technical curriculum, in which Technical and Superior Architecture and Engineering are included. Although these degrees have a greater influence in the region than in the rest of Spain, the formation is not particularly adapted to the productive fabric of the region since university instruction is usually more academic than technological. Even in entrepreneurial training, the contents do not fully relate to the productive and technological level. Moreover, scientists and researchers have not developed embedded technologies for industries already existing in the regional economy, but rather they focus on sectors not implanted in the area.

Finally, specific training is rarely oriented toward the sectors in the economic system of the area and, when it is, it is often related to manufacturing technologies and does not go into depth in production or the engineering of machinery construction (Sweeny, 1993).

This failure is possibly the greatest obstacle to firms choosing a long-term path of product and process quality upgrading since the incorporation of improvements into the productive system is seriously limited. The involvement of technicians in the process is decisive for improving goods and services produced by the firms. Integrating technical staff in SMEs, however, is often a Herculean task.

In the productive systems of footwear, ceramics, textiles and toys, entrepreneurs generally consider that their human capital is adequate to handle the machinery. They feel that they are able to respond to technological changes arising in the sectors, in spite of the fact that the study shows there is still only modest presence of personnel with graduate degrees in the firms (4.6 percent; Vázquez-Barquero et al 1999). Only the most innovative leading firms, mainly in the ceramics sector, are aware of the lack of skills and knowledge of available human resources and encounter difficulties in contracting skilled staff in the area.

The Role of Technological Institutes in the Region of Valencia

Since 1985, the year when the transfer of authority in all areas in Spain was completed, there has been a regional industrial policy in the Region of Valencia. The stimulus, promotion and coordination of this industrial policy is the responsibility of the IMPIVA (the Institute of Valencian SMEs). This public organism offers various services and activities to firms, supplied through the IMPIVA network, which consists of 15 Technological Institutes, four Business Innovation Centres (BICs), the Technological Park in Paterna and the IMPIVA itself. The Technological Institutes are intermediate organizations created jointly by the regional administration and entrepreneurs in the various sectors, where the firms participate in the funding of the Technological Institutes.

At present, the four Technological Institutes studied here are self-funded at an average rate of 60 percent, corresponding to income received for dues and services rendered, in comparison to an average rate of 15 percent in 1986/1987. They attempt to improve the competitiveness in industrial SMEs in the Valencian productive system through the axes of IMPIVA policy: technological modernization of the firms and diversification of activities (Table 9.3).

Table 9.3 Services offered by the technological institutes in the region of Valencia

• *Information and documentation*, through publications and information events, on any topic related to the sector and official grants for innovation. They also inform and advise in matters of design, fashion and CAD-CAM services.

• *Promotion of technical studies*, of materials, raw materials, productive processes, machinery and finished products, automation, CAD-CAM, computer applications, energy saving, plant optimization environmental pollution and R&D.

• *Testing and laboratory analyses*, on raw materials, materials and finished products. In this field, standardization, homologation and certification are included with the goal of improving the firms' industrial quality.

• *Consulting on technology transfer*, a well-developed field due to the knowledge of technicians, R&D projects carried out in their own laboratories or as a result of participation in European projects, transnational missions or collaborations with universities and investigative centers.

• *Human resource training*, through scholarships and recycling and specialization courses for technical personnel and professionals.

Source: IMPIVA. 1997.

The 15 Technological Institutes can be classified according to the characteristics of their sectors. The first group is directed toward activities that are already established in the territory, providing services in the most important mature sectors of the productive structure of the region. The second group attempts to promote new activities in the industrial system of the region: bio-mechanics, optics, computer science and electrical technology.

The four Technological Institutes selected from among the first group were: INESCOP in the footwear industry, AICE in ceramic activities, AITEX in the textile sector and AIJU in toys. These institutes adopt two approaches, one territorial and the other sectoral, corresponding to the basic characteristics of the productive and entrepreneurial structure in the industrial districts of the regions. Territorially, centres are located in the most important productive nuclei in the region, that is, in those territories where there is a concentration of industrial activity and related tacit knowledge. On the other hand, the sectoral specialization of the centres is instituted in response to difficulties encountered by the firms in mature sectors. Finally, the Institutes are aimed at satisfying the needs of SMEs, the type of firm prevalent in the region.

In fact, the establishment of the Technological Institutes responds to a feeling of urgency on the part of politicians advised by economists and technology experts as to the importance of technology in economic growth. The legislators' objective is to promote technical change and innovation in the firms in their territory in order to achieve the technological frontier already existing abroad, particularly in EU member states. It is hoped that the historic industrial nuclei would not only persist but ultimately become vehicles for social improvement and welfare by empowering their technological possibilities. But, what have the Institutes done to improve their performance over the years of their existence?

According to official IMPIVA data, arrived at through various indicators that attempt to quantify to what use the firms put their services, since 1985 the number of firms using the services provided by Technological Institutes increased considerably (Table 9.4).

In the initial phase of the Technological Institutes, the demand for support services by firms in traditional sectors showed the need to increase the quality of their products, not only to meet growing competition, but also to comply with standards imposed on goods in the context of European integration. Thus, the Institutes' testing laboratories have constituted the principal function of the Institutes since the second half of the eighties, and particularly since 1992, when the European Common Market was consolidated. In general, these organisms have developed experiments and quality testing on raw materials and intermediate and final products, as well as the accreditation, certification, standardization and homologation of products manufactured by firms. Thanks to this situation, the great majority of client firms can now be characterized as 'adapting SMEs' whereas in the past they were imitators.

The results show that training services are used as a platform for technological transfer among firms in all the Institutes and, from the beginning, these services are in great demand. Nevertheless, technological assessment and R&D projects services, the most important services for future progress because of their greater component of knowledge, are used less in comparison with the rest of the services rendered. Only AICE stands out due to its close ties to the university and its determined research policy within the ceramics sector (Table 9.4).

Table 9.4 Services provided by technological institutes: INESCOP, AICE, AITEX and AIJU, 1996

Technological Institute	Lab. Testing %	Tech. Assess. %	R&D Projects %	Training %	Info. %
INESCOP	24.25	12.38	42.13	18.18	91.94
AICE	13.73	77.18	32	4.81	3.33
AITEX	10.88	0.47	12.92	52.42	0.015
AIJU	51.12	9.95	12.92	24.57	4.7
TOTAL (100%)	77140	1898	178	2991	45454

Source: IMPIVA, 1997.

This demonstrates that little attention has been paid to R&D activities and creation of technologies by these centres. Thus, until 1992, there were practically no research projects developed in the Institutes. It is only after 1992 that this type of project begins to appear due to a greater collaboration between Institutes and firms in the field of technologies and innovation and in the growing participation of the Institutes in regional, national and, particularly, European programmes.

The Technological Institutes and Their Accomplishments

As seen above, the Technological Institutes have fulfilled their role as transmitters of information in the service of mature industrial sectors that have an adequate level of tacit knowledge. These Institutes, in symbiosis with the entrepreneurs, constitute the major reason industries in the case study regions absorbed ideas from outside sources. The Institutes cooperate in the adaptation of those ideas with the entrepreneurial fabric in order to "match" the labour factor (through the providing of services and formation) to the capital factor (through the diffusion of embedded knowledge) in the Valencian productive system.

It has also been seen that these Institutes have been unable to 'create ideas' in order to attain greater endogenization of the productive process. Why has this institution failed to create codified, that is, embedded knowledge? We believe there are three main causes, related to the nature of the sector, the lack of public design and Europe as a nation.

The nature of the sector One of the characteristics of a mature sector is that its technical level is relatively less sophisticated, the diffusion of its technology will tend to be more accessible at less cost. Therefore, most firms are not sufficiently motivated to invest in R&D because acquiring the technology and design that is standard in the sector is not a barrier. Most firms feel they can continue being successful in the market without having to innovate. Many of the Spanish firms analysed reported that, while investment in R&D may be 'chic', they do not see the need to do so in the light of their relatively good economic results. In the long term, this thinking is in error.

Lack of public design The Institutes were created in imitation of those existing in various European countries. However, this type of institution was established in Germany, The Netherlands or Italy, for example, to implement industrial policy oriented toward applied research while in Spain there has not been clear public support for applied research. Due partly to industrial backwardness in the country which, with exceptions, was delayed until well after the Second World War, engineering schools did not carry out research activities. Their objective was to simply transmit, albeit with a lag, theory and machinery from abroad. Spain's political isolation until the seventies only worsened the situation and, consequently, the technological level with respect to the rest of Europe suffered a considerable lag. Moreover, the increasing presence of multinational firms in Spain, although a modernizing factor in the economy, inhibited national research efforts both concerning R&D in large firms and in their subcontracting networks with SMEs. The knowledge chain flowing from basic and/or applied research to the industrial fabric did not exist and barely exists even today.

Europe as a nation Since it was incorporated into the EU in the eighties, it has been increasingly difficult to distinguish Spain as a country from Spain as a region of Europe. The industrial integration achieved among member states is particularly significant. Industry was, and probably still is, the most integrated market in the EU: no tariffs, no technical barriers, and no quotas or contingencies. Because of this lack of barriers, uncertainty is not generated among the agents nor is there significant deviation among the respective markets. Valencian firms in the target sectors have little need for a national manufacturer of capital goods, as would have been advisable in earlier periods in which commercial costs were high. Consequently, local firms, once again, will not be sufficiently motivated to demand national capital goods. This lack of incentive is reinforced by the risk that the capital goods nationally produced may be inferior to those of other European regions.

Conclusions

Firms in the mature industrial districts of the Region of Valencia are neither knowledge-capacitated nor are they capable of generating knowledge at the rate necessary to approach the technological frontier. The result is a low level of R&D activity in the local system and a lack of incentives to patent their products, except in the case of the small group of leading firms that are beginning to carry out research and protect the results obtained. However, considering the general situation of the systems analyzed, if firms on the whole wish to remain in the market, they must implement responses that avoid loss of competitiveness.

These responses are twofold. First, old capital goods have been substituted for new generation equipment and, second, changes in product design have been affected. Both ways involve processes of incremental innovation through the adoption and adaptation of external technologies. Both alternatives act simultaneously, making it possible for many firms in the mature sectors to show an adequate presence in the international arena.

In this context, the Technological Institutes of the Valencian Region significantly contribute to the perseverance of firms in the market and to their technological improvement. The services provided by the four Technological Institutes examined in this study have helped to improve product quality and to upgrade the skills of the work force thus facilitating the modernization of productive processes within the firms. This has brought about an increase in sales and greater access to foreign markets. In short, competitiveness has been improved. The Institutes have mainly emphasized support and diffusion of basic standards among firms within their areas of influence, thus playing an important role in the diffusion of technological practices in order to integrate firms into existing European industrial usage.

However, the present innovation process, based on the acquisition and subsequent adaptation of exogenous machinery or designs, implies that firms have resigned themselves to not generating technology and, therefore, to dependence on embedded knowledge created by others. The Institutes were not prepared to meet the needs of the firms once they were inserted into the European industrial environment and routine. In general, firms demand changes and improvements in services provided by these centres. The most innovative firms demand support for manufacturing of machinery and R&D projects, research and technological assessment services on productive machinery, processes and R&D and more advanced training services.

If there is insufficient attention to these issues, the industrial sector is forced to depend on the purchase and/or imitation of foreign technology. Since at present there is only a small percentage of leading local firms capable of creating technologies, the Technological Institutes would do well to increasingly direct their activities toward joint research programs with firms, perhaps leading firms in particular. If such alliances can be forged, the rest of the industrial and labour fabric within the territories studied may indirectly benefit from the knowledge created. Therefore, R&D projects should receive priority attention from the Institutes in the immediate future. Moreover, these Technological Institutes of the Region of Valencia, specialized in mature sectors, must also be capable of encouraging the production of knowledge embedded in machines. They must act to create awareness and provide support for the establishment of firms producing capital goods.

Acknowledgements

Javier Alfonso and Ana I. Viñas thank the Universidad Autónoma de Madrid for providing travel funds to attend the IGU's Meetings in Dongguan where this paper was presented. Antonio Vázquez Barquero and Ana Isabel Viñas Apaolaza thank the IGU for travel support to attend these meetings. We also thank the DGXII of the EU for funding our research.

References

Alfonso, J. (1999), 'Sobre la Dinámica del Cambio Socio-económico', in Innovación, Competitividad y Política Tecnológica. Ed. V Congreso RICTES (Ciencia, Tecnología y Sociedad), Santiago de Compostela, España.

Arrow, J.K. (1962), 'The Economic Implications of Learning By Doing', *Review of Economic Studies*, vol. 29, pp. 155-173.

IMPIVA (1997), *Memoria de Actividades*, 1996/97.

Jones, C.I. (1998), *Introduction to Economic Growth*, W.W. Norton & Company.

Lucas, R. (1988), 'On the Mechanics of Economic Development', *Journal of Monetary Economics*, vol. 22, pp. 3-42.

Marshall, A. (1964), *Economics of Industry*, MacMillan & Co Ltd., New York (1st edition 1892).

North, D.C. (1990), *Institutions, Institutional Change and Economic Performance*, Cambridge University Press, Cambridge.

Ottati, G. dei (1996), 'El Distrito Industrial y el Equilibrio entre Cooperación y Competencia', *ICE*, vol. 754, pp. 85-95.

Romer, P.M. (1986), 'Increasing Returns and Long-run Growth', *Journal of Political Economy*, vol. 5, pp. 1002-1037.

Romer, P.M. (1990), 'Endogenous Technological Change', *Journal of Political Economy*, vol. 5, part II, pp. S71-S102.

Sweeny, R. (1993), *Analytical Study of the Knowledge Acquisition and Technology Transfer Networks of the Universities and Economy of the Community of Valencia*, SICA Innovation Consultants Ltd., Dublin.

Vázquez-Barquero, A. (dir.), Alfonso, J., Sáez, A. and Vinas, A. (1999), *SME Policy and the Regional Dimension of Innovation: The Spanish Report*, SMEPOL report no. 6. DG XII, European Commission.

10 The Internalization of Knowledge in Manufacturing Value Added: The Geography of Ericsson's Mobile Systems, and the China Connection

Claes G Alvstam

Introduction

The issues of the optimal location, as well as of the local, regional and national economic impact of manufacturing production, have constituted the core of neoclassical theorizing about location in Economic Geography ever since the classical contributions of Launhardt, Weber, Lösch and their early successors. Accordingly, there is a dominance of physical and visible parameters in location studies. Relatively less attention has been paid to the deeper understanding of the spatial behaviour of 'invisible' components within the manufacturing sector, such as research and development (R&D), technical innovation capacity, technical cooperation synergies, and supporting management functions in general. Despite several attempts, at multilateral and national levels, to improve the general scope and quality of information on independent service production and trade, there are still many conceptual problems and shortcomings, as well as lack of a widely accepted organization for the gathering of primary data, regarding these new phenomena (Bryson and Daniels, 1998; Daniels, 2000; Harrington, 1995; Illeris, 1996).

With the rise of high-technology, and service-intensive industries in general, Industrial Geography has experienced a vast growth of research on the themes of knowledge and learning as essential parts of the basic understanding of the spatial behaviour of manufacturing firms. A particularly interesting topic is the growing share of R&D in total manufacturing value added. Traditionally, the cost of knowledge was internalized in the compound gross output of manufacturing, and only recently has it been distinguished as an essential part of the transaction value of physical production, comparable to other cost factors. Simultaneously, there has been a parallel growth of interest in the general spatial aspects of economic systems, both from contemporary mainstream economics (e.g. Krugman, 1998), and from business economists and policy-makers (e.g. Porter, 1990).

The growth of high-technology manufacturing that, as a result of new innovations in the late 1940s, gave rise to a new branch of industry and a range of completely new products and applications, the electronics sector, marked a significant change in the role of technical and intellectual knowledge within the firm. The importance of keeping pace with the new technical revolution was widely felt. Indeed, a massive building up of research laboratories and production development units was initiated within almost all electrical and non-electrical engineering sectors. The first steps were also taken in the direction of creating independent units in order to maintain a better control of the ever increasing research costs. The picture that we observe today is that of R&D gradually becoming independent from manufacturing, geographically as well as structurally. Consequently, there is an even greater need to distinguish between different types of R&D taking place within the firm and between firms.

This chapter focuses on how to describe, measure and analyze the geography of intellectual capital in physical and in invisible production. It is based on the assumption that the intellectual capital within the firm accounts for an increasing share of the gross output value of the production within traditional manufacturing as well as in pure services. As long as the investments in innovation capacity, technical and intellectual skills contributed to a less significant part of the added value than costs for raw materials, energy, land and capital, the direct costs for R&D capacity were seen to be, if not negligible, relatively marginal for understanding the spatial behaviour of the firm, at least until a few decades ago.

Furthermore, the desire of firms to form various strategic alliances and cooperative research endeavours with complementary organizations and other forms of business networks, assumes value can be added to products beyond that achieved by 'go-it-alone' strategies (Hakansson and Snehota, 1995). In these cases, the compound added value is shared between many different actors, not necessarily owning or controlling each other, but rather taking advantage of mutual benefits (Mowery et al 1996). One important consequence of these trends is the growing gap between various descriptions of company size, in which the traditional measures of the size of revenues, profits, assets and the number of employees are overtaken by its market capitalization. While the ratio between unit price of stocks to earnings per share (p/e ratio) in traditional industry sectors usually varies between five and 20, the corresponding quota in sectors that have been – a little too carelessly – labelled 'the new economy' occasionally exceed 80-100. This is despite the large-scale 'market adjustment' that has taken place since the peak of demand for knowledge-based firms in early 2000.

Even though recent developments in international stock markets have highlighted the vulnerability and over-valuation of 'dot.com' companies, it is quite clear that 'traditional' and 'new' industrial sectors are valued differently. From the New York-based Nasdaq exchange there are still numerous examples of recently formed high-valued companies that lack profits, but nevertheless attract venture capital. The logic behind such an apparent paradox is that investor speculations about future profits within the firm are not necessarily deducted from active items in the balance-sheet, but from the estimation of the intellectual capital that is continuously upgraded within the firm and within various types of combinations of competence between the firm and its current and future research cooperation with strategic alliance partners, (Feldman, 2000; Sveiby, 1997; Stewart, 1997; Zucker and Brewer, 1998). Needless to say, these links are awkward to trace and map.

Empirically, the chapter focuses on the Swedish firm, Telefonaktiebolaget L M Ericsson. Comparisons are also made with its close rival, Finnish Nokia, in order to demonstrate that two similar companies can generate different types of new knowledge, and thus to represent various types of intellectual value added. In order to reveal the shortcomings of conventional methods of describing and analyzing the geography of intellectual capital and knowledge generation, Ericsson's activities in China are given special emphasis. China is the largest long-term market for Ericsson, and at the same time the Chinese state is actively pushing for the establishment of local R&D centres between foreign investors and domestic research units. In this context, China is a good pilot case for similar hopes in other large, potentially emerging markets for mobile communications in developing countries, including India, Indonesia, Pakistan, Bangladesh and Vietnam. The information base for the study includes open interviews and telephone conversations with representatives of Ericsson during the period, March, 2000 to September 2001, as well as the annual reports of Ericsson and Nokia, 1997-2000.

New Geographies of The Production of Commercialized Knowledge

Seven general observations regarding the new organizational and spatial contexts of technical and commercial knowledge activities in manufacturing production can be summarized:

a) The new creation of goodwill: Until recently, the physical good, be it a final consumer product, an intermediate input, or a machine, or an apparatus aimed at creating new added value in other manufacturing sectors, represented the image of the company. The costs of production were predominately internalized in the physical product, together with its brand name and its goodwill (Urry, 1987). These assets have gradually become less tangible and more related to the image and reputation of highly sophisticated production and technological pole-position.

b) More open distinction between procurement of goods and services respectively: In cases where firms acquired knowledge, patents or technical skills from other firms, the costs involved were included in the aggregated expenditure of industrial inputs and/or semi-manufactured goods, and were not seen as an exogenous variable in the location decision of the acquiring agent. These patterns have now broken off in many branches of industry (Coffey, 1995).

c) Externalization of supporting functions to manufacturing: After having been located in close contact with physical production, and organized as a wholly-integrated and completely internalized part of the business, many of the staff functions surrounding manufacturing have during recent decades tended to become separate, independent, specialized, externalized and outsourced activities, not more closely related to the principal company than any other supplier or subcontractor (Allen and Pryke, 1994; Bryson, 1997; Walker, 2000).

d) The visualization of the revealed overhead costs in the manufacturing enterprise: Cost-chasing, downsizing and the creation of 'lean manufacturing' have also highlighted the costs of labour-intensive administration and general management, whose contribution to manufacturing value added is more subtle than pure manufacturing. The extremely high taxation of labour compared to capital have, in some European countries, further contributed to cutting off staff within activities such as secretarial services, security, invoicing, factoring, computer maintenance etc. Outsourcing of these activities and the creation of an independent business services sector has thus been a means to implement better cost-control, as seen from the manufacturer's point of view (Lindahl and Beyers, 1999).

e) New technological opportunities: The development of information and communication technology (ICT), particularly the linkage with the organization of economic space, and the emergence of entirely new forms of relations between the firm and its external suppliers of various business services. The long-term effect of ICT is increased decentralization, growing spatial dispersal of various productive activities within the firm, and the creation of new 'vassal' service enterprises offering specialist competencies in multi-client networks (Alvstam and Jonsson, 2001; Castells, 1996; Park, 1996; Warf, 2000).

f) Geographical dispersal of previously hidden assets: In the past, the intellectual capital of company R&D units were most often tightly connected with general management functions and located in close vicinity to the corporate headquarters. Accordingly, there was little need to treat the innovative capacity of firms as independent quantities. The tendency for sectoral clustering within limited areas of high skill and competence, sometimes held together by forces of inertias despite emerging opportunities is still evident (Daniels, 1995; Feldman, 2000; Gertler, 1995; Illeris, 1994; Malmberg et al 1996; Markusen, 1996). However, geographic concentrations of industry are no longer seen as the dominant spatial form, as they were once were.

g) Bringing out of previously hidden assets and liabilities: The physical good, be it a final consumer product, an intermediate input or a machine or an apparatus aimed at creating new added value in other manufacturing sectors, represented the image of the company, and the costs of production were predominately internalized in the physical product, together with its brand name and its goodwill. Now, the notion of a value-chain has become more explicitly linked with the rise of high-technology manufacturing and independent professional business service production (Porter, 1990; Lindahl and Beyers, 1999). The R&D costs, as they are recorded by public corporations within the manufacturing sector, seem to only partially cover the real investments, due to the large number of externalized intra- as well as extra-corporate research units. These observations potentially apply to research-intensive manufacturing, where the pace of technical development is high, such as electronics and pharmaceutical industries, but also to knowledge- and research-intensive business services companies (Almeida, 1996; Almeida and Kogut, 1997; Bryson, 1997; Daniels, 2000).

Geographers have in general been reluctant to study or even to observe the logic of cost accounting within the business firm, leaving this field to business economics in general and to the sub-fields of corporate finance, accounting and cost-benefit analysis in particular. Consequently, the distribution of different types of costs within the firm has been treated as a black box. Alternatively, the official records published by firms in annual reports have been accepted as a point of departure in analyses of the spatial distribution of costs. However, the book keeping of types of costs and cost centres has not aimed at identifying their geographical structure. Rather, such accounting has reflected concerns over the optimization of profits within the limits and restraints of tax legislation in different countries, and these accounts (and cost centres) may not reflect the actual spatial pattern of costs. Thus, an independent research company within a MNC, operating directly under the Group Executive Board, may be legally treated as a part of the Group Headquarters, and the compound R&D costs accordingly reported in the home country of the MNC. In reality, there may be many small research units dispersed around the world. In another case, the recorded research costs may be decentralized to independent companies in various companies, wholly-owned by the Group, while the 'core and brain' of the research efforts is in one place.

The greatest challenge for analyzing the spatial behaviour of firms within 'the new economy' is the identification of the geographic domain of R&D cooperation between companies with related activities, each searching for new consumer applications created by the continuous development of the new information and communication technology. Such strategic alliances, single-purpose joint ventures and more or less temporary agreements of research cooperation are motivated by a common insight that no single company is able to afford the costs of developing new knowledge completely alone, and that no single company possesses all technical capabilities that are needed to compete and to survive. The implicit metaphor is a mosaic of fragmented knowledge amalgamating to form new applications. If valid, this observation has important implications for research on industrial clusters, the geography of competitive advantage, and, in general, on firms in networks. It may even be argued whether it is relevant to apply conventional geographical thinking, expressed in relation to forces of concentration versus dispersal, to such a volatile and elusive quantity as the intellectual capital of the firm.

Furthermore, special attention should be paid to the role of the state regarding its role to stimulate the creation of new knowledge and its aims to contain newly generated knowledge within its own territory. It may be argued that the role of the state has been strengthened rather than weakened, despite numerous predictions of the opposite. Thus, state policy is shaping the new context of information technology as illustrated in the telecommunication industry by the national auctioning of third generation (3G) licenses to domestic and foreign net operators in several European countries. The supranational level also has an important role to play in order to understand how knowledge and intellectual capital are spatially concentrated or dispersed. The UN body, the International Telecommunications Union, for example, sets the rules in establishing common standards for mobile communications, such as the WCDMA (Wideband Code Division Multiple Access) standard adopted in November 1999 under the label of IMT-2000 Direct Spread.

The telecommunications industry provides a good example of the new challenges facing the spatial identification of added value in manufacturing. Indeed, the physical products of telecommunications, such as the handset phones, the pylons and the terminals, comprise only minor shares of the compound value added in the industry. The current introduction of the third generation of telecommunications, the mobile Internet and the mobile data communications, the rising shares of invisible service value creation will become even more enhanced.

The remainder of the chapter elaborates on these ideas with specific reference to Ericsson, in comparison to Nokia.

Ericsson: Towards a Geographical Interpretation of Value Chains

Short Historical Background

LM Ericsson, the Swedish transnational giant in designing and supplying mobile-telephone networks for operating companies, was founded by the engineer Lars Magnus Ericsson in 1876, and initially manufactured physics instruments and (since 1892) table-top telephones. Due to severe competition in the domestic market, mainly from Bell, Ericsson expanded abroad quite early, and was particularly successful in Russia, China, Mexico and Argentina during the last decade of the 19th century and the first decade of the 20th century. For a while Russia was so attractive that L M Ericsson considered moving his head office to St. Petersburg during the years around 1910. The World War and the revolution stopped this plan. During the inter-war years and the first decades after WW II, Ericsson continued to grow abroad within its core business to provide national telephone companies with the necessary infrastructure, and to supply telephones. Due to a number of financial crises, Ericsson was in reality controlled by the American ITT for a couple of decades. In the 1960s and 1970s telephony seemed to be a mature business with falling profit margins, and the company attempted to broaden its activities into the fields of office automation, radio systems and radar electronics. These initiatives were commercial disappointments, but created research competence in radio technology that later proved to be valuable.

What saved the company was the successful launch in 1978 of a completely new technology for telephone exchanges, the AXE system, that grew to an enormous commercial success, achieving a world market share in the middle of the 1990s of about 10 percent. A few years after the launch of the AXE, the development of mobile phones marked a new revolutionary period in Ericsson's history. An early leading position in technology based on the AXE system's flexibility and capacity, and its acquired know-how in radio technology, combined with the early development of a Nordic standard (NMT 1981) and a rapid growth in the demand on the domestic market, paved the way for international success. Thus, in 1991 Sweden accounted for less than four percent of its globally diversified sales (Figure 10.1). In 2000 Ericsson accounted for an annual turnover of US$30 billion and had 105,000 employees in 140 countries. The pace of market growth is immense. The number of mobile users world-wide will surpass fixed-line users by

2002 while the mobile Internet is expected to overtake fixed line Internet in 2005. It is predicted that the number of mobile phone subscriptions will exceed 1 billion in 2002, and that there will be roughly 1,000 million users of the 3G mobile Internet by 2005 (www.ericsson.com).

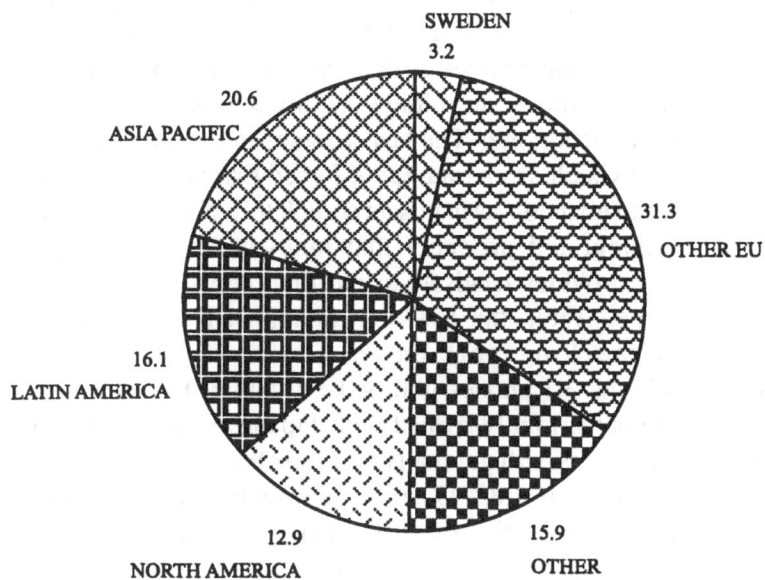

Figure 10.1 Ericsson's sales by geographic regions, 2000

Future Strategies – From Manufacturing to Knowledge Management

Like its main Nordic rival, Finnish Nokia, Ericsson adopted an integrated strategy, in which it was able to offer the final physical product as well as the compound infrastructural systems network for mobile telecommunications for business as well as for private customers. While Nokia developed a true excellence in the consumer products segment, Ericsson's main strength lies within the mobile systems sector, in which it reports a 30 percent global market share. Within the new segments of GPRS (General Packet Radio Service) and 3G, Ericsson's worldwide market share exceeds 50 percent. On the other hand, its global market share of sales of handsets is only 11.5 percent, compared to Nokia's 28 percent and Motorola's 16 percent (1999 figures). Roughly, in 2000, 21 percent of Ericsson's global turnover is accounted for within the consumer goods sector (i.e. the mobile handsets) and the share is gradually decreasing. Nokia's corresponding share, on the other hand, is 72 percent.

While Nokia recognized that it needed to focus on creating a consumer-oriented brand, Ericsson has been more concerned about the technology behind the phones. Ericsson's focus on technology rather than on simple logistics has recently caused delays in delivering products to the market. With typical product cycles of 18 months, missing the first six months makes it much harder to cover the initial R&D investments. Nokia has tighter control over its production line. The heavy losses in the consumer products segment, that were visible from mid-1999, forced Ericsson to take the tough strategic decision to abandon its handsets completely. Such a transformation from being an integrated producer that innovates product, constructs infrastructure and provides handsets to the final consumer, will likely be taken in two steps. First, the low-end models will be abandoned and outsourced to the Singapore-based global assembly company, Flextronics, while the 3G phones will be kept within the family for a few years. A new company, jointly owned with Japanese Sony, Sony Ericsson Mobile Communications, will gradually take over the entire consumer sector.

With the move towards 3G networks, which will enable much more data to be sent via mobile phones, Asian manufacturers, such as Korea's Samsung, will be preparing an assault on the European market. Ericsson's present strategy is to streamline manufacturing, and establish cooperation with other partners in R&D.

The strategic decision to transform the company to concentrate on research, infrastructure, systems development and business-to-business solutions may have been accelerated by the heavy losses. On the other hand, the explanations behind the losses revealed the vulnerability of outsourcing and on relying on independent suppliers. Ericsson admitted that a fire in March 2000 at one of its component factories in New Mexico, owned by Philips of the Netherlands, had caused the big losses in its handset business. The fire left Ericsson with large stocks of unfinished mobile phones awaiting a key component, a silicon semiconductor. Indeed, this extraordinary event revealed a larger problem of vulnerability from dependence on one single supplier. There were strong objections to phase out the production of mobile handsets, even though losses have continued. Thus, the mobile operators who buy infrastructure from Ericsson need a guaranteed supply of new generation handsets, and Ericsson will still need to keep abreast of trends in the handset market to understand how the industry is developing. Even though having a mobile division helps a company to win infrastructure contracts, in practice, most mobile operators do not regard it as vitally important that their network infrastructure provider makes handsets as well. The purchasing decisions are made separately. The network infrastructure business, moreover, has long term relationships and big long term contracts driven as much by technical as against cost, considerations. The mobile phone business involves more suppliers and shorter contracts, led by consumer demand.

In effect, Ericsson reinforced its worldwide leading position in infrastructure for mobile phones. Two opposite tendencies are implicated. First, the gradual externalization process of physical manufacturing, described in Figure 10.2, and second, the gradual internalization of the creation of knowledge (Figure 10.3). It is the latter process that is of particular interest in this study.

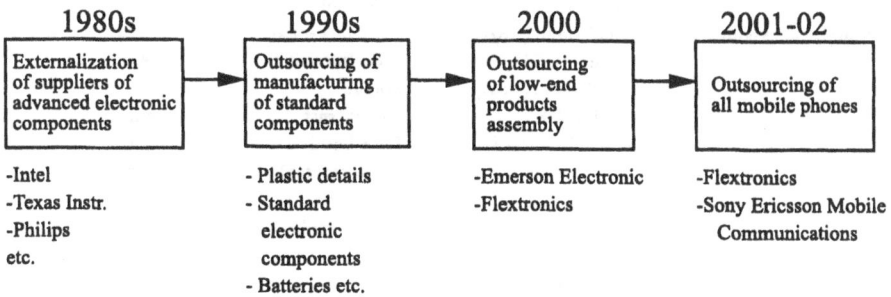

Figure 10.2 The gradual externalization process of physical manufacturing within Ericsson

The internalization of knowledge takes place in a variety of ways, and there is no natural taxonomy for finding a common structure. The consumer laboratories may be seen as a core activity of the future, acting as innovation centres, and market intelligence networks. Until now these have been mainly based in the Stockholm area, symbolizing mental proximity to the previous Group HQ, even though the operational top management has moved to London. The strategic role of the consumer laboratories is in times of rapid technical development, challenged, and in many ways out-manoeuvered, by centres of creation of new technical knowledge. This type of an internalization process takes place mainly through acquisitions. Recent examples are the takeover of Quallcomm's infrastructure division in order to speed up the standardization work, and the acquisitions of the American companies Torrent and TouchWave, as well as Danish Telebit, in order to strengthen Ericsson's position within the IP and data communication solutions. Telebit is the technical world leader in the next generation of the Internet Protocol, IP6.

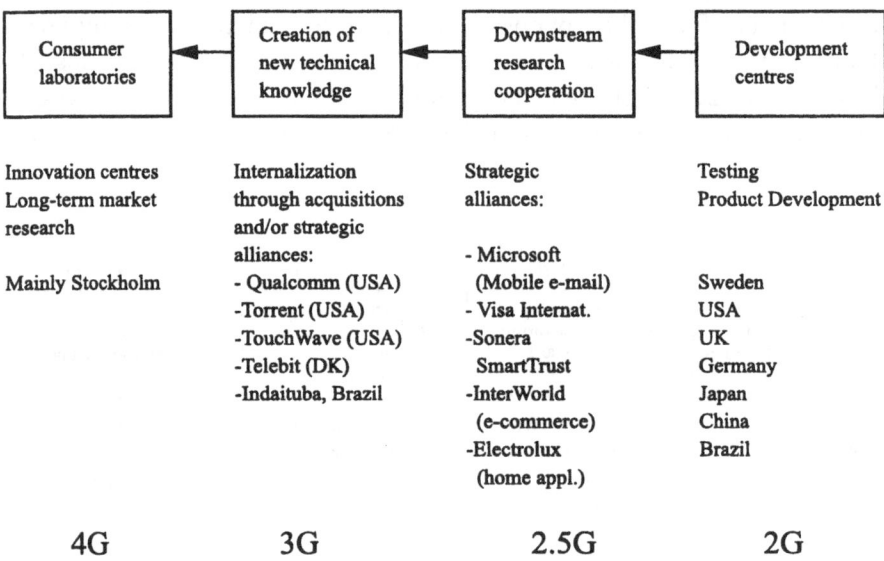

Figure 10.3   The gradual internalization of the creation of knowledge within Ericsson

An even more crucial part of the internalization of new knowledge within the company involves the creation of a large number of strategic alliances and research cooperation deals, aimed at uniting efforts between companies in different fields to find common solutions benefiting all parties. This kind of downstream research cooperation is illustrated by Ericsson's cooperation with Microsoft to create a compound solution for wireless Internet-access and mobile e-mail, and its joint ventures with Visa International and Sonera SmartTrust to assist market leaders to find technical solutions for safe e-commerce transactions. It is also notable that the three main competitors, Ericsson, Nokia and Motorola, have together initiated a common working group in order to develop a general standard for safe mobile e-commerce (MET). The cooperation with the world's largest whiteware appliance producer, Electrolux, aims for a common development of the 'smart home'. Together with the net-based investment broker Charles Schwab, Ericsson has formed a strategic alliance to develop applications for mobile investment services.

It is a typical development that the research units created by these alliances do not have a natural home base. They are characteristically geographically mobile, and are located to points where the involved partners are already based, mainly in North America. In June 2000, Ericsson announced an alliance with the US software group InterWorld to develop wireless e-commerce platforms. InterWorld designs robust, ready-to-use software packages for Internet sales. The wireless solutions, to be used in business-to-consumer and business-to-business sites, can be integrated into existing e-commerce platforms as well as new ones.

The development centres, finally, have the strategic role in being responsible for applied research, testing and consumer adaptations. These will be given an even more important role in the future as the number of application platforms continues to soar. Until now, the technology itself has been the driving force behind growth, while it is expected that the sector will enter a new phase of development characterized by an application-driven growth, ultimately less dependent on technologies and more centred around the users. The development centres are also elements in the corporate strategy to create a sense of 'locality' at distant markets, like Japan and China. More than 26,000 persons (25 percent of the total number of employed) were directly involved within various R&D functions in 24 countries in the year 2000 (www.ericsson.com).

The Power of Mobility

The technological development has been labelled 'the power of mobility'. The velocities in the 3G mobile network will be able to reach up to 2 Mbit/second, largely overtaking today's capacities of 9.6 Kbit/second (Figure 10.4). The long-term objective is mainly to amalgamate global and local solutions derived from the IP (Internet Protocol) to reach a platform of uniform complete applications.

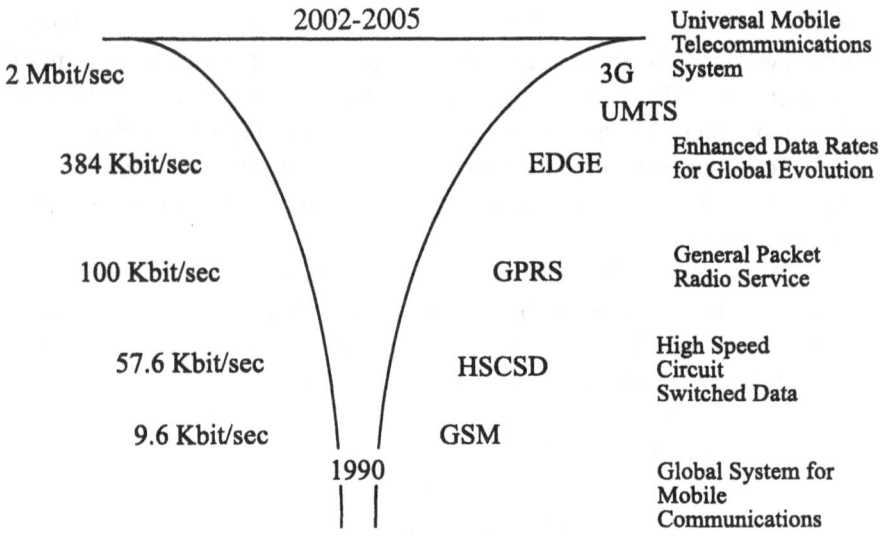

Figure 10.4 The development towards third generation of mobile communications

Source: Nokia, annual report 1999.

With the insight that applications will become less dependent on technology, and gradually become more market-driven, all types of downstream cooperation in finding new 'seamless' solutions for the customer will come into focus. The recent wave of giant consolidations among net operators, exemplified by Deutsche Telekom's acquisition of VoiceStream, and similar mega-deals involving Mannesmann-Orange, France Telecom-Orange, Vodafone-Airtouch-Mannesmann, show that very large companies are required to meet the demands for capital investment in the new applications. The net operators generally aim to spread their upstream cooperation agreements and orders among all large infrastructure and systems suppliers in order to reduce their vulnerability and curb market dominance. Even so, Ericsson plays a leading role in the introduction of the new technology, having captured more than 50 percent of the GPRS contracts, including Vodafone and Turkcell. Ericsson has also been chosen as supplier to seven out of ten operators within the 3G-segment, including customers of Vodafone, NTT DoCoMo and Japan Telecom (Ericsson, annual report, 2000).

The state plays a crucial role in relation to national and international net operators in controlling the auctioning of 3G-licenses. The common logic between the rapid consolidation and transnationalization of the net operators is the logical effect of expectancies of a huge market potential as a result of future demand for new applications. Knowledge is anticipated in all stages of the production cycle, from innovation centres through product development and manufacturers of the new generation of mobile phones to net operators and providers of the infrastructure behind the new types of services.

It is therefore misleading to treat R&D investments as costs in the balance sheet, while the intellectual capital generated is not treated as an asset. In order to decentralize the costs of R&D to the operating activities of the transnational firm, a larger share of the total costs for R&D has been concealed under other headings. Accordingly there is a tendency in many large firms, including Ericsson and Nokia, for the visible and officially recorded R&D investments to stagnate or even decline. In Ericsson's case the share of R&D in relation to gross income has declined from 17 percent in 1993 to an average of 14 percent over the 1997-2000 period (Figure 10.5). In Nokia the same share has stagnated to a level of around nine percent since 1996 (www.nokia.com).

It is a complicated task to estimate the actual value of generated R&D value by using only public data. From interviews in Ericsson, it has been possible, however, to establish an idea of how to measure the compound indirect R&D expenditure by estimating the R&D in various business areas. Even though this method is blunt, and vulnerable to criticism, it is nevertheless obvious that the gap between open and invisible R&D has increased sharply during the second half of the 1990s (Figure 10.5). The next step would be to allocate the value added from revenues in systems sales back to the knowledge centres in the company, and thereafter to estimate the added value in a given geographical location. This step remains to be taken, but preparations have begun to initiate such a measurement.

Ericsson in China

China was Ericsson's largest single market in 1998, accounting for 14 percent of its global sales. Due to the ongoing restructuring among the Chinese operators, and the introduction of a new market regulation, there was a clear stagnation in the activities in 1999 (see Figure 10.1). China lost its top-position in Ericsson's sales to the USA, while its market share decreased to nine percent. Nevertheless, during the 1990s, China has been the fastest growing market for mobile phones as well as for mobile systems infrastructure, and the one with the largest long-term potential. Even though Ericsson's share of the Chinese market for cellular phones in China has tumbled from 35 percent in 1997 to seven percent in 2001, the absolute number of sales has grown from 1.75 million units to 3.15 million units between 1997 and 2001. Nokia on the other hand has been able to maintain a market share at roughly 27 percent (12 million units), while Motorola holds 35 percent (Ericsson, China, phone interview, September 2001).

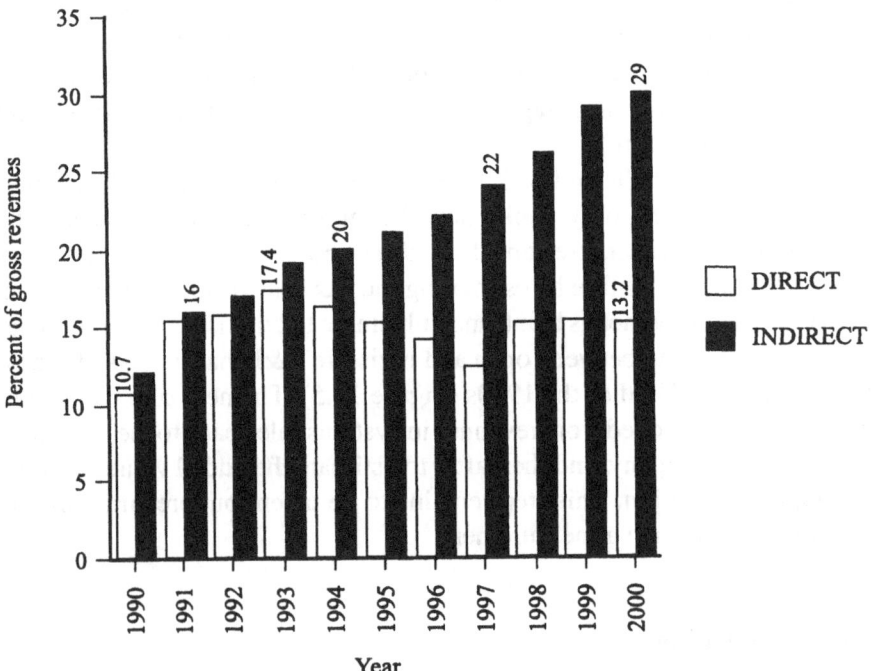

Figure 10.5 Direct and indirect research and development at Ericsson 1990-2000

Consequently, a location in China is necessary for all global actors to effectively access this market, and to gain deep knowledge of customer behaviour, local preferences, and the ability to adapt to local conditions. Through two production and sales companies, Beijing Ericsson Mobile Communication Co. Ltd., and Nanjing Ericsson Mobile Communication Co. Ltd., the company has gradually developed from pure sales of imported products to a higher level of local content. This development occurred in three initial phases: 1) local assembly of components; 2) local assembly of the final product; 3) building up of a local infrastructural system. The next steps were taken in the second half of the 1990s when development centres were established in cooperation with universities in Beijing and Shanghai. These centres deal mainly with applied customer-related R&D, and with technical adaptations to local standards and systems design. At present, there are preparations to contribute to the establishment of science parks specialized in mobile data design and mobile telecommunications, taking advantage of the huge research potential and intellectual capital that is available in China. It is predicted that, if successful, these science parks will develop into the exclusive group of 7-10 centres of basic mobile communication research in different parts of the world that will become the innovation centres for the 4G. It should also be noted that Ericsson's main competitors, Nokia and Motorola, have chosen similar strategies in China. It is expected that Sony Ericsson Mobile Communications, that will be responsible for the China sales as well, will integrate their previous acitivities within the new company. Sony has hitherto concentrated on the high-end segment, while Ericsson has rather worked in the lower price ranges.

The largest peculiarity of Ericsson's business strategy in China related to other parts of the world is the ultimate dependence on domestic state-owned partners in all stages of production, marketing, and sales. Formal cooperation with local companies is thus necessary, and good relations to national, regional and local authorities are even more essential than elsewhere. It is evident that the process towards establishment of special development centres, research cooperation and technology transfer in general has been accelerated by requests from Chinese authorities to proceed in such a direction, and that demands to transfer new technology are part of the negotiation process for new contracts and licenses. From the Chinese side there is a general attempt to spread out its cooperation to as many foreign actors as possible, thus thwarting any single company's ability to build up a too large strength. Examples of agreements concluded in 2000 are Ericsson's US$630 million contract with Guangdong Mobile to expand a GSM network to handle more than 18 million subscribers. The equipment supplier Lucent signed equipment contracts worth US$80 million with CITIC Pacific, and US$65 million with Guangdong Easter FiberNet. French Alcatel also won US$87 million in GSM contracts. These companies and rivals, such as Nortel, may gain from the introduction of Code Division Multiple Access (CDMA) technology to compete with GSM, now that delays in CDMA deployment seem to be resolved. The major domestic CDMA developer, China Unicom, is reported by the China Business Times to have won agreement on terms for ending joint ventures with US Sprint and others accounting for half of the US$1.4 billion in foreign funds that it was ordered to divest. Unicom insisted on progress in this effort before proceeding with CDMA construction.

It should be particularly noted that Ericsson's research activities in China are not recorded according to Chinese value added, but within a central business area for technical development, based in Stockholm. Neither are the Chinese staff involved in long-term technical cooperation and product development in China accounted for in China, but in Sweden, UK and USA, where the legal domiciles of the subsidiaries in question are to be found.

Conclusions

China is a good example of the new type of R&D locations carried out by large MNCs within the rapidly changing telecommunication industries. From the headquarters' point of view, R&D activities are becoming gradually more independent and located in concentrated archipelagos all over the world, and particularly to large, technically advanced nations in North America, European Union as well as in Japan. China, India, Brazil and Mexico are the most likely locations of advanced R&D centres among the emerging markets. It becomes increasingly essential to distinguish between different types of location behaviour between different types of research not always possible to identify for an outsider. A generalized pattern of the relation between the dimensions of concentration/dispersal on one hand and proximity to the headquarters in the home country is sketched in Figure 10.6. Strategic long-term consumer research within Ericsson is still concentrated to the Stockholm area close to the previous group HQ. This is where entirely new ideas are born, and long-term consumer trends and new technical applications are assessed and evaluated.

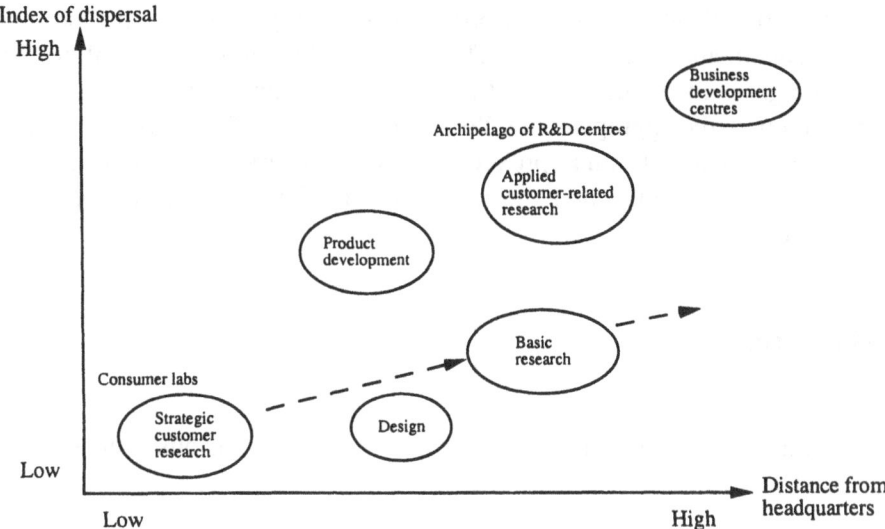

Figure 10.6 Spatial dynamics of research and development creation within the telecommunications industry

The special product development centres are today located at five clusters outside Sweden: USA, UK, Germany, Japan and China. Basic research in mobile communication technology has recently become so specialized and expensive that no single company can afford to bear these costs alone. Accordingly, there is a trend of externalizing this research to a couple of highly specialized centres, often in cooperation with related companies, and in some cases more or less directly together with the main competitors. It is expected that basic research units will continue to disperse, and continue to move away from the headquarters. The most dispersed R&D functions located at the longest distance from home are today various centres of applied customer-related research, as well as business development centres, operating in close vicinity to the customers, solving problems and identifying new opportunities and applications. There were in the year 2000 roughly 2,700 'mini-centres' operating within Ericsson Business Consulting in 36 different countries.

When it comes to developing a theory for the spatial pattern of R&D in rapidly changing high-technology sectors like telecommunications, it is evident that new measures have to be developed in order to better trace the location of the intellectual capital of the firm. This study demonstrates that there is a possible way to treat this problem by starting to identify the indirect R&D investments in companies. Thereafter, added value can be allocated from the gross revenues back to the R&D centres proportionately to their contribution to the aggregated intellectual capital of the firm. One advantage of such a step would be an improved and deeper understanding of the spatial distribution of income, and the preconditions for regional and national competitiveness.

Acknowledgement

I am grateful for the help and insights provided by various managers at Ericsson.

References

Allen, J. and Pryke, M. (1994), 'The Production of Service Space', *Environment and Planning D*, vol. 12, pp. 453-75.

Almeida, P. (1996), 'Knowledge Sourcing by Foreign Multinationals: Patent Citation Analysis in the US Semiconductor Industry', *Strategic Management Journal*, vol. 17, pp. 155-65.

Almeida, P. and Kogut, B. (1997), 'The Exploration of Technological Diversity and the Geographic Localization of Innovation', *Small Business Economics*, vol. 9, pp.21-31.

Alvstam, C.G. and Jonsson, A. (2001), 'Hyper-Footloose Business Services: The Case of Swedish Distance Workers in the Mediterranean Sun Belt', in D. Felsenstein and M. Taylor (eds), *Promoting Local Growth*, Avebury, Aldershot.

Bryson, J. R. (1997), 'Business Services Firms, Service Space and the Management of Change', *Entrepreneurship and Regional Development*, vol. 9, pp. 93-111.

Bryson, J. R. and Daniels, P.W. (eds) (1998), *Service Industries in the Global Economy: Volume 1, Service Theories and Service Employment*, Edward Elgar, Cheltenham.

Castells, M. (1996), *The Rise of the Network Society*, Blackwell Publishers, Oxford.

Coffey, W.J. (1995), 'Producer Services Research in Canada', *Professional Geographer*, vol. 47, pp. 74-81.

Daniels, P.W. (1995), 'Services in a Shrinking World', *Geography*, vol. 80, No. 347, part 2, pp. 97-110.

Daniels, P.W. (2000), 'Exports of Services or Servicing Exports?', *Geografiska Annaler*, vol. 82B, no 1, pp. 1-15.

Feldman, M.P. (2000), 'Location and Innovation: The New Economic Geography of Innovation, Spillovers and Agglomeration', in G.C. Clark, M. P. Feldman and M. Gertler (eds), *The Oxford Handbook of Economic Geography*, Oxford University Press, Oxford, pp. 373-94.

Gertler, M. (1995), 'Being There: Proximity, Organization and Culture in the Development and Adoption of Advanced Manufacturing Technologies', *Economic Geography*, vol. 71, No. 1, pp. 1-26.

Hakansson, H. and Snehota, I. (1995), *Developing Relationships in Business Networks*, London. Routledge.

Harrington, J.W. (1995), 'Empirical Research on Producer Service Growth and Regional Development: International Comparisons', *Professional Geographer*, vol. 47, No. 1, pp. 66-69.

Hotz-Hart, B. (2000), 'Innovation Networks, Regions and Globalization', in G.C.Clark, M. P. Feldman and M. Gertler (eds), *The Oxford Handbook of Economic Geography*, Oxford University Press, Oxford, pp. 432-50.

Illeris, S. (1994), 'Proximity Between Service Producers and Service Users', *Tijdschrift voor Economische en Sociale Geografie*, vol. 85, pp. 185-96.

Kogut, B., Shan, W. and Walker, G. (1993), 'Knowledge in the Network and the Network as Knowledge', in G. Grabher (ed), *The Embedded Firm. On the Socio-economics of Industrial Networks*, Routledge, London, pp. 67-94.

Krugman, P. (1998), 'What's New About the New Economic Geography?' *Oxford Review of Economic Policy*, vol. 14, no 2, pp. 7-17.

Lindahl, D.P. and Beyers, W.B. (1999), 'The Creation of Competitive Advantage by Producer Service Establishments', *Economic Geography*, vol. 75, No. 1, pp. 1-20.

Malmberg, A., Solvell, O. and Zander, I. (1996), 'Spatial Clustering, Local Accumulation of Knowledge and Firm Competitiveness', *Geografiska Annaler*, vol. 78B, pp.85-97.

Markusen, A. (1996), 'Sticky Places in Slippery Space: A Typology of Industrial Districts', *Economic Geography*, vol. 72, pp. 293-313.

Mowery, D.C., Oxley, J.E. and Siverman, B.S. (1996), 'Strategic Alliances and Interfirm Knowledge Transfer', *Strategic Management Journal*, vol. 17, pp. 77-91.

Park, S.O. (1996), 'Networks and Embeddedness in the Dynamic Types of New Industrial Districts', *Progress in Human Geography*, vol. 20, pp. 476-93.

Porter, M. (1990), *The Competitive Advantage of Nations*, Macmillan, London.

Stewart, T.A. (1997), *Intellectual Capital: The New Wealth of Organizations*, Doubleday, London.

Sveiby, K.E. (1997), *The New Organizational Wealth: Managing and Measuring Knowledge-based Assets*, Berrett-Koehler Publications, San Francisco.

Urry, J. (1987),'Some Social and Spatial Aspects of Services', *Environment and Planning, D*, vol. 5, pp. 5-26.

Walker, R.A. (2000), 'The Geography of Production', in E. Sheppard and T.J. Barnes (eds), *A Companion to Economic Geography*, Basil Blackwell, Oxford, pp. 113-32.

Warf, B. (2000), 'Telecommunications and Economic Space', in E.Sheppard and T.J. Barnes (eds), *A Companion to Economic Geography*, Basil Blackwell, Oxford, pp. 484-98.

Zucker, L.G. and Brewer, M.B (1998), 'Intellectual Human Capital and the Birth of U.S. Biotechnology Enterprises', *American Economic Review*, vol. 88, No. 1, pp. 290-306.

11 Environmental Knowledge, the Power of Framing and Industrial Change

Dietrich Soyez

Introduction: Issues at Stake and Line of Arguments

Current processes of the 'greening of industry' are pervasive. They vary considerably, depending on place, economic imperatives, political context, industrial sectors or corporate strategies. There is resistance within industry, rearguard actions, even hollow eco-rhetoric or fake eco-marketing. But ongoing change towards the greening of industry is indisputable. In what follows, these processes are called 'eco-modernization', originally 'ecological modernization' (Hajer, 1996).

Eco-modernization, often triggered by a number of distinct environmental and environment-related pressures, forcing corporate decision-makers to re-appraise the consequences of their actions, is a most important challenge for industry (Schot and Fischer, 1993; Newell, 2000). Moreover, the complexity of eco-modernization processes has stimulated a wide variety of scientific approaches to understanding the nature and implications of this challenge (cf. Gladwin's ironic parable, 1993, pp. 37-8). The point of departure for this chapter is the 'disciplinarity' of these approaches to the greening of industry. Conceptual and empirical lines of inquiry typically reflect the different perspective of individual scientific (natural as well as social) disciplines and the specializations within them, each with their traditional core contents and current fads. A problem with disciplinarity, however, is the fragmentation of knowledge. A crucial theme of my argument is that such fragmentation is barely avoidable but can be partially compensated by building bridges between different islands of knowledge. This argument requires a closer look into principles of knowledge construction. For the purposes of this chapter, the term knowledge is used to designate information embedded in specific experience and cognitive contexts, with its significance stemming from a given system's rationality and problem-solving heuristic (Willke, 1998, p. 13).

Disciplines have a disciplining impact not only on their practitioners but also on knowledge production itself (Messer-Davidow et al 1993). A closer look into the genealogy of disciplines, for example, illustrates the historically contingent and even adventitious way in which ideas and practices came to be grouped into disciplines – and what is taken for granted within a discipline occurs by both pure accident and conscious construction. The maintenance of a discipline, as well as its scientific legitimacy, however, is secured by what is called 'boundary-work', which aims at demarcating core areas, limiting methodological approaches and excluding what is designated as 'non-disciplinary', and even non-scientific. Indeed, it is contended here that similar processes have led to the almost complete exclusion of environmental topics in Economic and Industrial Geography, an issue that will be briefly addressed in the concluding paragraph.

These processes shaping the content, objectives and life of an academic discipline bear a remarkable similarity with what forms the essence of other organizations, such as firms or corporations (Morgan, 1986; Willke, 1998). They too are exposed to constant tensions caused by efforts to stabilize the corporate 'organism', without losing touch with its environments, including the larger societal "value environment" (Fredriksson and Lindmark, 1979). They further rely heavily on processes of disciplining their employees, be it with regard to simple knowledge 'elements', elaborated organizational 'processes', sophisticated 'context' adaptation or fundamental entrepreneurial 'paradigms', keywords according to the 'pyramid of organizational learning' (Willke, 1998, p. 44), as well as in defining the boundaries of responsibilities.

In this chapter, the imperatives of disciplinarity are challenged by a consciously 'un-disciplined', three-fold focus on boundary-permeating work or activity. First, concepts and findings of non-geographic disciplines are integrated in order to contribute to the reconstruction and explanation of specific greening processes in industrial production systems. Second, boundary-spanning capabilities – or inabilities – of important actors are assessed and, third, knowledge issues pertinent to the geographic discussion

– and its deficits – as regards the 'greening of industry' are addressed. A goal of the chapter is to build bridges between Industrial Geography and environmental issues.

With its focus on boundary-permeating work, this chapter pursues an eclectic rather than a unitary conceptualization, with concepts stemming from various fields, including Economics, Economic Geography, Political Sciences, Political Geography, Sociology, Social Geography. These concepts, ranging from 'stakeholder theory' (Freeman, 1984), 'transnational advocacy networks' (Keck and Sikkink, 1998), 'knowledge management' (Willke, 1998) to 'space-producing lobbies' (Soyez, 1997), help grapple with issues that transcend traditional disciplinary boundaries:

The line of argument is as follows. First, a tentative synopsis of eco-modernization in industrial countries since the 1960s is presented, leading to questions about the nature, role and diffusion of 'knowledge' in this context. Second, ensuing discursive struggles and the emergence of new frames are addressed, while the focus is narrowed to non-sovereignty- and non-academia-bound actors, notably non-governmental organizations (NGOs). These actors, seemingly of ever increasing importance, are understood not only as agents of both social learning and space-construction, but also of boundary-permeating knowledge transfer, thus triggering and/or influencing industrial change. Third, this latter is briefly presented with reference to recent case studies. In a concluding paragraph, the findings are generalized in an attempt to overcome the current, almost complete exclusion of eco-modernization issues in Economic and Industrial Geography.

Post-1960s Eco-modernization of Industry: A Synopsis (Based on Western European Experience)

Most observers place the start of post-war industry's eco-modernization, or the rise of modern environmentalism, in the late 1960s and early 1970s (McCormick, 1989; Fischer and Schot, 1993; Jamison, 1996). This chronology was triggered by rapidly increasing industrial production with a corresponding rise of pollution levels, on the one hand, and value shifts with regard to participatory political processes and quality of life concerns, on the other. In a rough periodization four distinct phases can be differentiated (Figure 11.1):

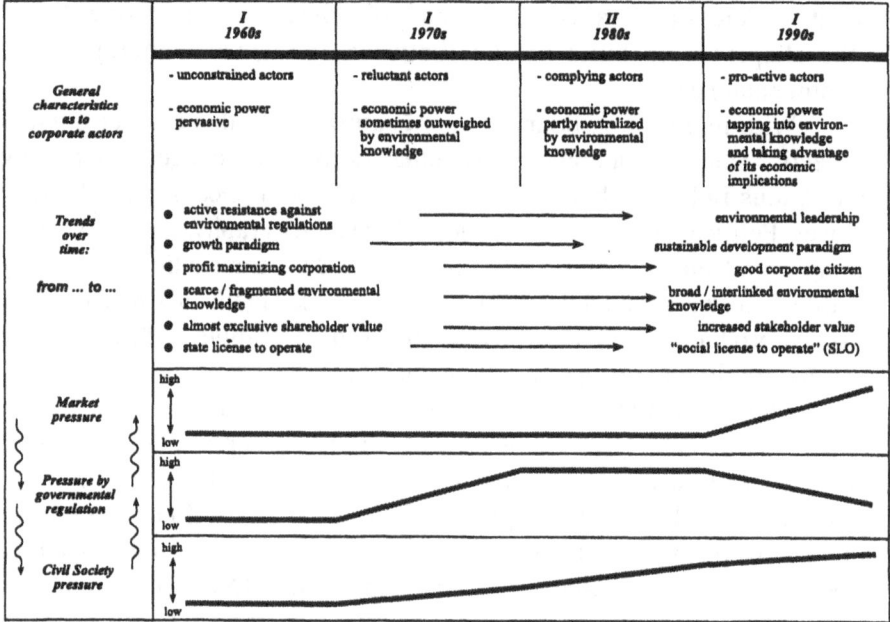

**Figure 11.1 The 'greening' of industry 1960-2000 in Western
industrial countries – a tentative synopsis**

a) prior to the late 1960s, industrial corporations can be regarded, almost without any exceptions, as powerful, un-restrained actors, imposing their will – and their environmental impacts – on both governments and communities. Regulatory frameworks, such as licenses to operate, did exist. More specific regulations, however, were only introduced when serious health or property risks were obvious. The growth paradigm was unchallenged;

b) during the 1970s the regulatory web was rapidly tightened as both ecological impacts, general environmental awareness and specific environmental knowledge increased. Pollution-abatement was imposed step by step, but almost exclusively in form of end-of-pipe technology. Industry's response was slow, reluctant, at times outright resistant;

c) the 1980s are marked, on the contrary, more often than not, by industry's growing compliance. Apparently, the industrial sector's economic and political clout to resist environmental concerns became outweighed by an indisputable record of disquieting impacts at local, regional and increasingly at national and international levels. Furthermore, the environmental movement and its allies, such as dedicated governmental authorities, research institutions, segments of the general public and the media, were increasingly able to form a counter-balance to industrial interests, sometimes by going international and adopting new concepts (such as 'sustainable development', see below);

d) during the 1990s, the concerns caused by unsolved old issues (such as acidification) and unexpected new issues (such as the ozone hole) become so high that many corporate decision-makers have turned into eco-pioneers, going 'beyond compliance', also beyond the 'end-of-pipe' philosophy, by designing environmentally more compatible processes and products. Global environmental issues are widely perceived. Sustainable development, broadly conceived as the will to balance economic, social and environmental factors in an inter-generational perspective, is adopted by many individual and institutional actors, if sometimes only as a catchword. The new power of advanced environmental knowledge is now often stronger than the traditional economic and political power of industrial (and union) interests as it helps to reach more ecological understandings of economic efficiency.

I contend that this evolution has resulted from processes of organizational learning that have been pushed by various kinds of pressures. These pressures most notably refer to: (a) regulatory pressures (for example, by state agencies); (b) market pressures (for example, by competitors, customers, suppliers 'going green'); and (c) civil society pressures (for example, by grassroots NGOs). These pressure types are not independent entities as they overlap and influence each other, for example, NGOs that organize a consumer boycott or state agencies that cooperate with either industry, civil society actors or both (cf. vertical arrows in Figure 11.1). In a recent World Bank report, these types of interactions among governments, markets and communities are even heralded as the 'new model' for the greening of industry (World Bank, 2000).

Over time, there are distinct patterns in the role of the leading actors in environmental initiatives, currently featuring a decreasing state commitment, whereas both market and civil society pressure seem to constantly increase (Figure 11.1). These changes are knowledge-driven or knowledge-dependent. Consequently, the underlying processes of knowledge production and transfer must be discussed, before the role of civil society actors can be analysed.

Knowledge Construction and Knowledge Transfer

Specifics of Knowledge Construction

Traditionally, the perception of knowledge construction and transfer was quite simple. Knowledge, at least advanced scientific and technical knowledge, was mainly regarded as expert knowledge. It was perceived as objective and true, although necessarily incomplete, as it reflected reality. Further, knowledge was seen as easily accessible in form of large bodies of literatures and other types of documentation, ready to be picked up by anyone interested in a particular field and readily to be transferred to scholars, students or employees. Current views are much more complex (cf. Haraway, 1991; Nonaka, 1994; Jamison, 1996; Hajer, 1996).

In particular, it is now argued that there is lay knowledge, as well as expert knowledge, and a person may be an expert in fields professional experts ignore. Consequently, depending on context, there is no clear answer as to which type of knowledge is more important. Further, knowledge is not simply a true reflection of reality but is socially constructed and influenced, for example, by interests, power or self-referential processes. From this perspective, the notion of objectivity is dubious, if not false, because knowledge is also situated, thus reflecting different worlds and world views, based on, for example, place, race, class or gender. Some knowledge becomes explicit or codified knowledge, easily picked up or transferred, while tacit knowledge and 'know-who' knowledge often remains inaccessible or extremely difficult to adopt.

Thus, there is not just 'knowledge' in a particular field, but multiple 'knowledges'. There are also different knowledge claims, politics of knowledge, contested and competing knowledge(s) as well as regions, sectors, layers, periods and cycles of knowledge. Furthermore, while some knowledge diffuses globally at an astonishing speed, other knowledges meet insurmountable barriers or need a lot of time to filter down. All this is particularly true for environmental knowledge.

Environmental Knowledge and its (Non-) Diffusion

Traditionally, professional environmental knowledge construction was confined to the academic domain of Natural Sciences, and to rapidly emerging highly specialized fields, in particular Ecology. This discipline, or set of perspectives, is at the origin of modern environmentalism in the early 1970s. It rapidly substituted for traditional conservationism and increasingly influenced a wide array of academic disciplines, Social Sciences included, to address environmental dimensions of their specific fields of inquiry. Specific ecological issues were subsequently made the subject of discussion by many neighbouring disciplines, such as Geo-Sciences, in particular Physical Geography, and Physics. In the 1980s, many other disciplines increasingly addressed a large variety of ecological dimensions, thus developing more or less influential fields of inquiry, such as Ecological Economics (Costanza, 1991) or Industrial Ecology (Allenby and Richards, 1994). There were also efforts to bridge existing gaps and lack of interaction, while providing overarching academic fields of knowledge, such as Environmental Sciences or Resource Management.

At the same time, with constantly increasing professionalism, another important strand of environmental knowledge construction developed within the environmental movement. In particular, grassroots activists, independent research institutions and environmental organizations took up ideas and concepts from academia, in order to 'translate' them into the discourse of science and technology policy (Jamison, 1996, p. 242), channel back alternative viewpoints, and develop new – sometimes radically new – approaches, such as the sustainable development paradigm (e.g. McCormick, 1991). Some of these elements and overarching concepts, predominantly those with a technological orientation, influenced state regulation (Phase I, II and early III in Figure 11.1). However, the direct reception of these ideas and findings by the corporate world was extremely limited (arguably with the exception of some sub-fields of Economics and Engineering, such as Marketing, Organizational Theory and Industrial Ecology, that is, domains close to the corporate world and therefore less prone to the barriers listed below (Allenby and Richards, 1994; Banerjee, 1998). Thus, there is no contiguous field of environmental knowledge; rather, there are largely unconnected islands of knowledge (or fragments thereof) in an archipelago of knowledges.

Recent findings as to knowledge construction and knowledge management provide a framework for explaining this hesitant, seemingly fortuitous pace of eco-modernization. Thus, for a long time, corporate actors and environmental actors (many scientists included) lived in clearly demarcated worlds of their own, where, more often than not, the *other* was either not perceived, not accepted in his or her 'expertship', or even regarded as the main *adversary*. Both groups of actors were influenced by largely differing "cognitive structures" and "cognitive praxis" (Jamison, 1996, p. 239) preventing them from finding a common ground, that is, shared meanings necessary for a meaningful exchange. Put differently, both groups constructed their respective environments almost solely for their own "sensemaking" (Weick, 1985), thereby filtering out other 'senses' of potential relevance for their own organizational behaviour. Both groups' specific knowledges were thus deeply "embedded knowledges", often specifically "situated knowledge" (Haraway 1991), far from being easily transferable. Indeed, possibilities to meet or to exchange findings, viewpoints or opinions were rare, that is, "communities of interaction" (Nonaka, 1994, p. 15) or "communities of practice" (Willke, 1998, p. 105) barely existed (with the possible exception of technologically oriented fields, such as engineering improving pollution abatement equipment). Moreover, "knowledge links" (Badaracco, 1991) did not exist and were difficult to establish. As there was no relevant previous knowledge enabling actors to appreciate new ideas, no community of practice, no openness, no trust, no routines, no autonomy, the whole gamut of specificities necessary for a functioning learning environment was lacking. What is more, in both groups of actors, "organizational defences" (Argyris, 1990) developed to screen out alternative knowledge perceived as threatening to the sensemaking of their own organization.

These patterns of mutual non- and mis-understandings, as well as organizational defences, were successively broken up. As a result, important initiatives and boundary-permeating endeavours become typical of the 1980s and, even more so, of the 1990s. In this context, civil society actors became especially influential.

The Power of Framing and More: New Capabilities of (Transnational) Civil Society Actors

Background and Strategies

Civil society actors that transcend international borders have become increasingly important in a wide range of policy areas in the international system (environment, human rights, disarmament). Most observers of the international scene are convinced that international politics in general has changed tremendously since this new type of actor in the international arena has gained more strength. In this view, NGOs,[1] transnational advocacy networks (TANs), transnational epistemic communities (Haas, 1992) or transnational social movement organizations (TSMO), play influential roles in a number of different policy areas (Smith et al 1997; Cohen and Rai, 2000), including environmental issues (e.g. Wapner, 1996).[2] In terms of the general background for the evolution of these entities, two points can be made.

First, the rising power of NGOs is related to recent changes within the traditional state system, marked, for example, by a growing erosion of power of the nation states in important policy areas and by the increasing fragmentation of formerly homogeneous interest or issue areas (Rosenau, 1990). Second, NGOs are occupying important niches which are not – or cannot – be filled by either the traditional nation states or the new supranational institutions, due to immense deficits in awareness, determination, and room for manoeuvre. Princen (1994, p. 41) gives an excellent illustration of NGO activities regarding environmental issues: "[...] it is influence achieved by building expertise in areas diplomats tend to ignore and by revealing information economic interests tend to withhold".

Following Risse-Kappen (1995, p. 312), the new transnational actors use primarily communicative rather than instrumental rationality and prefer the logic of conviction to that of cost-benefit analysis. This may be the main explanation for their ability to form 'winning coalitions' in certain contexts, an ability that is, according to concepts taken from political theory, an important prerequisite for becoming effective agents of political change by capitalizing on the relational power of underlying networks. Thus, with the appearance of new civil society actors constituting the environmental movement, new agents of social learning and, at the same time, new actors triggering industrial change, have entered the scene (Soyez, 1998). Whether they are active domestically or internationally, or linking all arenas by skilfully applying a consistent politics of scale (Swyngedouw, 1997), they have presented new counter-narratives, in the sense of 'critical geopolitics' (Ó Tuathail, 1998), undermined predominant paradigms, and thereby changed the rules, contributing to new power geometries (Massey, 1993) – and new Geographies of the corporate world.

This newly gained power of formerly powerless actors in discursive struggles can be explained by several factors and abilities, notably framing and agenda setting, networking, and the politics of leverage and shaming.

Framing and agenda setting "Environmental issues and problems are not present in nature: ...they are [...] the products of a collective, instrumentally dependent and institutionally circumscribed professional activity that we call science or scientific research" (Jamison 1996, p. 224) – and, one should add, products of other forms of social construction.

In order to be appropriately perceived by the general public, the media or politics, ensuing expert discourses have to be translated and recombined, transforming expert knowledge(s) into policy-oriented packages. This can be done by the scientist themselves – or any other actor, NGOs included. Thus, it is a necessarily selective, contingent process, influenced not only by the translator's own situated knowledge as well as his or her intentions and objectives, but also by the resonating capacities or peculiarities of potential receivers. Every message is given, or comes with, a meaning, a factual or symbolic content, a process called framing (Snow and Benford, 1988).

Social movements, or loosely knitted advocacy networks, have not only developed strong capabilities of framing, but also of putting skilfully framed issues on the agenda of politics and media (see Princen, 1994; Wapner, 1996). More often than not, such 'winning' frames result from innovatively interpreting generally available, but not generally perceived (or differently framed), scientific knowledge or original research done in what could be called 'green think tanks' (cf. Jamison, 1996).

As frames are necessarily selective interpretations of reality, however, they can be – and are – contested: scientific, public or movement discourses are arenas for competing frames. Thus, framing is a specific type of knowledge management, aimed at convincing those making decisions that a certain perspective on reality is more appropriate than others.

Networking Neither networking nor framing are strategies used solely by movement actors, but NGOs have only learned in recent years to apply it as skilfully as their main adversaries in politics and economics have done traditionally. To network and to form ('winning') coalitions and alliances can prove a powerful tool, not least in order to impose one's own frames. Networking is another form of boundary-permeating work: distances (spatial and others) are overcome, spatially dispersed capacities and competencies are bundled, fragmented domains of knowledge or authority are connected and, more recently, scales – from the local to the global – are transgressed and linked. Ironically, in these respects, the transnational corporation, the main adversary of environmental movements has served as a model for the transnational movement organization or the transnational advocacy network.

The power of networking has been dramatically increased by recent developments in communication technologies. Special importance must be attributed, in the context of NGO activities, to the internet, in particular e-mailing (e.g. the resistance against the Spring Meeting 2000 of multilateral development banks in Washington, D.C., cf. http://www.a16.org/).

Politics of leverage and shaming The environmental movement has proved to have the power to not only develop influential frames but also to impose them, using a wide array of 'eco-tactics', from banner-hanging to boardroom negotiations, from traditional lobbying to going transnational. In the latter case, it has become increasingly normal to take advantage of what Keck and Sikkink (1998) call the 'boomerang effect'. The thrust of this effect is, to mobilize, in situations of domestic deadlock, foreign allies (from grassroots groups to governments) to put pressure on domestic adversaries. This strategy was recently also adopted by indigenous movements resisting the industrial world's need for resources and ensuing ingressions into peripheral world regions (Soyez and Barker, 1998).

In these different contexts, the power of an argument or of a fact alone is often not sufficient to exert enough pressure on those who are indifferent or opposed. Instead, specific knowledge is used to hit adversaries' weak-spots. This can be done by exposing embarrassing facts (such as non-compliance to self-established rules) or by staging 'ethno-dramas' or 'action-events'. These politics of shame are particularly successful when actors with a high capital of respectability in society, or in the marketplace, are targeted.

Targets

In order to put pressure on the decision-makers of the corporate and governmental worlds, the main objective of civil society actors is to target highly visible actors with highly visible actions. The most important of these targets are (see Soyez, 1998):

Direct actions aimed at central actors of the industrial system, usually companies The direct action against central actors (individual business leaders, plant sites or headquarters) who are accused of real or perceived adverse effects caused by their industrial activities is one of the classic action strategies of lobbies at all levels. It has become evident, however, that banner-hanging on smoke-stacks and blocking-actions around international meetings might be enough to raise short-lived media attention, but do not suffice to induce a pervasive change of industrial behaviour. Thus, while these actions have not been completely abandoned, a clear change in strategy is apparent.

Influencing the value environment of important actors in the spheres of economics, politics, media and public opinion In order to induce pervasive change of actors and institutions. lobbying, in the original sense of the word, becomes the rule. In this view, lobbying is inconspicuous, discreet, sometimes even hidden interactions with representatives of the economic, political and public spheres, more often than not in closed board-rooms, meetings, conferences and negotiation rounds, both domestic and international. To succeed in these contexts, high professional competence is needed in those fields covered by the representatives of the corporate, institutional or governmental worlds, in particular technology and economics.

Changing international regimes A specific ability of dedicated transnational movements is to influence international institutional settings. More often than not this leads to both formal and informal new international 'regimes', involving rules of environmental governance, control and conduct that determine important actors' environmental behaviour. One illustrative example is the above mentioned 'World Conservation Strategy', the forerunner of the ensuing 'sustainable development' paradigm as extended and popularized by the Brundlandt Report.

Typical arenas of influence and pressure are large UN conferences. Agreements reached in these arenas influence the environmental behaviour of nation states, filtering down in form of new regulatory frameworks, subsequently affecting the regional and local levels. Other important targets for the implementation of new environmental regimes are powerful international institutions, such as Multilateral Development Banks (see below).

The transmission of consensual knowledge and the persuasion of discussion partners during informal as well as official occasions is crucial. Thus, 'communities of interaction and/or practice', as well as 'knowledge links' and 'trust' are built up. These are embedded in new "landscapes of knowledge" (Willke, 1998, p. 58), overcoming mutual 'organizational defences' that did not exist in the original archipelagos of fragmented knowledge, characterizing both the worlds of corporations and environmental movements.

Civil Society Actors and the Greening of Industrial Production Systems

To illustrate the processes addressed earlier, two case studies are outlined. Both cases feature a conscious change of perspective in which environmentalist and civil society pressures create responses from the industrial world. While the first case is abundantly documented in the international literature, the second one is taken from an ongoing study (Soyez, 2001). Both show crucial aspects of the intricate transnational interplay characterizing actions and reactions of the economic, political, societal and environmental spheres causing current industrial change.

Industrial and Governmental Response to Transnational Environmentalist and Media Pressure in the Forest Sector of British Columbia (Canada)

In the early 1990s exploitative forestry methods used in British Columbia's (BC) forest industries (and supported by government) increasingly became the target of harsh criticism (Hayter and Soyez, 1996). Initially, local environmental action groups as well as Canadian and American environmental NGOs (for example, Earth First!) formed loose coalitions with other NGOs, among them Greenpeace Germany. While this rapidly unfolding controversy became especially charged in Germany it involved a large number of transnational lobbies in many other countries. All groups involved soon realized the enormous potential that lay in the coordination of their efforts.

After intensified, and for Canada mostly negative, reporting in the German press throughout the summer of 1993, Greenpeace Germany started a large-scale forest campaign in the late autumn of 1993. In this campaign, Greenpeace used master-frames like 'biodiversity' or 'old growth forests' on Canada's West Coast as well as arguing that this region was a World Heritage site for humankind.

The accused were no longer the German end-users of too much paper, but the Canadian forest industry, with a special campaign focus on this industry's unsustainable large-scale clearcut practices – which were long thought appropriate in Canada. While escalating the direct action campaigns that were taking place in the forests of Canada's West Coast (with German participation and mass arrests of protesters), Greenpeace Germany used a different type of campaign to effectively 'encircle' the Canadian forest industry. Specifically, they lobbied German bulk users of Canadian paper (especially publishing houses like Burda, Spiegel, and Henri-Nannen) to buy only 'clearcut-free paper' and thought about calling for a boycott of all Canadian wood products, which to Canadian observers seemed to be a realistic danger.

These activities were framed by spectacular direct protest actions in front of the Canadian Embassies in Germany and Austria. They were chosen because there were no European offices of the criticized Canadian forestry corporations (such as McMillan-Bloedel, now Weyerhaeuser). In Canada's embassies in Europe, as a consequence, classical policy areas faded into the background for months while the ambassadors themselves and entire embassy departments were active in environmental diplomacy.

The pressure, simultaneously built up in Canada, Germany and other countries, such as in the UK, led to pronounced learning processes and ensuing adaptation measures in BC, both in forestry policy and in the forest industry in particular as well as with regard to land-use planning more generally. In the medium-term this will potentially cause (e.g. Hayter, 2000): pronounced changes in the industrial structure of the region, due to resource availability and accessibility as well as changed production technologies and product mix; a new competitiveness of B.C. products on sensitive markets, for example, in Europe; and new regional patterns of industrial and recreational land-use patterns in B.C.

The Financial Sector under Siege: The Role of NGOs and 'Green' Think Tanks in Washington, D.C.

Civil society actors today rely increasingly on professional knowledge and skills in order to cope with complex professional information held by their antagonists in highly specialized institutions. This is particularly true for any interaction with the financial sector. The slow change to 'green' financial institutions behaviour testifies to this (as to the World Bank cf. Rich, 1994).

Today, many important decisions with regard to the ongoing industrialization of the world are made in Washington, D.C., the site of the World Bank Group's headquarters. Many NGOs, such as FoE (Friends of the Earth) or BIC (Bank Information Centre), but also 'green' think tanks, such as World Resources Institute (WRI), act as knowledge-based coordinators of resistance against development projects that are contested by domestic and transnational social movement organizations or impacted locals. These Washington based actors combine 'indigenous' knowledge originating in the projects' host countries with advanced professional information – technical, legal, environmental – resulting from their embeddedness in the specific Washington, D.C. spatial context. This embeddedness furthermore provides: boundary-spanning relational power (linkages, coalitions and alliances with other NGOs, government agencies, transnational institutions, churches, universities and – not least – sympathizing informants in the targetted institutions); framing and agenda-setting power, and; a setting for rapid information diffusion and fast learning.

In other words, in Washington D.C. there is a unique combination of governmental and non-governmental actors, enabling NGOs to thrive in an environment that is as challenging as it is creative. As NGOs with a special focus on the environmental behaviour of financial institutions have recently been struggling with marked policy shifts, this local 'institutional thickness' is of crucial importance for appropriate strategy choices. Thus, Multilateral Development Banks increasingly tend to act no longer as major lenders as was originally the case, but as mobilizers of capital, thus taking a 'back seat' to private money. Development, in particular industrial development in the Third World, is increasingly financed by private corporations, very often complex consortia. The risks linked to these investments are taken over by new specialized institutions, such as MIGA (Multilateral Investment Guarantee Agency, a member of the World Bank Group, cf. West and Tarazona, 1998). NGOs now no longer can target only a handful of MDBs or governments. Rather, they have to deal with a multiplicity of actors, many of whom have not yet adopted – or do not care about – the rules of environmental conduct that the World Bank has gradually accepted, not without controversy.

The NGO answer to this new challenge has been to learn how the financial world works, where the important decision-makers are and why the circuits of capital flow take the directions they do (Chan, 1996; Ganzi et al 1998). This has enabled them to systematically identify different financial actors' weak spots and key personnel, and taught them how to channel relevant information to financial analysts, rating agencies, mutual funds managers, creditors and equity owners. Thus, environmental concerns are translated into environmental risks or liabilities, social and human rights concerns into political risks and labour and workers' rights into financial risks.

While these pressure tactics are increasing, ensuing impact patterns are far from clear, and more research in these fields is needed. Some events, however, seem to corroborate NGO actors' contention that these actions can have considerable effects, for example as regards project modification and mitigation effects, even if a complete withdrawal from major projects because of environmental or human rights issues seems to be an exception (cf. project documentation at BIC).

Conclusion

To conclude, two points are addressed, namely the *topical* and *conceptual* dimensions of the greening of industry as a research object for Industrial (and Economic) Geography and, the *disciplinary* dimension of the inclusion of this topic into future industrial (and economic) geographical enquiries.

From a *topical* and *conceptual* point of view, placed in the context of industrial production systems as a whole, the two cases presented in the last chapter address mainly processes connected to the procurement of resources and the financial system. If industrial production systems (or chains and webs of chains) are concepualized as the model presented by Dicken (1992), the cases thus fall into stage one and two of the model's inner core as well as into the model's outer shell. Indirectly, however, other stages or shells are affected, and for any of the model's elements different examples of societal pressure can be put forward (e.g. Newell, 2000; Soyez, 1998). Different types of pressures currently exerted on corporations heavily influence the greening of industrial production systems and, consequently, industrial change.

Many details of eco-modernization processes may appear purely technical, administrative or social. The common denominator, however, is their embeddedness in a societal value environment that is currently marked, more deeply and pervasively than in any decade before, by pervasive eco-modernization processes. To a large degree, these are pushed by a variety of civil society actors. Almost any industrial sector, any industry-related arena is affected. The impacted areas range from risk insurance to eco-consulting (see Chapter 12 by Schulz), from pollution abatement technology to product design, from resource management principles to marketing schemes, from environmental legislation to industry's voluntary initiatives and beyond-compliance strategies, from recycling strategies to research into options for possible industrial futures.

All these aspects are linked to knowledge, or to a variety of relevant knowledges, some of which were long marginalized, but are now filtering up and filtering down, predominantly due to the framing power of non-corporate and non-governmental civil society actors. To paraphrase Haas (1990, p. 11): environmental knowledge "infects the way" political and corporate actors think – and act.

This is not to say that every corporate or government actor has accepted the sustainability paradigm or that each Social Movement activist shares the same ideas about the 'greening of industry'. But it means that there is a myriad of impulses, bottom-up, top-down and obliquely zigzagging through all scale levels, which affect the industrial world as we know it, and that is worth studying also from a geographical perspective. Being only early witnesses of a potentially pervasive secular process, we cannot be sure about the nature of the 'greening of industry'. It may be that this trend is just a "new logic of business" (Freeman et al 2000) or the beginning of a new Industrial Revolution, as both claimed and predicted by visionaries such as Hawken et al (1999) or Flavin and Young, stating in the influential *State of the World*: "Once seen as a distraction to the real business of business, environmental concerns are becoming an engine of the next industrial revolution" (quoted in: Jamison, 1996, p. 233).

As Hayter and Le Heron (chapter 2) develop in more detail, ongoing processes of industrial greening can tentatively be conceptualized as elements of a new 'techno-economic paradigm'. Adherents of the Regulation School, on the other hand, should be tempted to evaluate the importance of civil society actors for regulation processes. These few comments should suffice to emphasize that ongoing processes of 'greening' can not only be linked, in Industrial Geography, with existing conceptual frameworks, but do, indeed, enhance their explanatory power.

Finally, from a *disciplinary* point of view, the traditional boundary-work within Industrial (and Economic) Geography has to be reassessed and re-shaped. In particular, the environmental implications of industrialization and industrial change far exceed the mainly technological or economic dimensions that to date have been the discipline's focus. A better understanding of other knowledge domains as well as an integration of other disciplines' findings would contribute to broaden and deepen existing approaches. Thus, more *boundary-permeating* work lies ahead for Industrial (and Economic) Geography.

Acknowledgement

The generous support of the German Research Council (Deutsche Forschungsgemeinschaft) is gratefully acknowledged.

Notes

[1] The term NGO is an ambiguous one, both in academic literature and everyday usage (Princen, 1994). In this chapter, NGOs are understood to include only normative, principle-based, non-profit groups which are financially independent of state or business actors.

[2] 'Transnational advocacy networks' (TANs) (Keck and Sikkink, 1998) are more or less loosely organized webs including actors such as: international and domestic nongovernmental research and advocacy organizations, local social movements, foundations, the media, churches, trade unions, intergovernmental organizations and finally parts of the executive and or legislative branches of governments. While being broader than 'epistemic communities' (Haas, 1992), both approaches focus heavily on scientific or other knowledge, conceived of as meaning shared by members of loosely knitted transboundary networks. Soyez (1997) has introduced the term 'spatially relevant or space-producing' lobbies, as opposed to 'space-producing' governmental actors, covering most NGOs and TANs as well as other non-sovereignty bound actors trying to lobby influential decision-makers of the corporate and governmental worlds.

References

Allenby, B.R. and Richards, D.J. (1994), *The Greening of Industrial Ecosystems*, National Academy Press, Washington D.C.

Amin, A. and Thrift, N. (1994), 'Living in the Global', in A. Amin and N. Thrift (eds), *Globalization, Institutions, and Regional Development in Europe*, Oxford University Press, Oxford, pp. 1-22.

Argyris, C. (1990), *Overcoming Organizational Defenses: Facilitating Organizational Learning*, Allyn, Boston.

Badaracco, J.L. (1991), *Strategische Allianzen. Wie Unternehmen durch Know-how-Austausch Wettbewerbsvorteile erzielen*, Wien: Ueberreuter.

Banerjee, S.B. (1998), 'Corporate Environmentalism: Perspectives from Organizational Learning', *Management Learning*, vol. 29, pp. 147-164.

Chan, M. (1996), *The Anatomy of a Deal: A Handbook on International Project Finance*, Friends of the Earth, Washington, D.C. (mimeo).

Costanza, R. (1991), *Ecological Economics: The Science and Management of Sustainability*,Columbia University Press, New York.

Dicken, P. (1992), *Global Shift: The Internationalization of Economic Activity*, Guilford Press, New York.

Fredriksson, C.G. and Lindmark, L.G. (1979), 'From Firms to Systems of Firms: A Study of Interregional Dependence in a Dynamic Society', in F.E.I. Hamilton and G.J.R. Linge (eds), *Spatial Analysis, Industry and the Industrial Environment*, 1, Wiley, New York, pp. 155-186.

Freeman, R.E. (1984), *Strategic Management: A Stakeholder Approach*, Pitman, Boston.

Freeman, R.E., J. Pierce and Dodd, R.H. (2000), *Environmentalism and the New Logic of Business: How Firms can be Profitable and Leave Our Children a Living Planet*, Oxford University Press, Oxford.

Ganzi, J., Seymour, F. and Buffet, S. (1998), *Leverage for the Environment. A Guide to the Private Financial Services Industry*, Washington, D.C.: World Resources Institute.

Gladwin, Th. N. (1993), The Meaning of Greening: A Plea for Organizational Theory,' in K. Fischer and J. Schot (eds) *Environmental Strategies for Industry: International Perspectives on Research Needs and Policy Implications*, Island Press, Washington, D.C., pp. 37-61.

Haas, E.B. (1990), *When Knowledge is Power: Three Models of Change in International Organizations*, University of California Press, Los Angeles.

Haas, P. M. (1990), *Saving the Mediterranean: The Politics of International Environmental Cooperation*, Columbia University Press, New York.

Hajer, M.A. (1996), 'Ecological Modernisation as Cultural Politics', in S. Lash, B. Szerszynski and B. Wynne (eds), *Risk, Environment and Modernity: Towards a New Ecology*, Thousand Oaks, London, pp. 246-268.

Haraway, D. (1991), 'Situated Knowledge: The Science Question in Feminism and the Privilege of Partial Perspective', in D. Haraway (ed), *Simians, Cyborgs, and Women: The Reinvention of Nature*, Routledge, London, pp.183-201.

Hawken, P., Lovins, A. and Lovins, H. (1999), *Natural Capitalism*, Little, Brown and Co., Boston, MA.

Hayter, R. (2000), *Flexible Crossroads: The Restructuring of British Columbia's Forest Economy*, University of British Columbia Press, Vancouver.

Hayter, R. and Soyez, D. (1996), 'Clearcut Issues: German Environmental Pressure and the British Columbia Forest Sector', *Geographische Zeitschrift*, vol. 83, no. 3/4, pp. 143-156.

Jamison, A. (1996), 'The Shaping of the Global Environmental Agenda: The Role of Non-Governmental Organisations,' in S. Lash, S. B. Szerszynski and B. Wynne (eds), *Risk, Environment and Modernity: Towards a New Ecology*, London, Thousand Oaks, pp. 224-245.

Keck, M. E. and Sikkink, K. (1998), *Activists Beyond Borders: Advocacy Networks in International Politics*, Ithaca, Cornell University Press.

Massey, D. (1993), 'Power-Geometry and a Progressive Sense of Place', in J. Bird, B. Curtis, T. Putnam, G. Robertson, L. Tickner (eds), *Mapping the Futures: Local Cultures, Global Change*, London: Routledge, pp. 59-69.

McCormick, J. (1991), *Reclaiming Paradise: The Global Environmental Movement*, Bloomington, Indiana University Press.

Messer-Davidow, E., Shumway, D.R. and Sylvan, D.J. (eds) (1993), *Knowledges: Historical and Critical Studies in Disciplinarity*, University Press of Virginia,Charlottesville.

Newell, P. (2000), 'Environmental NGOs and Globalization. The Governance of TNC', in R.Cohen and S.M. Rai (eds), *Global Social Movements*, The Athlone Press, London, pp. 117-133.

Nonaka, I. (1994), 'A Dynamic Theory of Organizational Knowledge Creation', *Organization Science*, vol. 5, pp. 14-37.

Ó Tuathail, G. and Dalby, S. (eds) (1998), *Rethinking Geopolitics*, Routledge, London.

Princen, T. (1994), 'NGOs: Creating a Niche in Environmental Diplomacy', in T. Princen and M. Finger (eds), *Environmental NGOs in World Politics: Linking the Local and the Global*, Routledge, London, pp. 29-47.

Rich, B. (1994), *Mortgaging the Earth. The World Bank, Environmental Impoverishment, and the Crisis of Development*, Beacon Press, Boston, MA.

Risse-Kappen, T. (ed) (1995), *Bringing Transnational Relations Back in: Non-State Actors, Domestic Structures and International Institutions*, Cornell University Press, Ithaca, N.Y.

Rosenau, J. N. (1990), *Turbulence in World Politics: A Theory of Change and Continuity*, Princeton University Press, Princeton.

Schot, J. and Fischer, K. (1993), 'Introduction: The Greening of the Industrial Firm', in K. Fischer and J. Schot (eds), *Environmental Strategies for Industry: International Perspectives on Research Needs and Policy Implications*, Island Press, Washington, D.C., pp. 3-33.

Snow, D. A. and Benford, R.D. (1988), 'Ideology, Frame Resonance, and Participant Mobilization', *International Social Movement Research*, vol. 1, pp.197-217.

Smith, J., C., Chatfield, C. and Pagnucco, R. (eds) (1997), *Transnational Social Movements and Global Politics: Solidarity: Beyond the State*, Syracuse University Press, Syracuse.

Soyez, D. (1997), 'Raumwirksame Lobbytätigkeit', in R. Graafen, R. and W. Tietze (eds), *Raumwirksame Staatstätigkeit, Festschrift für Klaus-Achim Boesler zum 65. Geburtstag*, Bonn, pp. 205-219.

Soyez, D. (1998), 'Globalisierung von unten': Transnationale Lobbies und Industrieller Wandel', in H. Gebhardt, G. Heinritz u. R. Wießner (eds), *Proceedings of 51st Deutscher Geographentag*, Bonn, pp. 55-65.

Soyez, D. (2001), 'Entgrenzt, aber situiert....Zur Politischen Geographie Transnationaler Bewegungsorganisationen', in P. Reuber and G. Wolkersdorfer (eds), *Politische Geographie, Handlungsorientierte Ansätze und Critical Geopolitics*, Heidelberger Geographische Arbeiten, vol. 112, pp. 117-32.

Soyez, D. and Barker, M.L. (1998), 'Transnationalisierung als Widerstand: Indigene Reaktionen gegen fremdbestimmte Ressourcennutzung in Ost-Kanada', *Erdkunde*, vol. 52, no. 4, pp. 286-300.

Swyngedouw, E. (1997), 'Neither Global nor Local: 'Glocalization' and the Politics of Scale', in K. Cox (ed), *Spaces of Globalization: Reasserting the Power of the Local*, Guilford Press, New York, pp. 137-166.

Taylor, M. (1995), 'Linking Economy, Environment and Policy', in M.J. Taylor (ed), *Environmental Change: Industry, Power and Policy*, Avebury, Aldersot, pp. 1-12.

Wapner, P. (1996), *Environmental Activism and World Civic Politics*, Albany: SUNY Press.

Weick, K.E. (1995), *Sensemaking in Organizations*, SAGE, London.

Willke, H. (1998), *Systemisches Wissensmanagement*, Lucius and Lucius, Stuttgart.

World Bank (ed) (2000), *Greening Industry. New Roles for Communities, Markets, and Governments*, Oxford University Press, Oxford.

12 Environmental Service-providers, Knowledge Transfer, and the Greening of Industry

Christian Schulz

Introduction

During the 1990s, environmental producer services emerged as a new segment of the so-called environmental industry. These activities, however, have received little attention from economic geographers. This chapter addresses this research lacuna.

Environmental producer services can be defined as knowledge intensive service firms, often highly specialized, which provide consultancy and technical assistance on environmental issues to manufacturing firms. These new markets owe their dynamic evolution to several factors. Thus, the broadly based 'flexibilization process' which is occurring across the industrial spectrum has included the externalization of production oriented services, a trend that has been reinforced by changing legal and political environments as well as a growing influence of different stakeholders. These latter influences have incorporated more and more voluntary initiatives in the manufacturing sector, further adding to the reasons for external advice from environmental services. A recent study of the OECD (1997, p. 7) notes this tendency:

> Environmental regulations and standards set the broad framework for demand for environmental goods and services. There is a shift in regulatory focus towards economic instruments, incentives and voluntary agreements which concentrate more on overall

environmental performance, and give greater flexibility in achieving environmentally satisfactory solutions. This has been coupled with greater emphasis on clean technologies and products. Both provide new impetus to the supply industry.

The rapid growth and growing importance of environmental service providers suggests a need for a better understanding of the economic geography of these activities, notably their specific location factors, ways of networking with other firms and labour market requirements. Beyond understanding their 'simple' dynamics, environmental service providers play potentially significant roles in the 'greening of industry'. Literature on producer services as whole reveals, service providers usually exert considerable impacts on the competitiveness and strategic planning of manufacturing firms. Coffey and Shearmur (1997, p. 404) emphasize that "through their role in investment, innovation, and technological change, high order services play a decisive role in the economic development process, particularly in facilitating overall economic change and adjustment". Florida (1996, p. 92) presents a survey in which 30 percent of the inquired manufacturing enterprises consider external consultants to be "key actors in pollution prevention efforts". In turn, this study recognizes that environmental producer services are one of the 'agents of greening' that strongly influence the reorganization and rethinking of environmental strategies by manufacturing firms (Figure 12.1).

To understand and evaluate the contribution of the environmental producer services for industrial change, it is necessary to highlight their spatial and functional organization and their ways of interacting with manufacturing firms. In addition, aspects of knowledge creation, knowledge transfer, problem-framing and agenda-setting need to be considered. This chapter, first, briefly overviews the different service activities that are linked to the environmental performance of manufacturing firms. Second, the theoretical framework of a recent survey (by D. Soyez and C. Schulz) is discussed at two levels. At a general level, the relations between changes in the political and social environment, the greening of industry and the emergence of new types of environmental producer services are analysed by means of a regulationist approach. Then, different theoretical concepts, notably related to 'actor networks' and inter-organizational learning processes, are discussed. Since this research project deals with two regional case studies in Germany and France, inter-cultural aspects, expressed as different framing patterns, regional discourse styles and constructivist issues, are also taken into account.

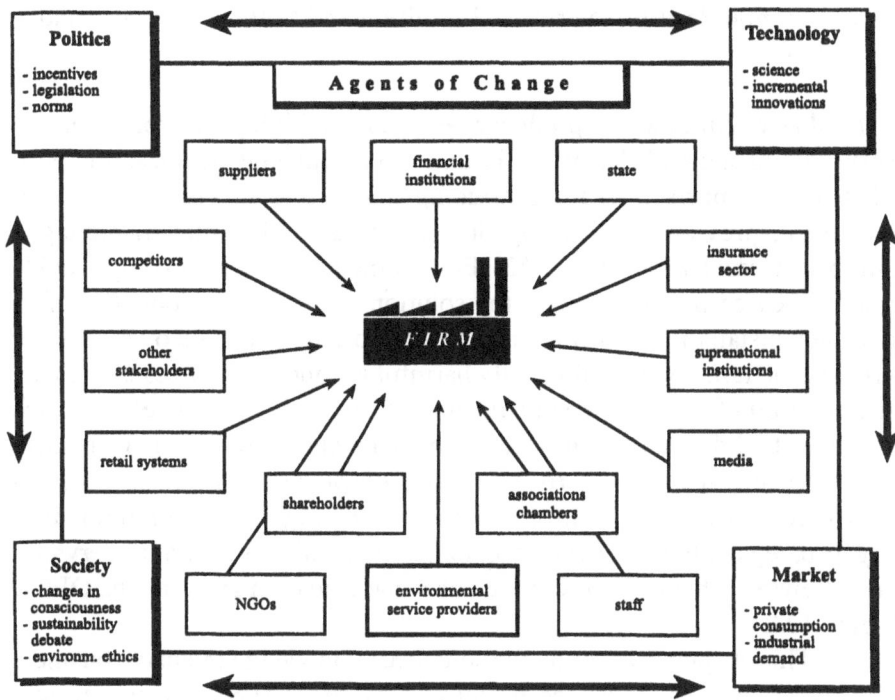

Figure 12.1   Factors in the 'greening' of industry

Finally, it should be noted that while the field work underlying the present study was not yet completed at the time of writing, information from these interviews has helped shape the views presented in this chapter. Indeed, a series of exploratory interviews with representatives of business associations, chambers of commerce, governmental organizations as well as service and manufacturing firms were completed in the summer of 1999 to help structure and prepare further steps of the study. The second phase in the survey, which is now complete although the data still have to be analyzed, involved a postal survey of about 400 environmental service firms – 200 identified in each area. The last survey step will involve in-depth interviews. The details of this survey and its results will be the subject of further publications.

Environmental Producer Services: Definition and Market Characteristics

Given that environmental producer services are highly heterogeneous, a useful, albeit general definition is that environmental business or 'eco-industries' comprise: "firms producing goods and services capable of measuring, preventing, limiting or correcting environmental damage" (European Commission 1994, p. 53). Environmental services not only provide punctual advice or assistance to the solution of single problems (e.g. fuel accidents, installation of a filter system), but often contribute over a longer period to the remedy of ecologically harmful production processes or to the incorporation of environmental strategies. Their activities, therefore, range from such 'traditional' tasks as waste water treatment, solid waste management or air pollution analysis to more contemporary functions such as environmental management consulting, auditing, environmental due-diligence, legal advice or even specialized financial and insurance services. The following scheme is partly based on a classification given by the OECD (Figure 12.2).

This chapter focuses on the so-called 'knowledge intensive business services' (KIBS) or 'advanced producer services' (APS), that is, services whose production requires an important knowledge input and is characterized by a high degree of individualization (see Daniels and Moulaert, 1991). Less knowledge intensive services or highly standardized services (for example, waste transportation) and services offered to other than industrial clients (public authorities, private consumers, other service firms) are not examined.

Despite inconsistencies in official statistics, some broad indicators of market share and economic importance are available. In 1993, about 7,600 companies in the USA offered environmental consulting/engineering services with a turnover of US$12 billion and a 16 percent growth rate, while the environmental business in the USA, had a turnover of US$132 billion with 11 percent annual growth (Kastner, 1993, p. 55). An analysis of the United Kingdom's environmental consultancy market in 1998 showed a growth rate of 10 percent (ENDS, 1999, p. 1). This market growth is close to that for traditional management consulting in Europe (15 percent), still one of the most dynamic sectors in the economy (Glückler, 1999, p. 21). All regional and sectional studies dealing with environmental services predict continued growth, at least for the next decade (e.g. Caldwell and Smallman, 1996; Sam, 1999).

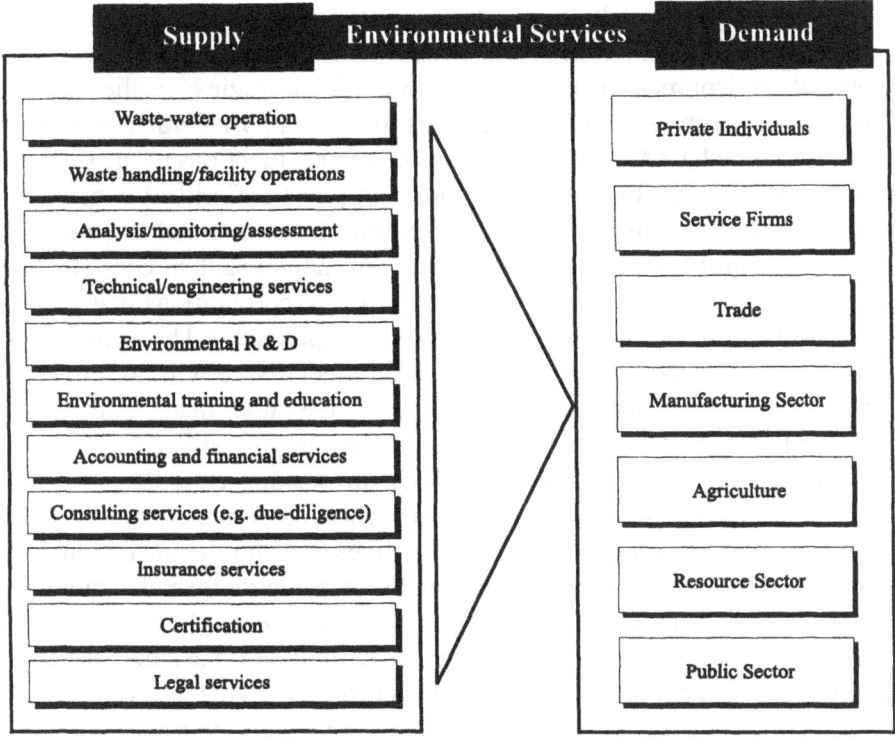

Figure 12.2 Providers and client types for knowledge intensive environmental services

Environmental Producer Services: Indicators of a Post-Fordist Regime?

According to Taylor et al (1995, p. 59), "the business enterprise is the political crucible within which social, economic and environmental issues and forces meet and are played out in particular and specific spatial and temporal contexts". Similarly, Fuchs and Mazmanian (1998, p. 201) stress the need for a "collective understanding of the greening of industry as a political, economic and organizational phenomenon". Given this advice, I will next outline the general political and economic environment relevant for examining the organization of producer-client interactions in the service economy.

During the last 30 years, growing awareness of ecological problems and more sensitive attitudes towards environmental issues has considerably changed environmental policy and economic strategies, in the leading industrialized countries. The tendency is towards a post-Fordist regulation system, in which traditional command-and-control functions of the state are more and more substituted by the emerging power of critical consumers, environmentally conscious stakeholders and governmental but market-oriented incentives (e.g. to foster voluntary initiatives such as eco-labelling and environmental management). Most industrial corporations have begun to fundamentally rethink their environmental policies. "The conclusions businesses are coming to are that the environment issue will not go away, that standards will only rise, and that a pro-active rather than a compliance strategy is for the best. This is the opportunity for consultants" (Caldwell and Smallman, 1996, p. 17). Enterprises are turning towards pro-active environmental policies, trying to become 'good corporate citizen[s]', with more and more agreeing to voluntary initiatives (e.g. the chemical industry's Responsible Care Programme) or passing through environmental audits, such as ISO 14001 or the European Union's EMAS. In these trends, the emergence of a new accumulation regime can be seen that implies the need for permanent innovation and adaptation in environmental matters. This generates a new demand for specific services that are rarely provided in-house.

From an organizational perspective, some authors consider the greening efforts in the manufacturing industry as a result of a general reorganization process. 'Does lean mean green?' was the title of a study conducted by the Massachusetts Institute of Technology (Maxwell et al 1993), which "suggests a relationship between lean production and innovative environmental manufacturing practices" (Florida, 1996, p. 82). In general, as Wood (1996b, p. 342) points out, "experienced companies actually use consultancies more than inexperienced companies for broader management skills". The MIT study provides empirical evidence showing that those enter-prises that are innovative on general organizational issues, are also leading in their environmental performance. Nonetheless, the new market for service providers seems to be far more than a simple side effect of an ongoing out-sourcing strategy. Banerjee (1998, p. 166) even argues the opposite by showing that in many cases the introduction of Total Quality Environmental Management Systems (TQEM) "has led to an enhancement of product quality and corporate performance as well', and that 'environmental issues facing industry today have been reframed as quality improvement issues". By analysing the externalization strategies and the quality of relationships between service providers and their clients, the following sections shed light on the real impact that outsourcing strategies may have on environmental matters.

Externalization Strategies and Modes of Interaction

Beyers and Lindahl (1996, p. 352) identify three main types of motivation for manufacturing firms to externalize producer services. First, there are pure cost-driven considerations, based on a neoclassical view of the transactions cost approach, driven by the perception that outside provision is less expensive, related to a general downsizing strategy. Second, there are quasi-cost considerations, where aspects of flexibility dominate, for example to reduce entrepreneurial risk, to manage an infrequent demand or to concentrate on core competencies. Third, there are non-cost considerations, that is, external services are purchased to compensate for a lack of expertise, to obtain unbiased third-party information, to adapt to the growing complexity of management by using more specialist know-how or, in the long term, to benefit from buyer-supplier dynamics.

These categories also reflect the importance of external service provision. While externalized services of the first category should be, in general, standardized routine services with no strategic impact on the client firm, the last category contains high order services of which quality or specificity is more important than their price. In terms of environmental producer services, it is necessary to analyse the out-sourcing decision process of each case of externalization in order to understand the strategic potential of the external input.

In a second step, the form of cooperation between a service provider and the client firm is examined. In this context, the differentiation made by Tordoir (1994, p. 328) helps explain the different intensity levels of producer-client-interaction (Figure 12.3). In particular, Tordoir distinguishes:

- sparring relations, where consultancy services are supplied to (top) management and communication takes place on the same level in terms of information exchanges and competencies, based on a good interpersonal 'chemistry';

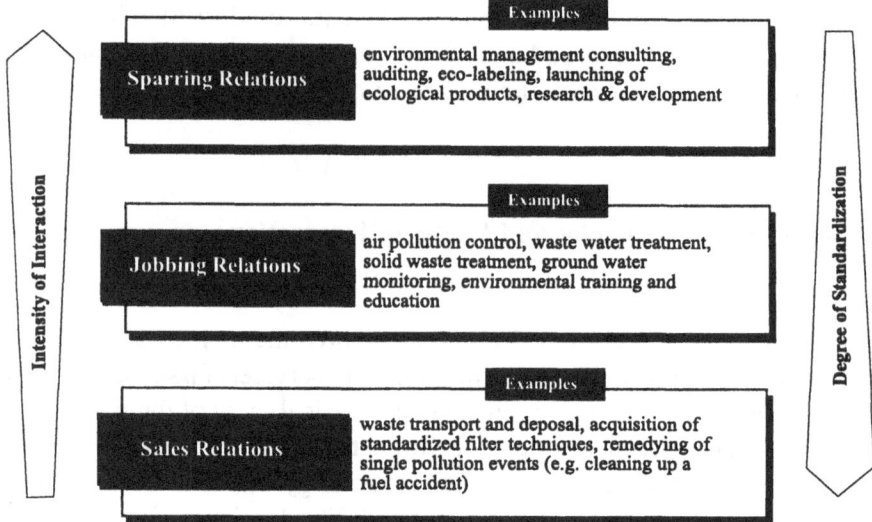

Figure 12.3 Types of relationships between environmental service providers and clients

- jobbing relations, that are most common for professional business services, that is 'typical in cases where the knowledge involved has a specialist and technical character in its widest sense, and where the client is a professional himself in the relevant field'. Jobbing relations feature less interaction than sparring relations;
- sales relations with a 'product-like character of services', which are developed beforehand by the provider.

In the case of advanced producer services, sales relations are unusual, whereas jobbing relations and sparring relations are most common. In this study, interactions with knowledge intensive environmental services are supposed to fulfil at least the criteria of jobbing relations. To be defined as a sparring relationship, they should be characterized by an intensive and continuous process of co-operation (or "co-production" in the terms of O'Farrell and Moffat, 1995, p. 121) with the client, working in the area of strategic decisions and the orientation of the firm's environmental policy. This could be the case, for example, when an external specialist is engaged to develop and install an environmental management system.

The above mentioned classification (Figure 12.3), based on Tordoir's (1994) proposal, is a first attempt to qualify the different environmental services in terms of their degree of interaction with the client. Admittedly, the dividing lines between these types are not clear and there are overlaps and real exceptions. For example, a training and education service could become a sparring relation when dealing over a longer period with the top management level or when developing a proper training concept for the whole enterprise. In contrast, eco-labelling can be a simple sales relation when it is a standardized certification act.

Networking Issues – Weak Ties and Strong Results?

At first glance, networking activities, not only between service producers and their clients, but also among service providers themselves (strategic networks, virtual organizations), are of geographical interest in terms of their functional organization, the building of clusters and their relationship to an institutional environment. The question of whether distance matters in the client-consultancy relationship, or whether there is a spatial concentration of service providers resulting in "clusters of specialist business service firms" (Wood, 1996a, p. 654), allows us to analyze the importance of these activities for regional development. In this context, backward linkages also have to be taken into account: "spatial proximity between producer services and the sources of creators of knowledge, information and technical ability is crucial. A given producer service establishment must therefore have linkages to specialised consultants, complementary producer services, research institutions, universities" (Coffey and Bailly, 1992, p. 864).

In the area of corporate environmentalism and environmental services, intermediary institutions such as chambers of commerce and industrial associations, play a significant role. Through their efforts to transfer information and to foster exchange and inter-firm relationships by offering conferences, workshops or simply 'network dinners', they contribute to the emergence of local or regional networks. Rasmussen, who conducted one of the few regional studies on environmental services, emphasizes this important opportunity for consulting firms to find partners for strategic cooperation during such informal meetings ('kaffeklubber'), with particular reference to an association of Danish firms ('Dansk Komite for fast Affald' or DAKOFA) handling solid waste (Rasmussen, 1992, p.194).

In addition to this spatial aspect of networking and its impact on regional economies, the quality of such networks needs to be analyzed, especially in terms of their potential for knowledge exchange and learning processes (Grabher, 1973, p. 8ff.). In this context, Grabher identifies the following key terms and questions:

- reciprocity: where lies the reciprocity or mutuality in the co-operation?
- interdependence: does this mutuality lead to interdependencies between networking partners?
- loose coupling: how are the ties bound, what is the organizational or contractual basis for cooperation?
- power: are there power asymmetries?

Thus, there needs to be more understanding of, first, the relationships between producer (environmental service provider) and client (manufacturing firm), where governance by the client may play a role in very sensitive areas (e.g. influencing the results of expert reports). Second, there needs to be more understanding of networks between different (specialized) service providers, where the coordinating role of one firm (broker) may result in a dominating position (Figure 12.4). Indeed, it is evident that 'a network organization extends the learning relationships to external sources of know-how' and the 'diversity of information and experience-sharing needs is a crucial determinant in the externalization of knowledge services from industrial firms' (Monnoyer-Longé and Mayère, 1994, p. 305). Consequently, the character of inter-firm learning processes helps explain the impact service providers may have on their network partners.

Inter-organizational Learning and Knowledge Transfer: Key Factors for the Greening Process?

Lundvall (1993, p. 53) criticizes the transactions cost approach on the basis that it neglects the importance of interactive learning as an element of innovation and interaction between firms: "the most important processes

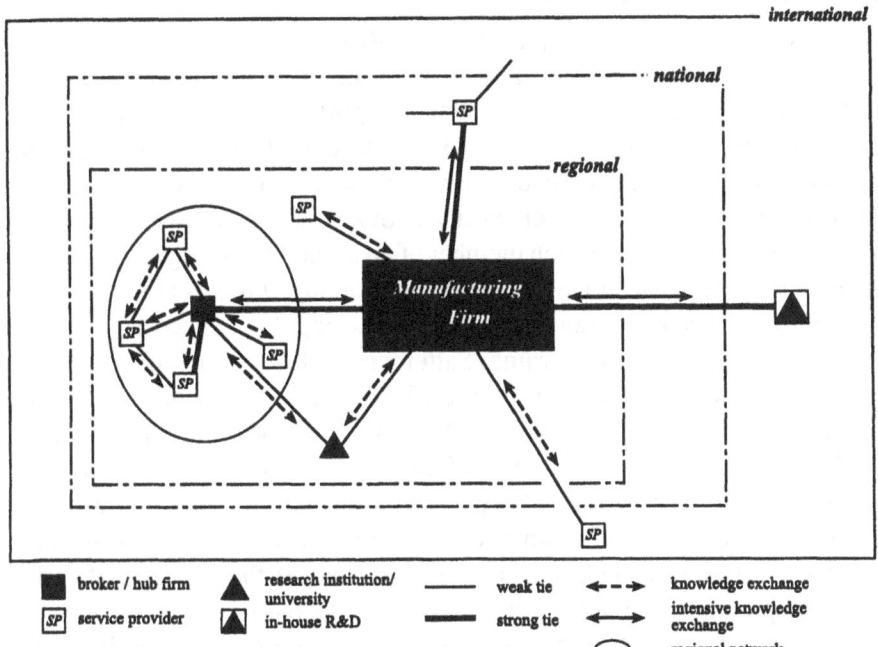

Figure 12.4 Networking between service providers, clients and research institutions

determining the dynamics of the modern economy are actually processes of learning and change". Inter-organizational learning is both a result of inter-firm cooperation as well as a precondition for it. In both cases, it should be seen as an element of strategic change that can considerably modify the identity of an enterprise. In a similar vein, in a discussion of "environmental co-makership", Fuchs and Mazmanian (1998, p. 199) recognize that inter-firm cooperation provides "the potential to foster greening through resulting technology transfer and increases in the pool of knowledge, interactive research activities and the development of international environmental standards".

Inter-organizational learning requires a particular attitude of the learning partner, a kind of openness towards new perspectives, meanings and solutions. Given this prerequisite, manufacturing firms may considerably benefit from external advice. Especially on environmental topics, this openness seems to be more frequent due to a need of expert assistance: "The lack of understanding and the uncertainty around environmental issues is a key factor that influences how firms learn to successfully integrate environmental issues" (Banerjee, 1998, p. 148).

To understand industrial change, the structure of a learning process and the role of its main actors are vitally important. In order to be able to evaluate the effect service providers have on the manufacturing industry (and vice-versa), it is necessary to distinguish types of inter-organizational learning and their different results. The simplest way of learning by knowledge transfer is the individual learning of each member of an organization (e.g. an employee of a manufacturing enterprise learning by interacting with an external service provider). Here, even the transfer of 'tacit knowledge' may become possible. Apart from this individual learning, Sattelberger (1991, p. 15) distinguishes five possible levels for learning processes. These are: a) representative learning of an elite (e.g. top management level); b) learning of other entities of the firm (divisions, specialist teams); c) change of the knowledge shared by all members of an organization ("organizational maps", "shared frames of reference", "community of assumptions"); d) modification of the organization itself by incorporating learning experiences; and e) the use, modification and development of the organization's knowledge base.

This classification suggests a certain hierarchy or quality level. Indeed, the levels a) and b) cannot only be the first steps in a corporate learning process. Levels c) and d) are more integral impacts of an external input: "Organizational learning occurs when it becomes embedded in the memory or in the collective consciousness of the organization as a whole such that this memory guides the behaviour of the members of the organization" (Banerjee, 1998, p. 149). Undoubtedly, e) has the farthest-reaching effect, more or less decoupled from the initial service performance. In the empirical research underlying this study, this relationship between different target groups and the way of valorization of new knowledge by the firm has to be especially respected.

On the other hand, there is the "paradoxical constitution of the consultant's body of knowledge", that is, consultants that automatically learn from the business that they are teaching (Willke, 1998, p. 123, cited in Glückler 1999, p. 10). The fact that inter-organizational learning rarely is a one-way knowledge transfer, leads to a high potential for backward learning effects, that is, for transfer of explicit or implicit knowledge from the client firm to the service provider. This is knowledge that the service firm will use on other occasions and that it may transfer to other clients. Thus "environmental learning" (Banerjee, 1998) can function in both directions in the relationship.

The "strength of weak ties" (Granovetter, 1973) is often considered the crucial element of successful networking. In the present case, however, strong ties – as shown in Figure 12.4 – may play an important role in terms of knowledge exchange and learning processes. Thus, "strong ties are not irrelevant to information flows, and in fact decision-makers appear to be influenced by their strong-tie network" (Bryson and Daniels, 1998, p. 269). We thus have to look very closely at the type of relationship and the level of confidence between the partners.

Does Green Mean Green? Intercultural Aspects and Regional Discourse Styles

The two regions studied in Germany and France are characterized by different national and regional regulation systems, cultural backgrounds and political influences. Although there is a strong tendency to harmonize environmental legislation in the EU (for example, air pollution control, waste management, eco-auditing), there are still quite obvious differences to be taken into account. On the one hand, the framing of environmental issues through the manufacturing industry depends considerably on the way ecological problems are perceived in the population, which themes are claimed by environmentalists and how they are presented by the media. In many cases, the political reactions are part of this framing and are strongly influenced by more or less arbitrary agenda-settings – often very fragmented. Further, as Hunt et al (1997, p. 10) found in their study of environmental attitudes and understandings of business owners or industrial managers, "shared concerns may be constructed and communicated through Trade Associations' and other bodies that 'produce an awareness of an issue coupled with a perception of the necessity to act". As the earlier discussion on networks indicates, the role of intermediary institutions can be crucial.

Additionally, environmental service providers as "agents of greening' are themselves part of a larger context (as management consultants in general are): 'consequently, management consulting is a locally, institutionally and culturally contextualised business" (Glückler, 1999, p. 35). Like other firms too, environmental service providers are embedded in a specific local or regional environment, are part of a social context that is continuously constructed and reconstructed during interaction' (Grabher, 1993, p. 5). In this study, contextualization is of great interest and is analyzed by looking at the different regional styles of discourse and simultaneously by verifying cultural and professional socialization trajectories of the actors. Regarding the latter, both sides – individuals in the service firms as well as members of the client firms – have to be investigated.

Conclusion

To illuminate the field of environmental producer services and their impact on industrial change through 'environmental learning', different theoretical concepts and perspectives are necessary. Organizational and, to a certain degree, transactional approaches help to understand the spatial functioning and the networking of this new branch. The analysis of inter-organizational learning processes with regard to inter-cultural and constructivist aspects is supposed to provide further insights.

Two other issues will be addressed in the future by the present study: the increasing importance of internationalized services and the question concerning the limits for externalization of environmental services in a flexibilization strategy. First, growing internationalization activities of service providers as can be observed in many other fields also matter in the examined branch. More and more environmental consultants serve transnational corporations or get in contact with foreign clients through their activities with international firms. This internationalization is supposed to imply a cross-border transfer of knowledge and corporate culture to client firms in other countries. Thus the greening of industry process in many countries may be strongly influenced by the attitudes and strategies of several 'leading' countries in terms of corporate environmentalism.

Second, there are certain limits for the externalization of environmental producer services, particularly in large corporations pursuing flexibilization and outsourcing strategies while focusing on their core competencies. In the chemical industry, for example, many core competences and necessarily include environmental issues. In other words, if too many services are externalized the firm risks to support an unintentional knowledge transfer because it becomes difficult to keep sensitive details of the production process secret. On the other hand, externalization of environmental services does not necessarily release the manufacturing firm from its responsibility for this sector. As the representative of an iron and steel plant in France emphasized, his firm is very aware of this problem and has identified several fields to be kept in-house and where an externalization shall not be possible in the future (e.g. waste water treatment).

Beyond these issues, further research should focus on aspects neglected in this study, such as the specific labour market requirements of this branch, or the role of 'epistemic communities' in the personal relationships between service suppliers and their clients. Hopefully, this study has shed some light on the dynamic and heterogeneous sector of environmental producer services. If the insights offered by this research focus are only one element in the far more complex process of industrial change in ecological matters, the environmental producer services are certainly not the least important agent in this greening process.

Acknowledgements

The generous support of the German Research Council (Deutsche Forschungsgemeinschaft) in funding the research on which this paper is based and for the travel expenses to the Dongguan conference of the IGU which is gratefully acknowledged. I am also grateful to Linde Kanther and Bret Karnatz for proof reading the English drafts of the paper.

References

Banerjee, S.B. (1998), 'Corporate Environmentalism. Perspectives from Organizational Learning', *Management Learning*, vol. 29, 2, pp. 147-164.

Beyers, W.D. and Lindahl, D.P. (1996), 'Explaining the Demand for Producer Services: Is Cost-driven Externalization the Major Factor?', *Papers in Regional Science*, vol. 75, 3, pp. 351-374.

Bryson, J.R. and Daniels, P.W. (1998), 'Business Link, Strong Ties, and the Walls of Silence: Small and Medium-sized Enterprises and External

Business-service Expertise', *Environment and Planning C*, vol. 16, pp. 265-80.

Caldwell, N. and Smallman, C. (1996), 'Environmental Consultancy in the UK: Structure and Implications', *Management Decision*, vol. 34, pp. 15-23.

Coffey, W.J. and Bailly, A.S. (1992), 'Producer Services and Systems of Flexible Production', *Urban Studies*, vol. 29, pp. 857-68.

Coffey, W.J. and Shearmur, R.G. (1997), 'The Growth and Location of High Order Services in the Canadian Urban System, 1971-1991', *Professional Geographer*, vol. 49, pp. 404-418.

Daniels, P.W. and Moulaert, F. (eds) (1991), *The Changing Geography of Advanced Producer Services. Theoretical and Empirical Perspectives*. Belhaven, London.

ENDS (1999) (Environmental Consultants Directory), *Environmental Consultancy in the UK: A Market Analysis*, 1997/98. http://www.ends.co.uk/consultants/maexec.htm (10.11.1999).

European Commission (1994), *Panorama of EU Industry 94*, European Commission, Brussels.

Florida, R. (1996), 'Lean and Green: The Move to Environmentally Conscious Manufacturing', *California Management Review*, vol. 39, pp. 80-105.

Fuchs, D. A. and Mazmanian, D.A. (1998), 'The Greening of Industry: Needs of the Field', *Business Strategy and the Environment*, vol. 7, pp.193-203.

Glückler, J. (1999), *Management Consulting – Structure and Growth of a Knowledge Intensive Business Service Market in Europe*, Institut für Wirtshafts- und Sozialgeographie, Universität, Frankfurt a.M.

Grabher, G. (ed) (1993), *The Embedded Firm. On the Socioeconomics of Industrial Networks*, Routledge, London.

Granovetter, M. (1973), 'The Strength of Weak Ties', *American Journal of Sociology*, vol. 78, pp. 1360-80.

Hunt, J., Parvis, M. and Drake, F. (1997), *Business Construction of the Environment*, Working Paper of the University of Leeds, School of Geography 97/12. Leeds.

Kastner, O. (1993), *Environmental Consulting in the USA in Reference to the Situation in Europeand Austria*, Institut für Ökologische Wirtschaftsforschung, Vienna

Lundvall, B.-A. (1993), 'Explaining Interfirm Cooperation and Innovation: Limits of the Transaction-cost Approach', in Grabher, G. (ed), *The Embedded firm. On the Socioeconomics of Industrial Networks*, Routledge, London, pp. 52-64.

Maxwell, J., S. Rothenberg and Schenck, B. (1993), *Does Lean Mean Green: The Implications of Lean Production for Environmental Management*, MIT, Cambridge, Massachusetts.

Monnoyer-Longé, M.C. and Mayère, A. (1994), 'Networks in Knowledge-Intensive Firms', *Tijdschrift voor Economische en Sociale Geografie*, vol. 85, 4, pp. 303-310.

OECD (1996), *The Global Environmental Goods and Services Industry*, OECD, Paris.

OECD (1997), *Environmental Policies and Employment*, OECD, Paris.

O'Farrel, P.N. and Moffat, L.A.R. (1995), 'Business Services and Their Impact Upon Client Performance: An Exploratory Interregional Analysis', *Regional Studies*, vol. 29, 2, pp. 111-124

Rasmussen, J. (1992), 'MiljOservicesektoren i StorkObenhavn: Lokal Forankring og Internationaliserin', in Illeris, S. and P. Sjoholt (eds), *Internationalisering af Service og Regional Udvikling i Norden*, NordReFO, Kopenhagen, pp. 188-202.

Sam, P.A. (1999), *International Environmental Consulting Practice: How and Where to Take Advantage of Global Opportunities*, New York.

Sattelberger, T. (1991), 'Die lernende Organisation im Spannungsfeld von Strategie, Struktur und Kultur', in T. Sattelberger (ed.), *Die Lernende Organisation: Konzepte für Eine Neue Qualität der Unternehmensentwicklung*, Gabler, Wiesbaden, pp. 11-56.

Taylor, M., M. Bobe and Leonard, S. (1995), 'The Business Enterprise, Power Networks and Environmental Change', in M. Taylor (ed), *Environmental Change: Industry, Power and Policy*, Avebury, Aldershot, pp. 57-81.

Tordoir, P.P. (1994), 'Transactions of Professional Business Services and Spatial Systems', *Tijdschrift voor Economische en Sociale Geografie*, vol. 85, 4, pp. 322-332.

Willke, H. (1998), *Systemisches Management*, Lucius and Lucius, Stuttgart.

Wood, P. (1996a), 'Business Services, the Management of Change and Regional Development in the UK: A Corporate Client Perspective',

Transactions of the Institute of British Geographers, vol. 21, pp. 649-665.

Wood, P. (1996b), 'An 'Expert Labour' Approach to Business Service Change', *Papers in Regional Science*, vol. 75, 3, pp. 325-349.

13 Competitive and Green? Determinants of Successful Environmental Management in the Manufacturing Sector

Boris Braun

Introduction

In the future, environmental performance will be an increasingly important factor in determining the competitive success of manufacturing companies. Economic and ecological goals do not necessarily contradict each other (Braun, 1998; Hurt and Ahuja, 1996; Porter and van der Linde, 1996). However, the environmental challenge will require considerable changes in the organization of firms as they respond to ever-tightening environmental legislation and to the needs of more environmentally aware consumers, investors and employees. The so-called 'greening of industry' will also have long-term consequences for technological change and regional development. This paper explores an increasingly important facet of these profound changes, namely, the implementation of environmental management systems (EMS) as a way to combine industrial and environmental knowledge.

The late 1980s saw a growing interest in the concept of corporate environmental auditing around the globe. In the early 1990s, the British Standard (BS) 7750 became the point of reference for international efforts to set internationally comparable standards for corporate environmental management. The progress that has been made over the years, resulted in the publication of the European Eco-Management and Audit Scheme (Council

Regulation, 1836/93) (EMAS) by the European Communities in 1993 and the ISO 14000 series (environmental management) by the International Standard Organization in 1996. Both standards are comparable in their major objectives. Both emphasize the importance of business ethics and a continuous improvement of environmental performance. EMAS and ISO 14001 differ in spatial scope (European versus global), organizational approach (site versus organization), and obligations for public information. Especially with respect to the latter aspect, EMAS is the more demanding standard, as it requires the publication of an environmental statement.

EMAS has been available for voluntary participation since April 1995. It basically works as an external control and validation of internal policies, programmes, management systems, and audit cycles. It sets no physical targets. It only requires compliance with existing national environmental standards. German industry has been particularly keen to adopt EMAS. In most other European countries the uptake of EMAS has been considerably slower (Braun and Geibel, 1999). These differences are partly a consequence of the varying pace of the member states in establishing the national prerequisites necessary to administer the system. Some member states have also implemented measures to promote small firm participation (notably, grants and supported pilot projects). Apart from these different regulatory and financial frameworks, there are other reasons for national differences in the adoption of EMAS by private companies. In the United Kingdom, for example, the national standard BS 7750 which later was converted into ISO 14001, proved to be more attractive to firms than EMAS. However, many European companies perceive the two systems as complementary rather than exclusive. The reformulation of EMAS (EMAS II) which became effective in April 2001 has further strengthened links between the two systems.

From a political as well as from a business perspective, a critical issue is how far environmental management systems can support efficient solutions to environmental problems. The answer to this question will undoubtedly influence the future acceptance of environmental management standards. EMAS as an instrument of environmental policy will only be successful if it leads to ecological, economic, and technological gains. From a long-term perspective, a more comprehensive question can be formulated: How far can industrial environmental management and environmental protection contribute to sustainable economic and ecological development? Until now, economic geography has been hesitant to answer this question.

The following sections of this chapter approach the theme of successful industrial environmental management in two steps. The first step investigates how far the central goals of industrial environmental management have been achieved by German enterprises until now. The second step is a statistical analysis of the factors which are significant for successful environmental management. On the basis of the empirical findings, I offer recommendations for individual enterprises and regional policies to improve environmental and economic performance.

The analysis is based on data from a representative sample of German manufacturing establishments validated according to EMAS. It covers 385 establishments; these were more than 40 percent of all validated manufactures in Germany by the beginning of 1998.[1] Thus, the sample is large enough to guarantee representative results. A comparison between the responding manufacturing establishments and the parent population on several control variables indicated that the structure of the respondents does not differ significantly from the parent population. The manufacturing enterprises in the sample represent very different ownership characteristics, size groups, industries, and location types. In addition to the quantitative information provided by this sample, qualitative experience was gained from more than 70 firm interviews conducted in Germany and the United Kingdom.

Success Dimensions of Environmental Management

Identification of the success dimensions of environmental management evokes the same problems as the evaluation of economic success in general. Difficulties arise especially because of the subjectivity of the term 'success' and the different indicators to measure it. Moreover, from the perspective of economic geography, it is important to incorporate criteria that assess regional or local effects. In broad terms, however, it is possible to identify four general success dimensions of environmental management, from which measurable success indicators can be derived. The four dimensions are: the economic dimension (cost-effectiveness); the ecological dimension (decreasing energy and raw material input); the technological dimension (innovation activities); and the communication dimension (enhancing the dialogue with stakeholders).

The Economic Dimension: Costs and Benefits

Environmental management standards are only generally accepted in the private sector if they lead to substantial cost savings or to new market opportunities. Yet, interviews with firms reveal that the chances of developing new markets or increasing market shares as a consequence of environmental management are still low. A process in which the fulfilment of environmental management standards becomes a necessity within supply chains is only developing slowly. At least until now, the cost side of environmental management is far more important for the firms. Cost savings, however, depend on the specific situation in each company. While some firms show net losses from having implemented an environmental management system (EMS), others show considerable gains. Thus, 58 percent of the surveyed manufacturing establishments claimed that they were able to achieve cost reductions within the first year after the system came into operation (Table 13.1). In most cases, cost-effectiveness had been achieved by optimizing the production process, mainly resulting in a considerable reduction of energy inputs and harmful by-products. For a majority of manufacturing establishments (61 percent), total expenses for establishing and running the system still exceed cost reductions.

The Ecological Dimension: Decreasing Resource Consumption

Cost reductions are often closely related to a more efficient use of natural resources. The issue of resource consumption is a decisive criterion to assess the quality of EMAS as an instrument of environmental policy. Such evaluations are difficult to accomplish, however, because in many cases there is a lack of reliable data on resource and energy consumption, even more so on emissions of specific pollutants. For this reason, the data presented here can only provide a rough indication of environmental effects of EMAS.

Overall, about one-third of the surveyed adopters of EMAS in Germany were able to reduce their total resource consumption between 1995 and 1998. This result applies to the reduced consumption of water (35 percent), fuel oil (31 percent), electricity (24 percent), natural gas (20 percent), and petrol (15 percent). Reductions in relation to output are even more important than such absolute reductions. The percentage of manufacturing establishments that achieve economies in relative terms is considerably higher because most plants increased their output in monetary as well as in physical terms (Table 13.1). However, it is almost impossible to determine or quantify the exact impact of EMS on the overall resource consumption because the relevance of other organizational and technological changes is often substantial.

Table 13.1 Selected indicators for successful implementation of EMS

	Establishments sampled		
	yes (%)	no (%)	N
Cost-effectiveness of EMS			
(after the first year of operation):			
Running expenses reduced by EMS	57.5	42.5	334
Cost savings higher than costs for running the EMS	38.6	61.4	202
Changes in resource consumption			
(during and after implementation):			
Total decrease of energy and raw material consumption (a)(b)	31.6	68.4	313
Total decrease of water consumption (b)	34.7	65.3	303
Total decrease of fuel oil consumption (b)	30.5	69.5	203
Total decrease of electricity consumption (b)	24.3	75.7	309
Total decrease of natural gas consumption (b)	20.0	80.0	235
Total decrease of petrol consumption (b)	15.2	84.8	191
Relative decrease of energy and raw material consumption a)(b)(c)	62.9	37.1	270
Relative decrease of water consumption (b)(c)	63.8	36.2	264
Relative decrease of fuel oil consumption (b)(c)	62.5	37.5	177
Relative decrease of electricity consumption (b)(c)	57.2	44.6	269
Relative decrease of natural gas consumption (b)(c)	53.8	46.2	204
Relative decrease of petrol consumption (b)(c)	52.0	48.0	171
Environmental innovations (as a result of EMS):			
Product innovations	17.2	82.8	378
Process innovations	52.1	47.9	378

Table 13.1 continued

	Establishments sampled		
	yes (%)	no (%)	N
Environmental communication with relevant stakeholders:			
Successful environmental communication (d)	30.4	69.6	368
'Very strong' or 'strong interest' in company's EMS			
(10 most important stakeholders) (e):			
Employees	73.5	26.5	362
Owners, shareholders	60.6	39.4	350
Government, authorities	55.4	44.6	368
Customers	51.4	48.6	358
Other companies and competitors,	36.7	63.3	346
Insurance companies	35.3	64.7	360
Non-governmental environmental organizations	34.3	65.7	362
Trade, industry associations	31.4	68.6	359
Media	30.4	69.6	363
Local community	22.1	77.9	362

(a) for selected inputs (water, electricity, gas, fuel oil, petrol), mean of input figures available
(b) decrease of more than 5 % between 1995-1997
(c) estimates based on input and output figures available
(d) profiles of environmental stakeholder demands and stakeholder interest in EMAS coincide
(e) self-assessment of companies on a scale from 1 (very little) to 5 (very strong)

Source: Survey of German manufacturing establishments 1997/1998.

The Technological Dimension: Environmental Innovations

Technological innovation is another important dimension of successful environmental management. The complex relation of environmental protection and technological innovations is still not fully understood despite substantial research activities in this field in recent years (for example, Hemmelskamp, Rennings and Leone, 2000). According to recent research findings environmental innovations tend to create first-mover advantages predominantly for those firms which are active in growth markets. In declining markets, traditional cost-oriented process innovations are often a better way to sustain competitiveness.

Due to the process-oriented nature of EMS, innovations are mainly process innovations (Table 13.1). In some cases the adoption of environmental management actually resulted in a substantial change of manufacturing processes. In contrast, product innovations have been relatively minor and mostly related to changes in the manufacturing process. It has to be stressed, however, that environmental innovations are not necessarily high-tech innovations. Rather, they are frequently an intelligent and consistent application of well-known techniques, machines, and materials.

The Communication Dimension: Environmental Dialogue with Stakeholders

From the beginning, hope for improvements in the communication between enterprises and the public has been one of the central elements of EMAS. The critical role of environmental communication was also stressed by authors who have based their work on the stakeholder approach (Donaldson and Preston, 1995; Freeman, 1984; Grafé-Buckens and Hinton, 1998). The stakeholder approach was initially a business management concept that emphasized the role of the environment in which firms operate. A stakeholder is defined as any person or organization who is affected or affects the way in which a company manages its business, for example, shareholders, employees, customers, banks, environmental organizations or the state. In order to be successful in the long term, enterprises have to find a sustainable balance of these different stakeholder demands. Despite its obvious normative character, the stakeholder concept can also be used as an empirical tool to analyse external demands on management decisions.

Table 13.1 lists the groups of stakeholders that were the most interested in the environmental management of the surveyed enterprises. The empirical results suggest that EMAS and the environmental statement have so far not significantly contributed to a functioning communication on environmental issues between companies and their external stakeholders (see Braun et al 2001). In particular, communication with so-called diffuse stakeholders (local communities, non-governmental environmental organizations, the media, etc.) is still in its infancy. Even according to comparatively low standards, not more than 30 percent of the surveyed enterprises could be classified as successful with respect to their environmental communication.

Determinants of Successful Environmental Management

If environmental management has the potential to combine ecological and economic gains, not all firms are able to profit from this 'win-win opportunities' or 'double dividends'. This potential remains untapped in many companies. Only a fraction of the EMAS-validated manufacturing enterprises in Germany (and in the United Kingdom) could be labelled 'ecological pioneers'. This leads to an important question: What are the internal and external determinants that are responsible for an ecologically and economically successful implementation of EMS within manufacturing establishments?

Basic Assumptions and Hypotheses

Little systematic research has been done on the success factors of environmental management from the perspective of economic geography. Economists and political scientists have carried out some relevant research since the early 1990s. Still, this work is only marginally concerned with spatial factors and relies mostly on rather anecdotal evidence or small firm samples (Freimann and Schwedes, 2000; Hillary, 1998; Sutton, 1997). Consequently, within geography, there is no comprehensive or generally accepted theory from which hypotheses could be derived. However, network and milieu approaches (Camagni 1991; Cooke and Morgan, 1993; Maillat, 1998), as well as work on regional innovation systems (Braczyk et al 1997), provide useful points of reference from a spatial perspective. Other promising lines of theoretical explanation are provided by two not explicitly spatial approaches: economic research which tries to identify critical success factors of private businesses (Steinle et al 1998) and the stakeholder approach (Grafé-Buckens and Hinton, 1998).

Based on the assumptions of these theoretical approaches, and experiences from firm interviews, it is assumed that successful implementation of environmental management is positively connected to both (favourable) firm-internal and external factors. The first set of (internal) factors is further differentiated as: a) environment-related and b) general internal factors. The second set is further differentiated as: c) regional and d) non-regional factors. A fifth intermediate category is defined as: e) network factors. Network factors by their nature hold a position between purely internal and external factors.

Network relations have to be actively pursued by the firm itself but they also rely on external conditions which cannot be significantly influenced by a single enterprise (for example, the existence of network partners within the region). Figure 13.1 represents this complex of different success dimensions and different success factors graphically. The following six specific hypotheses can be deduced from these relationships:

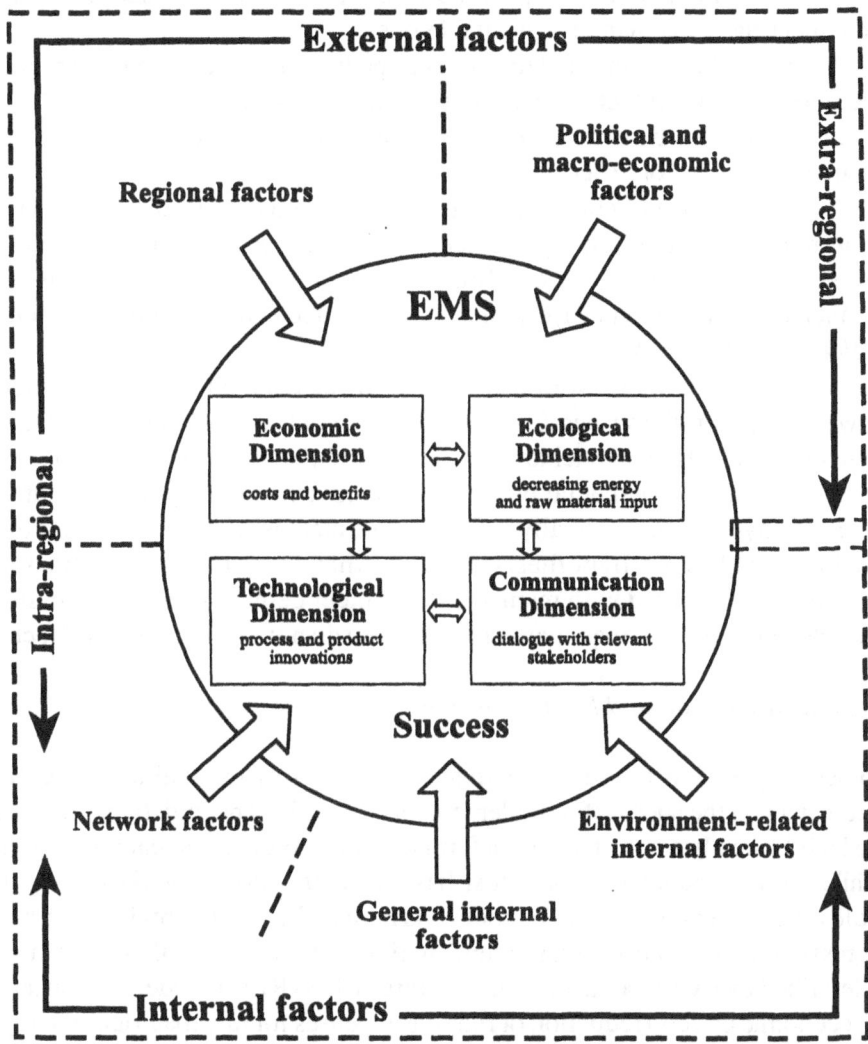

Figure 13.1   Determinants and dimensions of succesful environmental management in manufacturing

First, environmental management opens up considerable scope for individual implementation strategies. This hypothesis suggests that firm-specific efforts in environmental protection are more important for successful environmental management than general characteristics or the regional setting of an enterprise;

Second, economic and ecological benefits are closely related. According to this hypothesis, profitable enterprises are more successful in implementing EMS than less profitable ones;

Third, limited internal resources impede an adequate implementation of environmental protection measures on the company level. Firms, which have been assisted by external environmental consultants and supported by public funds, are more successful in implementing EMS;

Fourth, manufacturing establishments rely on their local and regional environment. Positive externalities (localization and urbanization economies) increase the chances of successfully implementing an EMS. This is an advantage for manufacturing establishments located in urban agglomerations and/or technologically dynamic regional settings;

Fifth, integration within environment-related inter-firm information networks supports the successful implementation of EMS. Apart from regional networks extra-regional exchange of information is important too; and

Sixth, environmental protection and environmental management at the company level are basically reflections of increasing pressures from the outside world. This hypothesis suggests that manufacturers, who see themselves confronted with intensive stakeholder demands, more successful in implementing EMS than manufacturers who do not yet feel the full effects.

Logistic Regression and Model Variables

Logistic regression is used to estimate the impact of internal and external parameters on the successful implementation of EMS. The logistic regression model is basically a non-linear transformation of an ordinary linear regression. While the dependent variable must have a binary format, variables of all scales are accepted as independent variables. In the following logistic regression models successful implementation of environmental management is described by five dependent variables. First, COSTRED (Model 1) measures the economic effects (reduction of running expenses for raw materials, energy, waste treatment, etc.). Second, RESOURCE (Model 2) measures the environmental effects (decrease of energy and raw material consumption) of environmental management. A third dependent variable, INNOVATE (Model 3), serves as an indicator for innovation processes initiated by environmental management. The fourth dependent variable, ENVCOMM (Model 4), indicates achievements in communicating with relevant stakeholders of the enterprise. The last variable, SUCCESS (Model 5), represents a more comprehensive overview. It is derived from the first four dependent variables and comprises different dimensions of EMS success.

In a second step 30 independent variables were selected for which explanatory potential can be assumed. Almost all variables were derived from the representative survey of EMAS-validated manufacturing establishments. There are only three exceptions. PATINT, a measure for the innovative character of the regional milieu, comes from the German Patent Atlas (Greif, 1998). ENVREG, an indicator for the state of the environment within the region, was taken from Korczak (1995). The classification of the variable AGGLO is based on the spatial typology of settlement patterns from the Bundesamt für Bauwesen und Raumordnung (1999).

The full set of selected indicators and variables is presented in Table 13.2. In most cases, clear causal effects (and signs) from the set of hypotheses can be expected. The effects of some general firm-internal factors such as industry (INDUSTRY), size (SIZECOM, SIZESITE), age (AGECOM, AGESITE), and market orientation (EXPORT) are more difficult to assess. Depending on the success criterion in question, different signs could be expected. For this reason, neutral effects were assumed for this set of variables.

In an analysis of direct effects between 'success indicators' and 'success factors', the majority of relations reveal the expected signs. On the other hand, only a relatively small number of potential success determinants indicates significant positive effects on all dependent variables. This points to the fact that success factors can have quite selective impacts on different aspects of environmental management. Thus, it is useful to calculate separate regression models for different dimensions of successful EMS implementation. The small number of significant relations also makes it necessary to limit the number of independent variables in the statistical models. Due to a stepwise backward exclusion of variables, with $p > 0.2$, only statistically relevant parameters are included in the logistic regressions.[2]

Table 13.2 Variables of logistic regression models

Dependent variables (success indicators):

COSTRED	Running expenses reduced by EMS (yes = 1, no = 0)
RESOURCE (a)	Relative decrease of energy and raw material consumption (yes = 1, no = 0)
INNOVATE	Product or process innovations as a result of EMS (yes = 1, no = 0)
ENVCOMM	Environmental communication successful (yes = 1, no = 0)
SUCCESS	Comprehensive indicator for EMS success (three or four success indicators = 1, less than three success indicators = 0)

Independent variables – internal factors related to environmental protection:

INNOPROD (b)	(+)	Environmental product innovations (yes = 1, no = 0)
INNOPROC (b)	(+)	Environmental process innovations (yes = 1, no = 0)
EARLY	(+)	Adoption of EMAS before July 1996 (yes = 1, no = 0)
ISO9000	(+)	Additional ISO 9000 certification (yes = 1, no = 0)
INTEGRAT	(+)	Quality and environmental management systems integrated (yes = 1, no = 0)
ISO 14001	(+)	Additional ISO 14001 certification (yes = 1, no = 0)
COMMIT	(+)	EMS initiated and supported by top management (yes = 1, no = 0)
EMPLOY	(+)	Interest of employees in environmental issues, scale from 1 (min.) to 5 (max.)
ENERGY	(+)	On-site usage of renewable energy (yes = 1, no = 0)
ECOPROD	(+)	Eco-aspects considered in product development (yes = 1, no = 0)
ADVERT	(+)	Environmental characteristics of products emphasized in advertising campaigns (yes = 1, no = 0)

Independent variables – general internal factors:

INDUSTRY	(o)	Dummies for different industries (yes = 1, no = 0)
GROUP	(o)	Subsidiary or branch plant (yes = 1, no = 0)
SIZECOM	(o)	Total number of employees (ln)
SIZESITE	(o)	Number of employees on the validated site (ln)
AGECOM	(o)	Age of company (in years) (ln)
AGESITE	(o)	Age of establishment (in years) (ln)

Table 13.2 continued

PROFIT	(+)	Profit situation, scale from 1 (very poor) to 5 (excellent)
EXPORT	(o)	Export as % of total production output (by tonnage)
RESEARCH	(+)	R&D expenditure as percentage of total turnover

Independent variables – network factors:

NETREG	(+)	Regular exchange of information with other companies on environmental issues, local and regional levels (yes = 1, no = 0)
NETNAT	(+)	Regular exchange of information with other companies on environmental issues, national and international levels (yes = 1, no = 0)
CONSULT	(+)	Assistance by external environmental consultant(s) (yes = 1, no = 0)
STAKE	(+)	Median strength of stakeholder demands, scale from 1 (min.) to 5 (max.)

Independent variables – external factors (regional and extra-regional variables):

WESTEAST	(+)	Location of establishment in Old or New Länder (West = 1, East = 0)
AGGLO	(+)	Location of establishment (urban agglomeration = 1, rural area = 0)
PATINT	(+)	Regional patent intensity (patent applications per 100,000 employees, annual average 1992-1994), source: GREIF 1998
ENVREG	(+)	Regional situation of natural environment, scale from 0 (very unfavourable) to 1,000 (very favourable), source: KORCZAK 1995
SUPPORT	(+)	EMAS process supported by public funds (yes = 1, no = 0)

In brackets: expected sign of regression coefficient
(a) also independent variable in model 1, (b) not tested in models 3 and 5

Results of the Logistic Regression Models

The logistic regression results are presented in Table 13.3. The table includes regression coefficient estimates (B) and their significance as well as partial correlation coefficients (R) to indicate the contribution of the respective covariate to the overall model. Positive B-coefficients indicate an increasing probability, negative B-coefficients indicate a decreasing probability of positive values of the dependent variable. All models are significant at the 1 percent level according to model chi-square statistic. The percentages of correct predictions are also acceptable (71 to 75 percent). The results of the logistic regression models can be summarized as follows:

Model 1 (COSTRED): The probability of actually achieving cost reductions is significantly increased if business establishments reduce their overall resource consumption (RESOURCE) and if they generate environmental product and process innovations (INNOPROD, INNOPROC). The environmental commitment of the employees is also important (EMPLOY). Experience already gained with environmental management (EARLY) is another, although less significant component which is positively related to the ability to cut costs. General internal factors are less influential. Neither industry and profits nor size seem to have a significant effect. The more complex impact of 'size' deserves a closer look. Additional chi-square tests show that small manufacturing establishments, especially those with less than 50 employees, have difficulties achieving cost-effectiveness. The role of regional and network variables is also contradictory. On the one hand, the regression results reveal that firms do not depend too much on their external environment. On the other hand, the innovation dynamics of the regional environment (PATINT), as well as the integration in regional information networks (NETREG), seem to increase the ability of manufacturing establishments to realize cost savings.

Model 2 (RESOURCE): Environmental gains (RESOURCE) are more difficult to predict from the given set of independent variables than cost savings. According to Model 2 only two variables influence the probability of achieving reductions in the overall resource consumption in a significant way. Decreasing resource consumption obviously depends on specific and very individual parameters (experience, market-opportunities, technology, etc.). These parameters are difficult to measure, due to their diverse and often problem-oriented character. Nevertheless, the positive sign for PROFIT tends to confirm the assumption that there is a close correlation between economic success and environmental progress. Again, the impact of 'size' proves to be more complex than the regression model suggests. As is clear from cross-tabulations and chi-square tests, reductions in the consumption of energy and resources significantly increases above a threshold of about 750 employees. Large manufacturing establishments still have an advantage in resource and energy efficiency. The roles of process and product innovations (INNOPROD, INNOPROC) and of progressive environmental technologies (ENERGY, ECOPROD) are surprisingly limited. The same is true for almost all of the external and network variables. The influence of external consultants (CONSULT) is inverse to the expected effect. An explanation for this could be that consultants are commissioned primarily by firms that have difficulties in establishing a working EMS. Experienced firms, on the other hand, are mostly able to implement an EMS without the help of external consultants.

Model 3 (INNOVATE): The ability to generate technological innovations seems to be primarily influenced by the degree of commitment to environmental protection (COMMIT, EMPLOY, ISO 14000), economic success (PROFIT) and the integration in national or international information networks (NETNAT). Moreover, the process is supported to a minor degree by an innovative regional environment (PATINT). Subsidiaries or branch plants are less successful when it comes to technological innovations (GROUP). The statistically irrelevant effect of R&D expenditure (RESEARCH) on environmental innovations is surprising at first sight, but the dominating process character of most environmental innovations can explain it. Process innovations – as is confirmed by other research work – are less dependent on expensive R&D activities compared to product innovations.

Model 4 (ENVCOMM): The relatively high percentage of correct predictions can be attributed to a small number of independent variables. Successful environmental communication is basically influenced by environment-related internal factors (ENERGY, ECOPROD) and – most importantly – by the intensity of stakeholder demands (STAKE). The more demanding the stakeholders, the harder will enterprises try to fulfill these demands. This can be interpreted as a particular form of a demand-pull situation. The negative sign of EARLY may imply the existence of learning-processes between 1995 and 1998.

Table 13.3 Logistic regression results

Independent Variable	Model 1 COSTRED			Model 2 RESOURCE			Dependent Variable Model 3 INNOVATE			Model 4 ENVCOMM			Model 5 SUCCESS		
	B	Sig.	R	B	Sig.	R	B	Sig.	R	B	Sig.	R	B	Sig.	R
RESOURCE (+)	0.411	**	0.114		not tested			not tested			not tested			not tested	
INNOPROD (+)	0.819	***	0.164		-			not tested			-			not tested	
INNOPROC (+)	0.381	**	0.112		-			not tested			-			not tested	
EARLY (+)	0.421	*	0.070		-			-		-0.410	**	-0.074		-	
ISO9000 (+)		-		0.261	n.s.	0.012		-			-			-	
ISO14001 (+)		-			-		0.274	*	0.058		-			-	
COMMIT (+)		-			-		0.557	**	0.106		-			-	
EMPLOY (+)	0.428	**	0.118		-		0.460	***	0.138		-		0.486	**	0.119
ENERGY (+)		-			-			-		0.402	**	0.075		-	
ECOPROD (+)		-			-			-		0.735	*	0.058		-	
ADVERT (+)		-			-			-			-		0.254	*	0.051
GROUP (o)		-			-		-0.310	**	-0.077		-			-	
PROFIT (+)		-		0.433	**	0.093	0.486	**	0.106		-		0.467	**	0.098
NETREG (+)	0.327	**	0.086		-			-			-		0.368	**	0.106

Table 13.3 continued

Independent Variable	Model 1 COSTRED			Model 2 RESOURCE			Dependent Variable Model 3 INNOVATE			Model 4 ENVCOMM			Model 5 SUCCESS		
	B	Sig.	R	B	Sig.	R	B	Sig.	R	B	Sig.	R	B	Sig.	R
NETNAT (+)		-		0.220	n.s.	0.011	0.319	**	0.082		-		0.252	*	0.048
CONSULT (+)		-		-0.457	**	-0.104		-			-			-	
STAKE (+)		-			-			-		0.887	***	0.215	0.459	*	0.049
PATINT (+)	0.007	**	0.091		-		0.005	*	0.059		-		0.006	**	0.082
Constant	-1.116	n.s.		-0.677	n.s.		-3.801	***		-3.850	***		-6.211	***	
No. of cases	223			194			241			281			254		
Model Chi² [df]	55.1 [7]	***		15.3 [4]	***		44.8 [7]	***		30.40 [4]	***		48.1 [7]	***	
Correct predictions (as %)	74.4			71.0			70.5			75.0			72.8		

In brackets: expected sign of regression coefficient
B: Regression coefficient; R: Partial correlation coefficient
Level of significance: *10 %, **5 %, ***1 %, n.s. not significant
Source: Survey of German manufacturing establishments 1997/9.

Model 5 (SUCCESS): The independent variable SUCCESS serves as a comprehensive indicator for successful environmental management which summarizes the findings of Models 1 to 4. About 30 percent of all surveyed manufacturing establishments were classified as successful implementers of environmental management. With the exception of the variable ADVERT, indicating environment-related product advertising, no new factor was added to the model equation. Some success factors, however, confirm their importance. According to Model 5, the environmental commitment of employees (EMPLOY), the profit situation (PROFIT), the dynamics of innovations in the regional environment (PATINT), the integration into information networks (NETREG, NETNAT) are the most important factors that are responsible for the overall success of EMS.

It has to be stressed that the logistic regression models are exploratory in character and that not all possible determinants can be depicted empirically. The models, for example, do not reveal any significant impact of the variable INDUSTRY. Still, some specific industries differ from this general picture. Manufacturers of medical instruments, for example, stand out positively in most of the success dimensions tested. Other industries, notably textile, clothing and furniture manufacturers, attract attention because of below-average results. The first two of these industries show considerable deficits in terms of energy and resource savings, or environmental innovations. The chemical industry reveals further differences. Chemical plants normally generate an above-average rate of product and process innovations. Yet, their achievements in cost cutting and the reduction of direct environmental effects by production are rather limited. These limitations probably do not reflect 'backwardness'. It is more likely that the chemical industry, which is very sensitive and politically controlled in its environmental effects, no longer enjoys easily accessible reduction potential compared to other industries.

Interpretation of the Empirical Results

If the findings of the empirical analysis with the basic hypotheses are confronted, a rather diverse picture emerges. While hypotheses 2 and 5 are basically confirmed by the empirical results, hypotheses 1, 4 and 6 are only partially confirmed and, despite its general plausibility, hypothesis 3 cannot be confirmed. Neither support by public funds nor external environmental consultants have a significant impact on the critical success dimensions of environmental management.

In general, the empirical findings reveal that there is a considerable firm-specific scope for achieving improvements in environmental protection (hypothesis 1). General firm characteristics, such as size, industry or product range, have only a limited impact on successful environmental management. Irrespective of these long-term structural characteristics, firms have enough room to find individual windows of environmental opportunity. Experience with management systems may help to find these windows but it does not guarantee that all possibilities of optimization become apparent. Environment-related internal factors in general tend to play an important role for costs, innovations and communication issues. Within the group of these environment-related factors the commitment of employees to environmental protection proves to be an especially influential factor.

Economic success and environmental performance show the expected close correlation (hypothesis 2). The reduction of costs and environmental gains are closely related and profitable enterprises are significantly more successful in environmental management than less profitable ones. These findings tend to confirm older empirical results on the firm-level (Fritz, 1995; Hurt and Ahuja, 1996) and more generally the optimistic win-win expectations of Porter and van der Linde (1996).

Hypothesis 4, expressing the traditional geographical focus on spatial criteria, has only been partly confirmed by the empirical results. 'Environmental pioneers' are not at all tied to a specific regional or local setting. Urban agglomerations are not necessarily a better breeding-ground for environmental innovations than rural areas. Manufacturing establishments in the periphery are not at a disadvantage, at least not with respect to the environmental performance. The same seems to be true for manufacturing establishments in East Germany. Obviously more important than general location factors is the regional innovation climate. Positive externalities and spill-overs within innovative milieus have a significant positive impact on cost savings and innovations. Regular exchange of information in environment-related networks has also proven to be a rather important success factor for environmental management (hypothesis 5). This result can be interpreted as an empirical confirmation of the 'network paradigm' (Cooke and Morgan, 1993), which emphasizes the role of cooperative linkages and the exchange of implicit or tacit knowledge.

Even though environmental management standards as such can be seen as a classic example of almost ubiquitous codified knowledge, at least some tacit knowledge appears to be necessary in order to implement these systems successfully on the firm level. Inter-firm cooperation seems to be even more important for environmental issues than for other areas of business. A closer look reveals that local and regional networks are particularly relevant for cost-related questions. For technological innovations, however, national or even international networking seems to be more important. These results are more or less in line with recent work on innovation-oriented inter-firm cooperation (e.g, Grotz and Braun 1997; Koschatzky 1997). Although these authors tend to be more sceptical with respect to the overall relevance of networks, they also stress the dominating role of extra-regional linkages for technological innovations.

Despite a highly significant impact of stakeholder demands on the effectiveness of environmental communication, it is too early to speak of stakeholder demands as a general success factor. External stakeholders are only marginally influencing the economic, ecological, and technological dimensions of environmental management. Therefore, hypothesis 6 is only partly confirmed. It should be considered, however, that internal stakeholders, and most prominently the employees of a firm, are among the most important success factors of environmental management.

Conclusions for Regional Politics and Private Enterprises

This analysis raises several specific conclusions for regional policy as well as for individual manufacturing enterprises. From the perspective of regional policy it is important to note that the environmental performance of enterprises is neither limited by the industrial structure nor by the settlement pattern of a region. For these reasons, regional economic policy that aims at environmental protection and economic growth is appropriate for peripheral areas with small-scale industries. In general, three points of reference for regional political initiatives are particularly promising.

The first group of initiatives relate to the promotion of environment-related networks which enhance a quick and cooperative exchange of experiences. These regional networks should remain open to the outside world to avert the danger of lock-ins. They should also be connected on a trans-regional level to similar networks in other regions. Especially the British experience with the so-called 'green business clubs' could be helpful in this context (Braun and Grotz, 2001). A second group of initiatives relate to the creation of open communications between enterprises and their external stakeholders (public authorities, politicians, neighbours, environmental organizations, regional customers and suppliers). If the immediate effects of this kind of regional communication should not be overestimated, the long-term potentials are high. A third group of initiatives relate to the promotion of environmental innovations in the public as well as in the private sphere. These initiatives will not be primarily instigated by environmental politics. Yet, they might serve environmental protection as well as economic interests.

With regard to manufacturing enterprises, favourable preconditions for environmental strategies depend upon the particular strategic goals desired by the firm (Figure 13.2). These goals relate to cost-effectiveness, reduction of resource and energy consumption, implementation of environmental process and product innovations, and improvement of environmental communication. In exploring the relationships between success indicators, strategies and management goals two aspects need to be underlined. First, success factors have a selective effect on different dimensions of environmental management. This means that the question pertaining to which factors should be especially promoted is dependent on the main strategic goals of an enterprise. Second, the success factors which could be identified in this study do not necessarily imply progress in environmental protection. They only make progress more probable. Within the list of success factors for environmental management two groups can be distinguished. The factors of the first group can be directly influenced by the management. These are classical environmental strategies. The factors of the second group either cannot be effectively influenced by the firm's management (especially regional aspects) or they rely on the general, non-environmental characteristics of an enterprise (e.g. profit situation, size). Thus, this second group of success factors describes the framework conditions for decisions aiming at environmental protection. The recommended firm-related management strategies (Figure 13.2) need to bear in mind the importance of these conditions. In practice, many of the success factors are in some way concentrated on communication aspects, both within the firm and between the firm and the outside world.

Goals of environmental Management	Success factors identified	
	Strategies	Favourable preconditions
Cost-effectiveness	• Reduction of resource and energy consumption • Promotion of process and product innovations • Fostering regular information exchange with other firms • Building-up understanding and environmental commitment of all employees	• Medium-sized manufacturing establishment • Innovative regional environment • Existence of partners for co-operation within a region
Reduction of resource and energy consumption	• Commitment of management to environmental protection • Fostering environmental competence in the company	• Good profit situation • Large manufacturing establishment
Implementation of environmental process and product innovations	• Commitment of management to environmental protection • Additional ISO-certification • Fostering regular information exchange with other firms • Building-up understanding and environmental commitment of all employees	• Good profit situation • Independent company or establishment • Innovative regional environment
Improvement of environmental communication	• Implementing publicly visible environmental protection measures (renewable energies, environmentally friendly products, etc.) • Implementing environment-related advertising strategies	• Good profit situation • Strong interest of internal and external stakeholders in environmental issues

Figure 13.2 Firm-related strategies to achieve central goals of environmental management

It is certainly too early to fully assess the environmental and economic effects of EMAS and the corresponding ISO 14001 standard. Manufacturers around the world still have a long way to go before they may call themselves and their products 'sustainable'. Nevertheless, the identification of critical success indicators and determinants of environmental management has given some answers to the question, how this 'long way' can be facilitated. It also demonstrates that manufacturing firms can be competitive and green!

Acknowledgements

The support of the German Research Council (Deutsche Forschungsgemeinschaft) in funding the research on which this paper is based and for the travel expenses to the Dongguan conference of the IGU which are gratefully acknowledged.

Notes

[1] The total number of EMAS-validated sites in Germany was 1,030 in December 1997. About 90 percent of these were classified as manufacturing. Until December 1999 the total number of validated sites increased to a level of 2,330. These were 74 percent of all registered sites within the European Union.

[2] The variables INTEGRAT, INDUSTRY, SIZECOM, SIZESITE, AGECOM, AGESITE, EXPORT, RESEARCH, WESTEAST, AGGLO, ENVREG and SUPPORT were excluded from all five models, due to their limited explanatory power.

References

Braczyk, H.-J., Cooke, P. and Heidenreich, M. (1997), *Regional Innovation Systems: The Role of Governance in a Globalized World*, ULC Press, London.

Braun, B. (1998), 'Locational Response of German Manufacturers to Environmental Standards', *Tijdschrift voor Economische en Sociale Geografie*, vol. 89, pp. 253-263.

Braun, B. and Geibel, J. (1999), 'The Implementation of Environmental Management: A Milestone in Firm Development?', in Van Dijk and P. H. Pellenbarg (eds), *Demography of Firms. Spatial Dynamics and Firm Behaviour*, Nederlandse Geografische Studies 262, Utrecht, pp. 285-302.

Braun, B., Geibel, J. and Glasze, G. (2001), 'Umweltkommunikation im Öko-Audit System – von der Umwelterklärung zum Umweltforum', *Zeitschrift für Umweltpolitik und Umweltrecht*, vol. 24, pp. 299-318.

Braun, B. and Grotz, R. (2001), 'Environmental Management in Manufacturing Industries: A Comparison Between British and German Firms', in L. Schäztl and J. Diez Revilla (eds), *Technological Change and Regional Development in Europe*, Physica, Heidelberg (forthcoming).

Bundesamt für Bauwesen und Raumordnung (1999), *Aktuelle Daten zur Entwicklung der Städte, Kreise und Gemeinden*, BBR, Bonn.

Camagni, R. (ed) (1991), *Innovation Networks: Spatial Perspectives*, Belhaven Pinter, London.

Cooke, P. and Morgan, K. (1993), 'The Network Paradigm: New Departures in Corporate and Regional Development', *Environment and Planning D*, vol. 11, pp. 543-564.

Donaldson, T. and Preston, L.E. (1995), 'The Stakeholder Theory of the Corporation: Concepts, Evidence, and Implications', *Academy of Management Review*, vol. 20, pp. 65-91.

Freeman, R.E. (1984), *Strategic Management. A Stakeholder Approach*, Pitman, Boston.

Freimann, J. and Schwedes, R. (2000), 'EMAS-experiences in German Companies: A Survey on Recent Empirical Studies', *Eco-Management and Auditing*, vol. 7, pp. 99-105.

Fritz, W. (1995), 'Umweltschutz und Unternehmenserfolg. Eine empirische Analyse', *Die Betriebswirtschaft*, vol. 55, pp. 347-357.

Grafé-Buckens, A. and Hinton, A.-F. (1998), 'Engaging the Stakeholders: Corporate Views and Current Trends', *Business Strategy and the Environment*, vol. 7, pp. 124-133.

Greif, S. (1998), *Patentatlas Deutschland*, -Deutsches Patentamt, München.

Grotz, R. and Braun, B. (1997), 'Territorial or Trans-territorial Networking:

Spatial Aspects of Technology-oriented Co-operation within the German Mechanical Engineering Industry', *Regional Studies*, vol. 31, pp. 545-557.

Hemmelskamp, J., Rennings, K. and Leone, F. (eds) (2000), *Innovation-oriented Environmental Regulation. Theoretical Approaches and Empirical Analysis*, ZEW Economic Studies 10, Physica, Heidelberg.

Hillary, R. (1998), 'Pan-European Union Assessment of EMAS Implementation', *European Environment*, vol. 8, pp. 184-192.

Hurt, S.L. and Ahuja, G. (1996), 'Does it Pay to Be Green? An Empirical Examination of the Relationship Between Emission Reduction and Firm Performance', *Business Strategy and the Environment*, vol. 5, pp. 30-37.

Korczak, D. (1995), *Lebensqualität-Atlas*, Westdeutscher Verlag, Opladen.

Koschatzky, K. (1997), 'Innovationsdeterminanten im Interregionalen Vergleich: Möglichkeiten zur Stärkung Regionaler Innovationspotentiale', *Geographische Zeitschrift*, vol. 85, pp. 97-112.

Maillat, D. (1998), 'Vom 'Industrial District' zum Kreativen Millieu: Ein Beitrag zur Analyse der gebietsgebundenen Produktionsorganisationen', *Geographische Zeitschrift*, vol. 86, pp. 1-15.

Porter, M. and van der Linde, C. (1996), 'Green and Competitive: Ending the Stalemate', in R. Welford and R. Starkey (eds), *Business and the Environment*, Earthscan Publications, London, pp. 61-77.

Steinle, C., Thiem, H. and Böttcher, K. (1998), 'Umweltschutz als Erfolgsfaktor – Mythos oder Realität?' *Zeitschrift für Umweltpolitik und Umweltrecht*, vol. 21, pp. 61-78.

Sutton, P. (1997), 'Targeting Sustainability: The Positive Application of ISO 14001', in C. Sheldon (ed), *ISO 14001 and Beyond. Environmental Management Systems in the Real World*, Greenleaf Publishing, Sheffield, pp. 211-242.

14 Globalization of the New Zealand Forestry Industry: The New Impact of Japanese Linkages

Christina Stringer

Introduction

Since the mid-1980s, the New Zealand forestry sector has become increasingly integrated into the Pacific Rim economy (Le Heron and Pawson, 1996). Prior to this time exports to Northeast Asia were intermittent. However, an extensive reorganization of production and a strengthening of trade linkages has occurred in response to national restructuring. The privatization of plantation ownership by the New Zealand government introduced foreign investors into the forestry sector and a reorientation within the production system has occurred in response to the flow of international capital and the subsequent establishment of production facilities by globally oriented forestry companies. New Zealand's integration into an Asian forestry production complex centred upon Japan has occurred, firstly, through low value exports and, secondly and more recently, through substantial investment by Japanese corporations in forestry ownership and processing facilities. This chapter addresses this integration.

Conceptually, New Zealand's integration into a Northeast Asian wood fibre import complex is examined within the commodity chain framework as originally articulated in world systems theory (Gereffi and Korzeniewicz, 1990). Empirically, the analysis focuses on the changing structure of commodity chain linkages from forestry to the end-overseas-user as a response to the state led commodification of the New Zealand forestry sector. The

253

analysis places special emphasis on the role of the *sogo shosha* and related inter-firm supply chain relationships in the mid- and late-1990s. The empirical analysis, which includes interviews with company and trade officials in Japan and New Zealand between 1995-1999, is in two main parts. First, I briefly outline the privatization of state owned assets, which laid the foundation for increased investment in the wood fibre regime. Second, the chapter analyses the strategic positioning of companies with respect to both resource ownership and non-resource investments. The commodity chain approach used in the chapter focuses on realignment within the forestry commodity chain, arising from company behaviour in the new environment that has emerged from restructuring, especially in New Zealand. The establishment of new and upgraded added value processing facilities is resulting in substantial readjustments along the supply chain and integrating New Zealand more fully within the Northeast Asian wood-fibre industry.

Commodity Chains

World systems theorists Hopkins and Wallerstein (1986, p. 159) define a commodity chain as a "network of labour and production processes whose end result is a finished commodity". The commodity chain was conceptualized in order to address a fundamental obstacle within the world-systems perspective, that of how to "depict and investigate the relationships that sustain and reproduce core-periphery relations over time and space" (Korzeniewicz and Martin, 1994, p. 68). In proposing the commodity chain as a means of analyzing the structure of the world economy, Hopkins and Wallerstein suggest that "concentration and decentralization, or shifts in the zonal location of nodes (e.g. from core to periphery), are associated with cyclical rhythms of the world-economy" (Gereffi et al 1994, p. 5).

Within global commodity systems research, the Gereffi and Korzeniewicz (1990) development of the Hopkins and Wallerstein model has perhaps gained the most prominence. Gereffi's (1994) industry specific work examines the relationship between the supply of materials, production, and consumption on a territorial scale (see also Gereffi, 1993, 1995; Gereffi and Korzeniewicz, 1990; Gereffi et al 1994; Korzeniewicz, 1994; Le Heron and Roche, 1996). Linking the commodity chain relationship to the concept of the core-periphery relationship Gereffi et al (1994, p. 2) accentuate the linkages between various geographical locations involved in the production of a single commodity.

Global commodity chains comprise a series of production nodes linked together in networks. It is suggested that profits accrue around specific nodes (where factors of production are assembled), rather than in nation-states as a whole, and that differential profits are created at various nodes within the commodity chain system (Appelbaum et al 1994). In particular, core-like nodes generate, and receive, a greater share of wealth than do commodities produced at peripheral-like nodes. While principally controlled by private economic agents (market forces and hence financial capital), commodity chains are also influenced by the state policies of both the producing and consuming countries (Gereffi, 1994; Dicken, 1998).

Two types of commodity chains are identified, namely 'producer-driven' and 'buyer-driven' chains (see chapter 20 by van Grunsven). Within producer-driven chains TNCs or similar large scale enterprises based usually in core localities coordinate the production system with the capacity to formulate backward and forward linkages and also to diversify. By way of contrast, within buyer-driven chains, large retailers and trading companies play a central role in "setting up decentralized production networks in a variety of exporting countries". Producer-driven chains "most characteristic of capital- and technology-intensive industries such as automobiles, computers, aircraft, and electrical machinery" (Gereffi, 1994, p. 97) also encompass the wood fibre industry. A central characteristic of producer-driven chains is the level of control the administrative headquarters of the TNCs exercise (Gereffi, 1994). Trade and production processes are linked across the zones of the world economy organized by large corporations involved in economic networks on a global scale. That the linkages are increasingly firm-specific as opposed to territoriality-based is leading to the "development of longer, more decentralized, and more flexible commodity chains" (Gereffi et al 1994, p. 8).

Edgington and Hayter (1997) apply the commodity chain framework to investigate the forestry sector. They conceptualize forestry, first, as an input-output structure consisting of a set of products, services and resources linked together in a value adding sequence. The added value sequence comprises six functions (decision making, research and development, production, distribution, marketing and business services). Second, they identify a governance structure consisting of authority and power relationships that determine how financial, material, and human resources are allocated and flow within a chain of forest products. Third, they examine the geographical dispersion and/or concentration of enterprises in different production and distribution networks and the desire of each region to control value added activity, an acknowledgment of territoriality. Finally, forestry is shaped at each stage of the chain through specific institutional relationships formed at different geographical scales. The chains are principally coordinated by enterprises through intra-firm and inter-organizational relationships.

Through the establishment of globally integrated networks, and hence nodes along the production chain, the Japanese *sogo shosha* play a fundamental role in Japan's resource related trade. Foreign affiliates and trade offices facilitate the procurement of raw materials and technology for the Japanese market. The *sogo shosha*, in turn, act as marketing agents in selling Japanese finished goods overseas. For overseas raw material suppliers the *sogo shosha* may be the only vehicle by which they are able to penetrate the Japanese market (Edgington and Hayter, 1997). In the 1980s the fastest growing segment of *sogo shosha* activity was third-country trade, where the trading houses facilitate trade between countries other than Japan (Edgington, 1990). In relation to the global timber trade, Marchak (1995, p. 4) reflects that "perhaps it is to Japan that we should attribute the current phase of globalization; certainly, it has been Japanese companies, urgently scouring the world for wood, who have pioneered global sourcing patterns...and Japan's Asian neighbours are not far behind". New Zealand has long been a site for the international activities of the timber and log sourcing divisions of the *sogo shosha* most of which have established branch offices in the country.

Realignment of the New Zealand Forestry Commodity Chain

Restructuring and New Interactions

Prior to the 1980s the state retained a strong presence in forest ownership; the rudiments of an afforestation programme in the form of experimental nurseries established in the late 1800s were followed much later by much larger planting booms in the early to mid-1900s (Roche, 1990, 1992). To increase forestry output and as a means of improving the economic performance and international competitiveness of the forestry sector it was deemed necessary by the state to privatize the state owned forestry assets.

Commercial principles, introduced into forest management, were followed by the subsequent sale of Crown forest assets.

Privatization of New Zealand's state owned plantations forest began in 1989 with the assets, (mostly the trees) associated with approximately 550,000 hectares of plantation forest (90 forests) and two state owned sawmills, being tendered for sale. Successful purchasers obtained assets on the land and land use rights for a minimum period of 70 years or two rotations, which ever is the longer. Land use rights remain in perpetuity except where otherwise notified, that is, in order to redress a successful Treaty of Waitangi claim, whereupon the land use period will be reduced. (This treaty is the framework for settlements between Maori and the New Zealand Government). By March 1991, approximately 45 percent of the forests originally offered for tender were sold. Fletcher Challenge Forests, controlled in New Zealand but 54.2 percent foreign owned, and Carter Holt Harvey, 51 percent controlled by International Paper of the USA, were dominant among the successful contenders, purchasing 60 percent of the forests sold. The remaining 40 percent was sold to Asian companies (39 percent) and domestic (one percent). In 1992, some 97,000 hectares of state-owned forests managed by New Zealand Timberlands Limited were sold to ITT Rayonier, an American owned transnational company, for $NZ366 million. ITT Rayonier, as a new competitor in the forest growing sector, was perceived as spreading New Zealand's forestry risk away from Asian markets (Bilek and Horgan, 1992).

Privatization of the state-owned forests concluded in 1996 with the $NZ2.026 billion acquisition by Fletcher Challenge Forests Limited (37.5 percent) in association with Brierley Investment (25 percent) and Citifor (China International Trust and Investment Corporation) (37.5 percent) of Forestry Corporation of New Zealand. CITIC New Zealand, a fully owned subsidiary of Citifor, was formed and in 1999, ownership of the Central North Island Forest Partnership became a 50/50 joint venture between Fletcher Challenge Forests and CITIC New Zealand. By 2001, the partnership was in dispute and legal mediation in effect.

Prior to the privatization of the forestry sector, Fletcher Challenge Forests and Carter Holt Harvey dominated the industry along with the New Zealand Forest Service (government owned). During the 1980s, an extensive offshore movement of capital occurred as part of New Zealand's ongoing internationalization process. Britton (1991) comments that the latest phase in internationalization occurred in order for New Zealand companies to penetrate overseas markets, to increase competitiveness, to escape the restrictions associated with a small market, and to diversify risk. During this period Carter Holt Harvey invested in forestry in Australia in order to fulfil demand by the Japanese pulp market. Additionally, Carter Holt Harvey invested in South America while Fletcher Challenge's investment strategy included Chile, Argentina and Canada. During the late 1980s increasing opportunities within the Northeast Asian markets became apparent. Trade linkages increased, in conjunction with the relaxation of regulatory regimes and the substitution of softwood logs for hardwood logs. During the 1990s, in order to develop forestry trade, which was becoming increasingly focused on the Northeast Asian markets, Fletcher Challenge Forests and Carter Holt Harvey established regional offices in Tokyo. The offices were established mainly to facilitate market entry and to promote marketing initiatives as the companies sought to develop and diversify into a range of higher-value products (Palmer, pers. comm. 1995). These trends highlight the degree of export diversification that is occurring in the search for added-value transformation of the industry.

Privatization of the former state-owned forests coupled with changes occurring in the global forestry complex proved to be a catalyst for change of forest ownership patterns in New Zealand. A redefinition of the wood-fibre industry in New Zealand occurred, accompanied by the further and different insertion of the New Zealand forestry sector into the global wood-fibre regime.

Governance, Coordination and Value

The changing internal regulatory environment of New Zealand and Japan, which allowed actors to begin to explore prospects, has aided interactions between these two nations. National reregulation, in response to global initiatives, enabled each country to reset themselves into the globalising economy. For the New Zealand forestry sector, the relaxation of import barriers has presented new market opportunities and challenges, requiring a re-articulation of production as producers adapt to new production-consumption relations.

Tariff structures have encouraged the importation of logs into Northeast Asia, as added value products warrant a higher tariff or non-tariff barrier. Under the Uruguay Round agreement Japan committed to eliminate tariffs on pulp and paper, to reduce the tariffs on solid wood products by 50 percent and to bind the tariffs on sawn timber (this applies to radiata pine) to

4.8 percent. Additionally, reductions were achieved on the tariff levels on panel products (Ministry of Forestry, 1996). The tariff reductions and eliminations have a ten-year phase-in period. Further developments occurred in November 1997 at the Asia Pacific Economic Co-operation (APEC) summit in Vancouver where APEC members agreed to fast track tariff reductions in forestry.

Traditionally, Japan has been a recipient of low grade timber exports from New Zealand; logs and flitches used in low value end uses such as the plywood and packaging industries influenced consumer perception of radiata pine as an inferior wood. While low value added logs comprise a significant portion of exports to Japan there is an increasing demand from Japanese customers for plywood and other added value products. New Zealand exporters, in an effort to reposition radiata pine as a higher value species, are seeking to establish niche markets particularly in the construction and appearance grade markets. Japan is a significant world market for sawn timber (principally for the packaging industry), fibreboard, plywood and panel products. Exports of panel products (fibreboard, plywood, veneer and particleboard) increased significantly in the mid-1990s in response to increased investment by Japanese corporations in production facilities. Imports of New Zealand log and timber are dominated by the Japanese *sogo shosha* (88.5 percent). This is in contrast to North America (54 percent) where the role of the trading house is to a degree being circumvented by building companies (Edgington and Hayter, 1997).

Companies, Networking and Strategic Positioning

Resource Investments: Fletcher Challenge Forests

Fletcher Forests' customer base in Japan has historically consisted of trading houses and sawmills, principally in the log and raw material trade. In particular, Fletcher Forests has a long established relationship with the Sumitomo and Nichimen Corporations. In 2000, nine percent of Fletcher Forests' exports, comprising logs for packaging and laminated lumber for the housing and construction markets, are destined for the Japanese market (Fletcher Challenge Forests, 2001).

With an increasingly global supply of logs coupled with finite markets in which prices are likely to become leviable due to market saturation, Fletcher Forests is evaluating how to succeed in an international manufacturing industry. The company is targeting the large-scale added value down-stream markets in order to control the commodity chain from forest through to consumer (Hayward, pers. comm. 1997). Historically, radiata pine has not been permitted to compete in the Japanese housing market due to consumer perception of radiata pine as a low quality packaging material. Considerable progress has been made to position radiata pine as a high-value structural product. With an increased volume of radiata pine being exported from both Chile and New Zealand, Fletcher Forests recognized a need to create high value markets and hence their entry into the Japanese housing market and other finished type applications for the Asian market. Within the Japanese housing market, Fletchers targeted the tatiuri (spec builders) segment.

In 1999, Fletcher Forests had a market share of approximately 4,000 housing starts out of a total of 1.4 million units per year (Fletcher Challenge Forests, 1999) and are seeking to develop further added value products in order to broaden their market share. The Kawerau laminating plant commissioned in 1996 was designed to meet Japanese specifications. Laminated lumber and finger jointed products are produced solely for the Japanese housing market using lower quality logs which are transformed into high-value laminated products. Production capacity is currently 40,000m³ per year with 90 percent of output destined for the Japanese market. Product development is occurring in conjunction with the Japanese company Sumitomo House, the trading arm of Sumitomo Corporation, in order to meet very rigorous Japan Agricultural Standards (JAS) and Approved Quality (AQ) certification. The Kawerau plant was the first non-Japanese wood manufacturing facility to receive AQ certification for treated dodai (laminated lumber ground sill) products. To date, Fletcher Forests have received JAS certification for laminated posts, medium density posts, and Douglas fir finger-jointed components and are awaiting approval for further products.

By moving further downstream, in order to meet the particular quality standard of the 'house makers', and thus circumventing the trading houses, Fletcher Forests have removed a former perceived barrier (Hayward, pers. comm. 1997) (Figure 14.1). A direct line operating between the Kawerau manufacturing facility and the Japanese 'house maker' consumers, that was not possible 10 years ago (Hayward, pers. comm. 1997), is reflective of the diminishing role of the *sogo shosha* as they are threatened by the movement

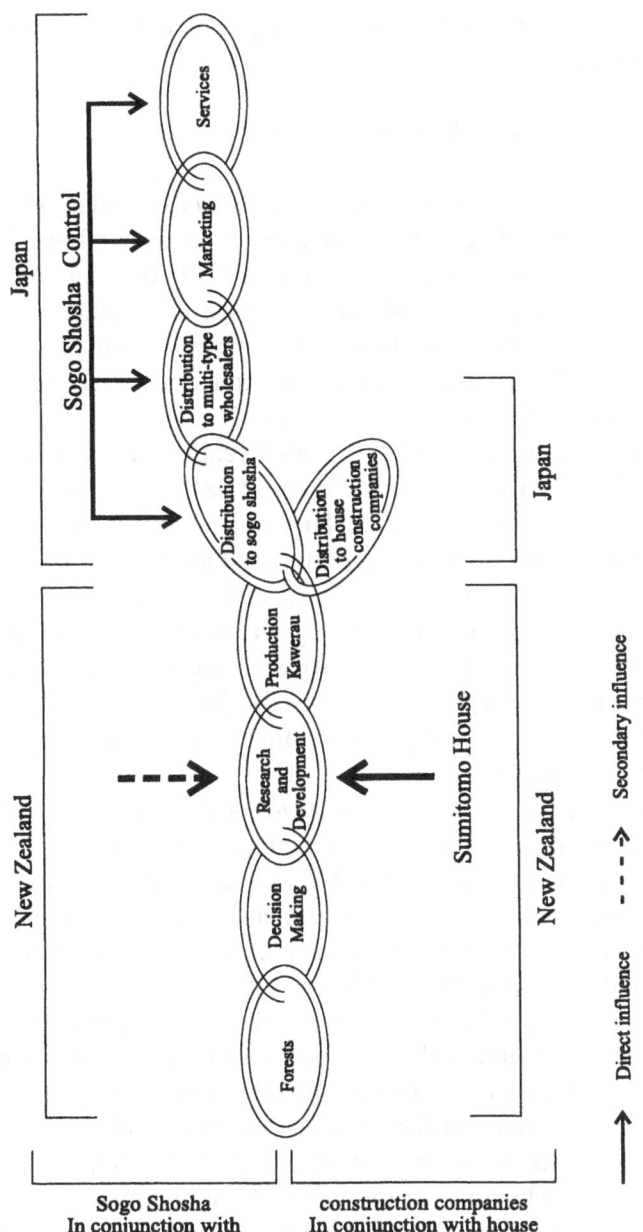

Figure 14.1 Fletcher Challenge Forests: Production chain components

of building companies into the timber production chain (Edgington and Hayter, 1997).

Resource Investments: Juken Nissho Limited

In 1990, Juken Nissho was established in New Zealand in order to purchase the cutting rights to five former state owned forests totalling 43,531 hectares for $NZ125 million: Ngaumu (Wairarapa), Patunamu (near Gisborne), Wharerata (East Coast), Aupouri and Otangaroa (Northland). Juken Nissho Limited is jointly owned by the Japanese general trading house Nissho Iwai (15 percent) and a major Japanese wood products manufacturer, Juken Sangyo Limited (85 percent). Nissho Iwai imports forest products: approximately 35 million cubic metres of wood is imported yearly into Japan of which Nissho Iwai handles 2.3 to 4 million cubic metres (Kase, pers. comm. 1996). Nissho Iwai has been present in New Zealand since 1967 when they established a branch office to facilitate the export of radiata pine to Japan. Through an established presence in the New Zealand market Nissho Iwai perceived a sustainable forest plantation as being an attractive investment in order to secure timber supply. Juken Sangyo were approached as a potential joint venture partner due, in part, to a long established trading relationship between the two companies. More importantly, Juken Sangyo has expertise in forestry management and operates several mills in Japan.

During the 1980s Juken Sangyo successfully developed techniques permitting the use of softwood in the laminated veneer lumber (LVL) process leading to an increase in demand for softwoods. Juken Sangyo is a leading Japanese manufacturer of "stepping boards for stairways using laminated veneer timber" and is a major producer of high value end uses such as flooring, fittings, and doors (Gresham, 1991, p. 14).

As owners of cutting rights over forests the joint venture partners determined that market sales of wood for processing was not a profitable venture and therefore processing should occur in-house. New Zealand is recognized as a source for fast rotation clearwood and the economic benefits of processing the wood in New Zealand versus shipping the logs (and consequently waste volume) are such that it is essential for profitability that primary processing occurs in New Zealand. Processing plants producing sawn lumber, LVL (laminated veneer lumber), and LL (laminated lumber) operate in Masterton and Gisborne sourcing logs from the nearby Juken Nissho owned forests. In 1991, Juken Nissho purchased the Kaitaia Triboard Mill and has upgraded and expanded the facility. With a maximum production capacity of 140,000 m³ per year the mill cannot keep up with demand (Kase, pers. comm. 1996). In response to increased demand from the Japanese market Juken Nissho has continued to expand the New Zealand based operations. In 2000, Juken Nissho announced the construction of a timber processing mill in the Kaitaia area. Five months after the opening of the mill Juken Nissho announced an up to $NZ50 million expansion to double the mill's capacity (Dominion 16 July 2001). Following the establishment of production facilities in New Zealand, Juken Sangyo began closing mills in Japan (Press, 1992).

While the second level of processing (advanced finished products) currently occurs in Japan at the Juken Sangyo mills, production facilities have also been established in China and the Philippines for the further processing of products from New Zealand. The Juken Sangyo wood panel plant in the Philippines is dedicated entirely to the advanced production of partly processed products from Juken Nissho's operations in New Zealand (*New Zealand Journal of Forestry*, 2000). The export of radiata pine products (via the Nissho Iwai shipping line), in semi-processed form, is principally to Japan while a small amount of triboard is sold on the New Zealand domestic market and small quantities of product are sold in Korea. Approximately 70 percent of Juken Sangyo's products are made from New Zealand pine and marketed under the brand "Jupino". Through backward integration Juken Sangyo and Nissho Iwai control the commodity chain from the natural resource through marketing and sales (Figure 14.2). As such the company is instrumental in repositioning radiata pine within the Japanese market where wood is valued for its aesthetic appearance.

Resource Investments: New Oji Paper Company

While the sale of New Zealand's state owned forests saw the entry of two major Japanese corporations into the production end of the industry, Japanese presence in the New Zealand industry dates back to 1972 with the establishment of an integrated sawmill and pulpmill. New Oji Paper, Japan's largest pulp and paper manufacturer formed in 1993 by the merger of Oji Paper and Kanzaki Paper, has substantial investments in New Zealand. The company's principal interest is in Pan Pacific Forest Industries (NZ) Limited. A combined pulpmill and sawmill was established in 1971 as a joint venture

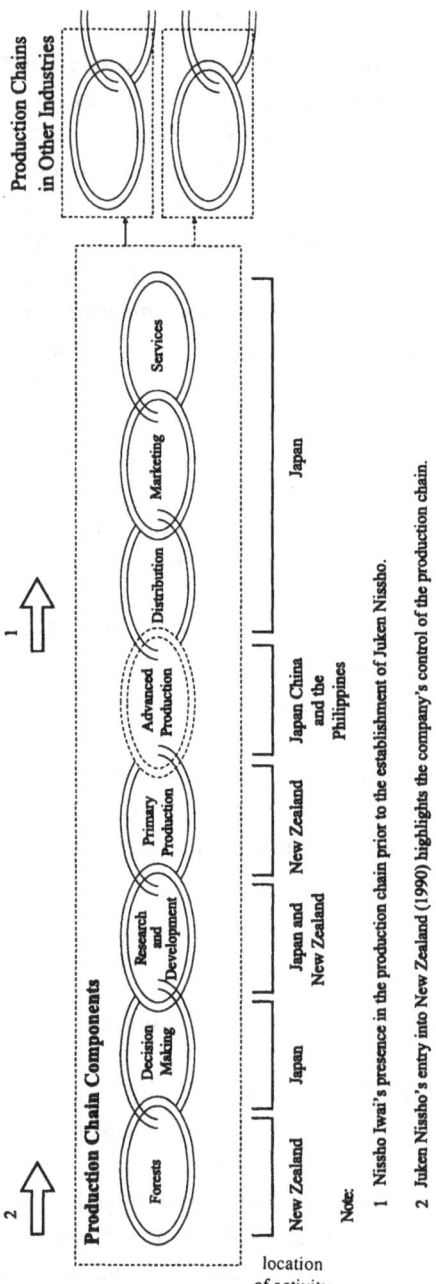

Figure 14.2   Juken Nissho: Production chain components

(Carter Oji Kokusaku Pan Pacific Limited) between Carter Consolidated Limited (now Carter Holt Harvey) and Oji Sankoku (Oji Paper and Sanyo Kokusaku Pulp Co.). By 1993, Oji Sankoku had obtained 100 percent ownership and the name of the company was changed to Pan Pacific Forest Industries (NZ) Limited.

The different capabilities of the companies were important in the subsequent development of a timber processing plant. While Carter Consolidated had the necessary expertise to establish the plant, Oji Sankoku provided technical expertise and guaranteed markets in Japan. An integrated operation was formed with the by-products from the sawmill being used in the mechanical pulp process. The sawmill mainly produced wood in flitch form for the Japanese packing industry. In response to changes in the Japanese paper market the pulp mill was converted from a refined ground wood process to a thermo-mechanical pulping process, a conversion dependent on Japanese technology. Semi-processed pulp is exported to Japan where it is repulped, further refined, and blended with other mechanical pulp to produce a paper suitable for the Japanese newspaper industry (Pan Pac, 2000).

The acquisition of assets and cutting rights to the former state forests (approximately 30,000 hectares) and the subsequent formation of Hawkes Bay Forest Limited removed concerns about a wood supply shortage and the company entered a new expansionary phase. Upgrading and expansion has occurred allowing the pulp mill to move further along the higher value added chain and examine other market opportunities. Currently there is a limited amount of product sold to Australia, USA and elsewhere in Asia. The sawmill processes approximately 500,000 tonnes of logs annually with an additional 100,000 tonnes of wood chips being sourced outside the region (Pan Pac, 2000).

In 2000 Pan Pac continues to be an integrated operation within an industry where the general tendency is to separate the solid-wood processing from the fibre/residual processing operations: "The nature of the forest has meant we have had to develop our solid wood side to remain profitable" (Pritchard, pers. comm. 1997). The $50 million upgrade announced at the beginning of 1996 involved the remodelling of the former sawmill and the introduction of new state-of-the-art technology in order to remain internationally competitive. Production capacity is over 250,000 cubic metres annually (Pan Pac, 2000).

Pan Pac's forestry division (formerly Hawkes Bay Forest) output is in the vicinity of 490,000 tonnes of logs a year; approximately 70 percent of which goes to Pan Pac where poorer quality logs are transformed into pulp and exported to Nippon Industries and New Oji Paper in Japan. Pan Pac supplies around 20 percent of New Oji's fibre requirements for their Tomakamai Mill in Japan. Medium quality logs, used for building and packaging crates, are semi-processed at Pan Pac with final processing occurring in Japan; high-grade logs are sold to domestic sawmills. Nippon Paper Industries, formed in 1993 by the merger of Sanyo Kokusaku Pulp and Jujo Paper, distributes approximately 50 percent of Pan Pac's export lumber. Pan Pac's control of the primary and secondary production stages has ensured the widespread integration of much of the forestry in Hawkes Bay into commodity chain linkages centred on Japan. Developments have occurred in the sawmill operation as the company has sought to move further down the added-value chain away from the constraints associated with production purely for packaging.

Strategic Positioning and Non-Resource Investment: Tachikawa Forest Products Limited

Tachikawa Forest Products Limited, a prominent end-user of radiata, has imported radiata pine from New Zealand for around 30 years. In 1988, Tachikawa obtained approval from the OIC to invest in a sawmill operation in New Zealand. The availability and cost competitiveness of log supplies coupled with the advantages of lower production costs made New Zealand an attractive investment location. In comparison to Japan, log prices are 50 to 60 percent less in New Zealand, likewise wood chip (used in the pulp and paper industry) and processing costs are considerably lower (Tachikawa, pers. comm. 1996). Tachikawa entered the New Zealand market in conjunction with the Nichimen Corporation (40 percent), a major importer of logs and lumber. Advantages to the joint venture included an established presence by Nichimen in the New Zealand market, as a long time buyer of wood and the operator of a shipping line allowing production to be shipped at cheaper costs.

A single site operation more central to the forests, was located in Rotorua. Tachikawa Forests own approximately seven hectares of land – half of which is zoned as residential and used as a buffer zone. Two state-of-the-art sawmills (vacuum kiln), kiln-drying facilities, and warehouses for dry storage have been established on the industrial site. A lack of space permits only a two-day log inventory in comparison to other sawmills, which generally hold one to two weeks' inventory. Hence the implementation of a just-in-time system has been necessary which Koji Tachikawa (pers. comm. 1996) perceives as making the operation "more efficient and more productive; otherwise it would be an impossible situation". The timber is dried and treated for stabilization and further strengthened for use as posts and beams. Between 35 to 40 percent of production is exported to Japan mostly as packaging lumber to make boxes, crates and pallets. Twenty to thirty percent remains in the domestic market for remanufacturing purposes, and the balance is exported to various markets within the Pacific Rim including the Philippines, Malaysia, China, Vietnam, Indonesia, Taiwan, Korea, Thailand and China (Tachikawa Forest Products, 2001).

A Discursive Conclusion

Economic restructuring pressures and measures in New Zealand have propelled the country's forestry industry into a more competitive international environment. This was achieved through the offshore movement and expansion of New Zealand domiciled companies coupled with the sale of former state-owned forests. Privatization measures aided in the restructuring of the industry, permitting the formation of more in-depth trading linkages with Northeast Asian markets. Changes that were occurring within the forestry industry in New Zealand occurred alongside changing supply-demand market forces in Asian markets and were aided by progressive changes occurring within the regulatory regimes at national levels in each country. Le Heron (1997, p. 64) argues that "the mix of companies (now mostly overseas owned) currently involved in New Zealand are attempting to align forest resources of the country into evolving wood supply and distribution networks centred mainly in Northeast Asia". This process has continued with the major forestry companies pursuing more in-depth investment in processing facilities in New Zealand.

Over the past few decades, forestry trade between New Zealand and Northeast Asia has been intrinsically and principally linked to Japan's increasing demands for timber. While unprocessed exports are still a significant component of forestry trade, there is a move away from low value products (periphery activity) into more end-application specific products. This trend highlights a move further down the commodity chain and towards increasing the share of economic surplus (core activity) in New Zealand, albeit due to substantial Japanese investment. Nevertheless, the marketing and business service links of the commodity chain, both of which are core activities, remain concentrated in Japan.

Due to increasing costs of production, coupled with an archaic, declining sawmilling industry (Hayward, pers. comm. 1997), Japanese entities are moving closer to sources of raw material and hence gaining control of the added-value chain from New Zealand to Japan. New entrants in the forestry processing industry in New Zealand, predominantly buyer-driven, have shaped production networks and shortened the commodity chain linkages through tighter involvement in more stages. The *sogo shosha* are significant investors in New Zealand, often controlling a vertically integrated operation from the forest to the end market, in order to secure the procurement of raw materials and by establishing processing activities for low added value products principally to supply the Japanese market. The *sogo shosha* have entered the New Zealand market principally through traditional joint venture arrangement highlighting the nature of the embedded inter-firm relations within Japanese corporations (Hayter and Edgington, 1997).

The most visible of Japanese investors is Juken Nissho whose control of the production chain ensures quality control and encompasses "the full production-consumption cycle: raw material supply, the design and manufacture of components and finished goods, exporting, distribution, and retailing. These products and services are connected in a sequence of value adding economic activities" (Gereffi, 1995, p. 47). In contrast to ownership of forest resources that Juken Nissho and New Oji have embarked on, Tachikawa Forest Products has not seen resource ownership as necessary for added-value transformation. Although the company is constrained from expanding, further downstream investment in New Zealand is being considered actively and in the interim Tachikawa has become an essential supplier to Panahome Industries in Rotorua which produces added-value products for the Japanese market. While Japanese companies are dominant within the production chain Fletcher Challenge Forests has developed forward linkages within the Japanese market through the establishment of a mill solely to meet Japanese consumer demand. Additionally, the development permits Fletcher Challenge Forests greater control of the added value chain in conjunction with established networks and highlights an alteration to the former *sogo shosha* dominated production chains.

While this chapter has examined the realignment of an historically and geographically specific component of the Asia-Pacific wood-fibre commodity chain from a New Zealand perspective there are wider processes at work. Within the past decade there have been substantial changes within the regulatory framework of many Asian-Pacific Rim countries impacting on supply options, exports, imports and processing options. The nature of the regulatory changes means that companies are accessing the wood-fibre value chain in order to meet both the domestic industrial consumption needs of the 'home' country and also those of consuming countries such as Japan and the United States. Through the reorganization of production and trade linkages the forestry sector in New Zealand is becoming increasingly embedded in a global commodity chain centred upon Japan. The distribution and marketing components of the commodity chain are imposed on the initial transformation stages of the chain as production is adapted to meet market demand with control dominated by Japanese interests. The production market is changing to become a customer-led system in contrast to a former resource-led system. Through financial linkages the production process is becoming an integrated system evidenced by the reduction in the number of nodes within the commodity chain process.

Acknowledgements

I gratefully acknowledge the contributions of Richard Le Heron and Roger Hayter in shaping this chapter, Igor Drecki and Jan Kelly for design and cartographic work, and the University of Auckland (Graduate Research Fund and the Department of Geography) for financial assistance.

References

Appelbaum, R., Smith, D. and Christerson B. (1994), 'Commodity Chains and Industrial Restructuring in the Pacific Rim: Garment Trade and Manufacturing', in G. Gereffi, and M. Korzeniewicz, (eds), *Commodity Chains and Global Capitalism*, Greenwood Press, Westport, Connecticut, pp. 187-204.

Bilek, E.M. and Horgan, G.P. (1992), 'The Challenges of Privatisation: New Zealand's Experience with Forestry', Proceedings from International Union of Forest Research Organizations, Moscow, pp. 1-37.

Britton, S. (1991), 'Recent Trends in the Internationalization of the New Zealand Economy', *Australian Geographical Studies*, vol. 29 (1), pp. 3-25.

Dicken, P. (1998), *Global Shift: Transforming the World Economy*, 3rd Edition, Paul Chapman, London.

Dominion (2001), 'Timber Giant May Outlay $50 million', Wellington, 16 July, p. 16.

Edgington, D. (1990), *Japanese Business Down Under: Patterns of Japanese Investment in Australia*, Routledge, London.

Edgington, D. and Hayter, R. (1997), 'International Trade, Production Chains and Corporate Strategies: Japan's Timber Trade with British Columbia,' *Regional Studies*, vol. 31 (2), pp. 151-166.

Fletcher Challenge Forests Limited (1999), Annual Report, Wellington.

Fletcher Challenge Forests Limited (2001), HYPERLINK http://www. fcf.co.nz.

Gereffi, G. (1993), 'International Subcontracting and Global Capitalism: Reshaping the Pacific Rim', in R. Palat (ed), *Pacific-Asia and the Future of the World System*, Greenwood Press, Westport, Connecticut, pp. 67-81.

Gereffi, G. (1994), 'The Organization of Buyer-driven Global Commodity Chains: How U.S. Retailers Shape Overseas Production Networks', in G. Gereffi, and M. Korzeniewicz, (eds), *Commodity Chains and Global Capitalism*, Greenwood Press, Westport, Connecticut, pp. 95-122.

Gereffi, G. (1995), 'Contending Paradigms for Cross-regional Comparison: Development Strategies and Commodity Chains in East Asia and Latin America', in P.H. Smith (ed), *Latin America in Comparative Perspective: New Approaches to Methods and Analysis*, Westview Press, Boulder, Colorado, pp. 33-58.

Gereffi, G. and Korzeniewicz, M. (1990), 'Commodity Chains and Footwear Exports in the Semiperiphery', in W. Martin (ed), *Semiperipheral States in the World-economy*, Greenwood Press, Westport, Connecticut, pp. 45-68.

Gereffi, G. and Korzeniewicz, M. (1994), *Commodity Chains and Global Capitalism*, Greenwood Press, Westport, Connecticut.

Gereffi, G., Korzeniewicz, M., and Korzeniewicz, R. (1994), 'Introduction: Global Commodity Chains', in Gereffi, G. and Korzeniewicz, M. (ed), *Commodity Chains and Global Capitalism*, Westport, Connecticut: Greenwood Press, pp. 1-14.

Gresham, S. (1991), 'Juken Nissho Limited', *New Zealand Forestry*, February, p. 14.

Hayter, R. and Edgington, D. (1997), 'Cutting Against the Grain: A Case Study of MacMillan Bloedel's Japan Strategy', *Economic Geography*, vol. 73, pp. 187-213.

Hayward, W. (1997), 'Fletcher Challenge Forests Limited', Auckland New Zealand, 30 June 1997.

Hopkins, T. and Wallerstein, I. (1986), 'Commodity Chains in the World-Economy Prior to 1800', *Review*, vol. 10, pp. 157-70.

Kase, Y. (1996), Nissho Iwai. Tokyo, Japan, 11 June.

Korzeniewicz, M. (1994), 'Commodity Chains and Marketing Strategies: Nike and the Global Athletic Footwear Industry', in G. Gereffi and M. Korzeniewicz (eds), *Commodity Chains and Global Capitalism*, Greenwood Press, Westport, Connecticut, pp. 247-265.

Korzeniewicz, R. and Martin, W. (1994), 'The Global Distribution of Commodity Chains', in G.Gereffi and M.Korzeniewicz (eds), *Commodity Chains and Global Capitàlism*, Greenwood Press, Westport, Connecticut, pp. 67-91.

Le Heron, R. (1997), 'The Fibre of Forestry: A Perspective on Structural Changes and Challenges Relating to New Zealand's Forestry Scene', *Proceedings of the 4th Joint Conference of the Institute of Australia and the New Zealand Institute of Forestry*, Canberra, pp. 63-71.

Le Heron, R. and Pawson, E. (eds) (1996), *Changing Places: New Zealand in the Nineties*, Longman Paul, Auckland.

Le Heron, R. and Roche, M. (1996), 'Eco-commodity Systems: Historical Geographies of Context, Articulation and Embeddedness Under Capitalism', in D. Burch, R. Rickson G. and Lawrence (eds), *Globalization and Agri-food Restructuring: Perspectives from the Australasia Region*, Avebury, Aldershot, pp. 73-89.

Marchak, P. (1995), *Logging the Globe*, McGill-Queen's University Press, Montreal.

Ministry of Forestry (1996), *Forestry Sector Issues*, Ministry of Forestry, Wellington.

New Zealand Journal of Forestry (2000), 'Juken Nissho plant boost', May, p. 20.

Palmer, G. (1995), New Zealand Forestry Corporation, Tokyo, Japan, 30 October.

Pan Pacific Forest Industries (N.Z.) Limited (2000), Company brochure, Napier.

Press (1992), 'Mills Planned for Northland', Christchurch, 24 September, p. 14.

Pritchard, B. (1997), Pan Pacific Forest Industries (N.Z.) Ltd. Napier, New Zealand, 10 September.

Roche, M. (1990), 'Perspectives on the Post-1984 Restructuring of State Forestry in New Zealand', *Environment and Planning A*, vol. 22, pp. 941-959.

Roche, M. (1992), 'Privatizing the Exotic Forest Estate: The New Zealand Experience', in J. Dargavel and R. Tucker (eds), *Changing Pacific Forests: Historical Perspectives on the Forest Economy of the Pacific Basin*, Forest History Society, Durham, North Carolina, pp. 139-154.

Tachikawa Forest Products (NZ) Limited (2001), http://www.tachikawa.co.nz.

Tachikawa, K. (1996), Tachikawa Forest Products (NZ) Limited. Toyko, Japan, 10 June.

15 Cultural Embeddedness, Corporate Strategy and Foreign Investment in Poland: A Tale of Two Firms

Jane Hardy

Introduction

In the conventional literature, foreign direct investment (FDI) is an 'engine of growth' which unproblematically brings benefits to host economies (Lipton and Sachs, 1990). For the lagging and peripheral economies of Central and Eastern Europe (CEE), these benefits relate to the upgrading of products and processes, the transfer of technological and managerial expertise and the stimulus for competition (Dobosiewicz, 1992; Dunning 1993). However, studies on the embeddedness of firms in CEE (Grabher, 1993; Grabher and Stark, 1997), and research into the transfer of managerial practices across national boundaries (Coller and Marginson, 1998; Elger and Smith, 1994; Marginson et al 1995; Smith and Elger, 1997), suggest that such processes and outcomes are likely to be much more complex as firms enter new markets, acquire new assets and operate in new institutional set-ups. While other literatures on the impact of firms and localities have primarily focused on the direct economic impacts of FDI on employment, supply chains and technology, institutionalist approaches have gone beyond conventional measures of embeddedness and emphasized the way in which economies are shaped by enduring collective forces (Hodgson, 1988). This approach accords great importance to habits, routines and norms, whereby economic agents are the products of past behaviours and learning as well as current incentives.

Elsewhere I have argued that the 'institution bending' effects of FDI in Poland greatly exceed their economic impacts. In a study of 12 foreign investments in Lower Silesia, managers of incoming foreign investments reported that replacing the managerial practices of the previous regime was central to the restructuring process. It followed that in the case study firms the impact of foreign investment on technological change had been incremental, while changes to management practices and organizational structures represented a quantum leap (Hardy, 1998). Although in some cases embedded tacit knowledge was regarded as a potential asset, the majority of managers viewed previous legacies and behaviour as a set of habits that needed to be circumvented or changed. Thus transnational corporations in CEE have been to the fore in instilling new material and discursive practices through deep restructuring of workplaces and workplace behaviour, in response to what were regarded as the behavioural constraints of the previous regime (Smith and Thompson, 1992; Swaan and Lissowka, 1996).

This chapter adopts an institutionalist approach to examine two foreign controlled case study firms that operate in Lower Silesia in Poland.[1] The first firm, Vehico, is a producer of commercial vehicles, which in 1998 had five factories in Europe. After a short-lived joint venture with the Polish bus and truck company they started production on a greenfield site in 1998. Enginco is one of the largest global engineering companies. In Lower Silesia they purchased a State Owned Enterprise (SOE) that produced power generators and engines through the privatization process. The main purpose of the chapter is to evaluate the process of instilling 'new knowledge' in Poland by these firms, given the constraints imposed by the embedded knowledge of workers and managers from the previous regime. While Enginco deployed 'shock therapy' tactics and used parent company blueprints to implant radically new behaviour and routines, Vehico took a more pragmatic and gradualist approach which purported to draw on local tacit knowledge. Further, the chapter explores the implications of these contrasting strategies for regional embeddedness.

Drawing on Zukin and Dimaggio (1990), the chapter begins by exploring elements that comprise the cultural embeddedness of a firm, focusing in particular on those aspects which have previously been neglected. Cultural embeddedness is viewed as that set of social conventions embracing behavioural norms, standards and customs and the rules of the game underlying social interaction within the firm (Schoenberger, 1994). The cultural embeddedness of the individual firm is an amalgam of a general consensus about management techniques (coloured in the short term by various management fads), institutional conditioning in the country of origin

(Kristensen, 1997), the idiosyncrasies of agents within the firm, and contestation by local workers and managers as they cross national boundaries. This chapter primarily focuses on the processes underlying the entry of firms into new institutional contexts, notably where incoming firms have to negotiate with or around existing established behaviours and understandings. In other words, as firms cross national boundaries they need to introduce and establish new material and discursive practices through restructuring management practices and changing workplace habits and norms. This aspect draws on the notion of 'soft' or 'knowledgeable' capitalism (Thrift, 1999) whereby it is necessary for managers to effect rapid change in the workplace.

The chapter begins by unpacking four dimensions that are a subset of cultural embeddedness: entry and acquisition, intra-firm governance, restructuring strategy and agents of change, and the informal institutions of workers and work practices. I then examine how these dimensions are reflected in, and influence, the way in which the case study firms deal with embedded knowledge from the previous (Polish) regime as they attempt to instill new knowledge. The implications of these are then explored in relation to the two case study firms in the remainder of the paper.

Dimensions of Cultural Embeddedness Within the Firm

The first dimension of embeddedness relates to the process of entry and acquisition. The mode of entry, whether by acquisition or greenfield investment, will influence the extent to which past legacies and linkages are preserved and built upon. Greenfield investments may circumvent existing linkages and insert new habits and norms while brownfield entry via privatization methods may develop old linkages (Grabher and Stark, 1997). Further, the process and outcomes of negotiations at a local and national level regarding entry (greenfield) or the acquisition of existing assets (brownfield) affect the nature of local embeddedness. The strength and bargaining position of the parties involved influences the nature of the bargain, for example, localities in competition with other regions for investment are more likely to make concessions to the firm. Further, there may be asymmetrical knowledge and learning which is disadvantageous to local negotiators or national governments, which allows firms who are experienced internationalizers to cherry pick assets.

The second dimension of embeddedness focuses on intra-firm governance. The structure of the firm and the place of the subsidiary in the corporate hierarchy influences the parameters of activity and its capacity for decision-making at a local level. A body of recent literature, often linked to the necessity of continuous learning within the firm, suggests that firms have become less hierarchical and headquarters are now architects of co-ordination rather than control centres (Cooke and Morgan, 1998). For example, it is argued that global mandates give a firm a wide range of autonomy and quality functions, in comparison with subsidiary status (Birkinshaw and Hood, 1998; Roth and Morrison, 1992), resulting from worldwide responsibility and strategic control for a product or product line. Related to this trend, is the degree of managerial autonomy with regard to critical functions such as innovation and reinvestment.

The third dimension of embeddedness is the restructuring strategy of the firm and the role of agents of change who institute these processes. To what extent do firms transfer existing models within the firm and how far do they consciously attempt to replace existing behaviours? What method is used to effect organizational change, and does the firm operate on a short or long time scale? How far do firms use outside external expertise, such as consultants, or are internal agents used to identify best business practice within the organization which is then cascaded to other parts of the company? Techniques such as benchmarking allow firms to quantify their efficiency in relation to rivals (and affiliated operations) and to create internal performance 'league tables' and the transfer of 'best practices'. Such benchmarking is likely to push in the direction of homogenous managerial and work practices. A related issue is whether firms use local personnel and draw on their local knowledge or whether they have a policy of using expatriate managers. Further, central management programmes of training, meetings and social events to build managerial networks are often used to inculcate all middle and senior managers with the corporate culture of the firm.

While the previous dimension of embeddedness is related to changing organizational culture and management style, the fourth dimension of embeddedness, focuses on the informal institutions of workers and work practices. These informal institutions relate to predominant habits, norms and social conventions within the workplace (Hodgson, 1988). The importance of local knowledge, and in particular the tacit knowledge of workers is viewed as increasingly important in gaining competitive edge in a global economy where knowledge is increasingly ubiquitous and codified

(Nonaka and Takeuchi, 1995). What is the attitude of the firm to the behavioural norms of the previous period? Is this embedded tacit knowledge regarded as an asset and a source of competitive advantage or a problem to be circumvented by transferring production methods from the home country?

I now turn to explore these dimensions of embeddedness with regard to the case study firms in the Lower Silesia region of Poland (Table 15.1).

Table 15.1 Selected features of Enginco's and Vehico's operations in Poland

Dimensions of Embeddedness	Enginco	Vehico
Entry and acquisition	• Circumventing existing institutions • Cherry picking assets • Leasing and low investment • Shallow roots in locality	• Institutional avoidance • JV to industrial centre • Significant investment
Intra-firm governance	• Matrix structure • Profit centres • Internal competition	• Heterarchical • Subsidiary discretion
Restructuring strategy and agents of change	• Transfer of existing models • 'shock therapy' • Creation of global managers • External agents of change	• Rotation of managers • Shared view • Ex-patriots as facilitators
Informal institutions and work practices	• Blueprints • Codified practices • Use of technical knowledge	• Pragmatism • Use of local tacit knowledge • Socialization

Entry and Acquisition

The first dimension of embeddedness relates to the motive for and method of entry. After discussions with the Polish national government, Vehico and Truckski, the leading Polish bus and truck producer, agreed on a joint venture. Vehico started assembly production on Truckski's site, in Lower Silesia in 1993. However, the joint venture with Truckski only lasted one year because of a failure to agree on employment levels. Subsequently, Vehico started production on a greenfield site 20 miles away on the edge of the largest city in the region. Between 1995 and 1998 Vehico invested US$20 million in developing bus and truck assembly on the new greenfield site. In June 1998, however, Vehico's head office in Scandinavia took a major policy decision and designated the production site in Lower Silesia to be the centre of Vehico's bus production in Europe, that is, for markets in Central, Eastern and Southern Europe, as well as Poland. By mid-1999 a new factory to build complete buses had been constructed on $40,000m^2$ of land. The additional investment amounted to US$50 million and was planned to create 1,500 new jobs by 2001. Additional employment would be created not only in production, but also by the location of the company's logistics and data processing functions on the same site.

In the case of Enginco, pressures in the power generation market and the need to internationalize came from both push factors in the domestic market and pull factors in international markets. The market for power transformers had matured, limiting demand in established markets while the need to modernize obsolete equipment through privatization had created a high level of demand in emerging market economies. Thus the transformation of SOEs and the deregulation of monopoly product markets had opened up new markets and intensified competition between big players. The high level of investment by Enginco in Poland was driven by both desires to access market and to lower costs.

When Enginco purchased Lodmel in 1992 it was one of the largest firms making power generators in the Soviet bloc. Negotiations took place between Enginco Power Ventures and the Polish government with Enginco purchasing 100 per cent of the shares. The purchase of the company was controversial because, according to Solidarity, Enginco used cherry picking tactics by splitting the former SOE into three parts; Lodmel (core area of generator production), Lodmel Drives (engine production) and Zemal which owned the land, buildings and social assets.

Enginco bought Lodmel and Lodmel Drives while Zemal remained as a State Owned Enterprise which leased the land and buildings to Enginco. This behaviour was entirely consistent with entry tactics used by Enginco in other parts of Poland and Central and Eastern Europe, namely to negotiate in advance precisely how much of the business it wanted to take on and cede control of the rest to the state.

The entry strategies and subsequent development of these two firms had implications for their embeddedness in the regional economy. In the case of Vehico, the decision to reject a joint venture and opt for greenfield production was based on a desire to avoid old institutional legacies and start with a clean sheet on which it could imprint the corporate culture of the firm. However, the pressure of international competition driving rationalization, rapidly led to changes in the Vehico industrial system with the subsequent designation of Westski as an industrial centre, that substantially increased its potential for local embeddedness. Favourable production conditions in terms of a skilled, reliable and relatively cheap workforce and a good relationship with local government meant that by 1999 Vehico's Westski site had changed from a truncated assembly site to an industrial centre with substantial additional investment to produce for the European market. This was associated with decisions related to sourcing and relationships with local institutions which would embed it in the local economy at a number of different levels. In contrast, Enginco's mode of entry through the privatization process, enabled the firm to gain access to skilled labour and embedded technological knowledge while avoiding the costs associated with search, greenfield development and ownership of land and buildings. Leasing rather than ownership of assets meant that it was much more footloose and could operate on shorter time horizons if it wanted to sell or switch production. That is, it had shallow roots in the locality. The implied vulnerability was reflected in the closure of Lodmel Drives in the 1999 after a series of redundancies and disinvestment.

Intra-firm Governance

The second dimension of embeddedness relates to the governance structure of the two firms and their place in the corporate hierarchy or heterarchy. Since its formation, Enginco has espoused a corporate strategy of being local and worldwide – 'a company with many homes'. Its organizational structure was designed to exploit the two-fold advantage of local specialization, on the one hand, and global resources and co-ordination on the other.

The matrix system meant that the senior management of Lodmel was answerable to one country manager in Warsaw and two Business Area Managers in Germany and Switzerland. The role of the Business Area Manager was to monitor the trajectory of business in general, and more specifically to decide the production range of each factory, which export market they would serve and how factories should pool their expertise and research for the benefit of the company worldwide.

Enginco's operation in the Westski region reflects the way in which most of the company's employees are organized in 1,200 autonomous small units with an average of 222 employees. Every month the Business Area headquarters distributes detailed information on performance measured by crucial parameters, such as failure rates, throughput times, and inventories as a percentage of revenues. These reports generate competition for outstanding performance and produce 'constructive internal competition'. The Business Area management board meets four to six times a year to shape the global strategy and monitor performance. This allows business area managers to benchmark plants against each other and to encourage the adoption of similar management practices without detailed hands on interference in plant operations. The Business Area has two additional tools to correct poor plant performance, and thereby encourage conformity with its programmes: export allocation and management promotion. Another mechanism to create competition was by Enginco's trade company at Mannheim. This company bids for contracts and then selects producers from factories within the Enginco network that demonstrate the highest quality and lowest costs. The trade company is in a better position to tender for large contracts than individual firms because it is able to offer financial packages for projects.

The organizational structure of Vehico was more flexible and multi-dimensional in comparison with the formal and rigid matrix organization of Engine and Headquarter functions were comparatively more diffuse. Vehico Management on the Westski site, as discussed in the next section, had a greater degree of autonomy regarding the organization of production, human resource management (HRM) policies and sourcing decisions. The plant was only responsible for production, with sales and marketing taking place elsewhere. Further, decisions regarding new investment and the expansion or closure of plants were taken in Scandinavia. Unobtrusive control was in evidence with the constant monitoring and measurement of quality according to criteria that were established to every last detail in Gothenburg. As revealed in the discussion of the labour process, the philosophy of the company was that the best way of meeting these externally determined criteria was to mobilize the tacit knowledge of workers rather than the prescriptive and restrictive approach of Enginco.

The conclusion is that global mandates and smaller units give a large degree of autonomy, but that this autonomy is highly constrained and not necessarily conducive to local embeddedness. In this case, freedom for manoeuvre from Headquarters is replaced by the despotism of the market and a different form of competition, which is an internal or self imposed coercion (Mueller and Loveridge, 1997). Internal competition for investment and repeat investments means that local actors have to employ radical change and learning is accelerated through the internalization of competitive markets which demand the replacement of existing institutional arrangements.

Restructuring Strategy and Agents of Change

The third aspect of embeddedness focuses on the restructuring processes through which firms instigate change, and the nature of the agents of change in this process. The two firms had very different ways of dealing with the inherited deep-rooted nature of rigidities of organizations and instilling new work and managerial practices. Former SOEs were characterized by centrally allocated vendors and buyers, enduring functional structures, domination by technocratic and engineering elites and strong inertia with little impetus for, or tolerance of change and uncertainty. Part of Vehico's decision to discontinue the joint venture and start greenfield production was motivated by a wish to circumvent existing institutions.

Vehico's corporate culture conformed much more closely to Hedlund's (1991) view of 'heterarchical' Scandinavian firm that relied much less on hierarchical authority, formal measurement and internal markets. Rather, Vehico relied on socialization, rotation of managers and the emergence of shared views of the company's strategy and identity. Whereas Enginco's managers were generally older and inherited from the previous company, Vehico tended to recruit young Polish middle managers. Scandinavian managers held key positions in the company, which included the CEO, the Finance Manager, and the Global Sourcing Manager. While the ultimate goal was to imbue middle and senior managers with the 'Vehico way of thinking', the route taken was very different from that of Enginco. There was much less evidence of the right way of doing things codified in manuals or procedures, particularly regarding HRM and supply management. The style was much less hierarchical and appeared, on the surface at least, to be much more flexible. Proximity, personal contact, coaching and a strong 'hands-on' approach on the part of the Scandinavian management were seen as preferable to formal training. The decisions of Headquarters were mediated through the Managing Director and there was a significant presence of Scandinavian managers in key positions such as sourcing and finance to drip feed and reinforce company culture.

If Vehico took a gradualist and pragmatic route to restructuring, the Enginco approach to the transformation of management culture was much more akin to 'shock therapy'. Enginco had given priority to creating a core group of global managers. These people were cultivated by rotating people round the world to give them line experience in three or four different countries to create a global perspective and to make sure their loyalty is to the company not individual countries or branches. According to the CEO, these global managers were required to understand the local culture and also, 'to sort through the cultural debris of excuses and find opportunities to innovate'. Therefore, shock troops comprising Enginco global managers and consultants were sent in to administer a short sharp shock and effect rapid change in the structure and culture of the organization.

One of the main tasks was to create a small group of Polish managers who could be left to run the firm once the 'agents of change' had moved on elsewhere. Previous management in SOEs were representatives of the Communist Party and powerful administrators. However, an unusual balance of power prevailed with top managers earning only as much or even less than the best performing workers. Workers had political clout, particularly since powers had been given to Workers Councils in 1981 to hire and fire managers. The process of creating new managers was threefold. First, as with other Enginco acquisitions in CEE, extensive training sessions were conducted to identify and select managers from the previous SOE who were most likely to adapt and change. These individuals were then put through intensive education and training to expose them to the principles and practice of the market economy in general and Enginco management system in particular. Second, the material incentives and prestige of managers were substantially increased with large wage increases, company cars, remodelled offices and foreign travel.

This intense restructuring took place in the four year period from 1990 and by 1994. Enginco Lodmel had all Polish management, the consultants having moved on and the expatriate managers transferred elsewhere. The matrix structure with constant financial reporting and manuals supplied to the HRM and supply managers regarding Enginco policy, meant that there was strict arms length control with little room for manoeuvre for incumbent managers. As we have seen, the Enginco Business Area decided which market the firm could sell in, and the level of investment and R&D continued to take place in Switzerland. The main area for which the factory workers and managers had responsibility was cost reduction and productivity, although targets for these parameters were handed down from the Business Area.

In the case of Vehico control of the company was centralized through the use of expatriates who have internalized corporate norms and values. Thus transferring expatriates who can be trusted to implement corporate policy and procedures are de facto centralizing mechanisms.

Informal Institutions and Work Practices

As we saw in earlier sections, Vehico and Enginco have adopted different policies to deal with the norms, habits and past behaviours of workers and incumbent management. While Enginco used a 'shock therapy' approach to restructuring and corporate strategy, Vehico's more gradualist approach was based on drawing on local tacit knowledge. These differences are manifest in how these two firms attempted to restructure production, as we shall see in this section.

Enginco's process of restructuring and changing the work process stands in sharp contrast to that of Vehico. In the early 1990s Enginco laid down the four principles according to which it was going to demonstrate the applicability of business and managerial reform in CEE. These principles are: first, the immediate reorganization into profit centres with well defined budgets, strict performance targets and clear lines of authority and accountability; second, to identify a core group of change agents from local management, and give small teams responsibility for championing high priority programmes; third, to transfer Enginco expertise from around the world in support of local change processes, without too much interference; and fourth, to keep standards high and demand quick results.

The Production Manager at Enginco had worked for the company for 19 years and outlined the main changes that had taken place since 1990 which included new technology, new processes, extensive training in business skills, cost measurement and people management. The major change in the organization of work was that in 1994 previously centralized control of quality control, purchasing and process engineering was devolved to five units that corresponded to the five different stages of production.

According to the Production Manager, visits to plants in other countries had inspired the new restructuring that was essentially an internal idea which had evolved through discussions with colleagues in other countries and workers in the Westski factory. However, there was a significant gap between company rhetoric and practice in Enginco. A senior Manager suggested that the Polish management were not consulted about the restructuring and that it had been imposed by the parent company. Further he (unofficially) suggested that Polish managers felt that they were treated as second class citizens by the parent company. Further, the foreman who gave me a tour round the factory emphatically denied any worker involvement in the restructuring of work within the factory.

The Polish Production Manager at Vehico explained that in the case of workers from previous SOEs such as Truckski (the truck and bus company with which Vehico had a joint venture) old habits and attitudes had to be unlearned. In particular, the motivation of workers by responsibility rather than by control was a new departure, in that Polish management under the previous regime had been based on a culture of hierarchy, discipline and order. Ideas and edicts came down from senior managers and workers were expected to comply passively and there was therefore little or no incentive to innovate.

The Production Manager identified a number of areas where past behaviour was a barrier to adapting to the Vehico culture and way of doing things. First, Polish firms were (and are) much more hierarchical and based on seniority, and therefore some of the older workers found it hard to accept orders or instructions from someone who was much younger, for example in the case of the Polish Production Manager. Second, another continuity with past behaviours that was deemed to be obstructive related to individuals who tried to improve their personal standing by proposing organizational changes on the basis of self interest, rather than the efficiency and development of the firm. Thus some individuals sought to mould the emerging organizational structure to give themselves a good position or a power base. Lack of pecuniary incentives in the past meant it was important to seek a position that provided status and/or the ability to access material benefits through unofficial or informal structures. Third, the Production Manager reported problems in trying to get workers to take responsibility for their work, and gave instances where quality was not up to standard and workers made excuses or blamed poor components rather than finding solutions. He suggested that it was difficult to get workers to see that it was not important to apportion blame or to attach guilt but look for processes or procedures to ensure that mistakes did not re-occur.

The details of work organization in Vehico evolved in a much more participatory and iterative way and there was no attempt to impose a blueprint of work organization that had been used elsewhere. However, the extent to which tacit knowledge from the locality has informed work design should not be overstated, in that management worked constantly to overturn institutional roadblocks to change. The young Polish Production Manager was a critical link in mediating between the company culture promoted by the Scandinavian management, and the habits and norms of workers.

In both cases the different corporate strategies are evident in the organization of work. In the case of Enginco this has largely been transferred and was in line with the restructuring that took place in other acquisitions. Vehico's production strategy was much more adapted to local circumstances both in terms of the scale of production and accessing local tacit knowledge of workers, and in particular that of Polish middle management. In the final analysis the whole process was predicated on getting workers to unlearn what were considered to be the unhelpful legacies of past behaviour.

Conclusion: Different Routes, Same Destination?

The summary in Table 15.1 suggests that Vehico and Enginco represent contrasting examples of cultural embeddedness that is reflected in markedly different ways of dealing with 'old inappropriate knowledge'. First, the brownfield investment via the privatization process in the case of Enginco by-passed local structures and involved laying shallow roots in the locality. The second dimension is the way in which organizational features of the two firms influenced their strategy for disembedding old legacies and instilling new managerial practices. The matrix structure of Enginco involved the subsidiary being answerable to two Headquarters outside the locality and the parameters for local learning were tightly set by cost constraints. However, learning was not a one-way street as the country Headquarters served to gather and disseminate information about institutional influences in the national economy, which framed their operations in local economies. By contrast, Vehico organizational structure was more heterarchical, providing for greater room for manoeuvre in local decision making regarding production and more open as to the possibility of learning from tacit knowledge in the locality.

A third difference between the two firms was evident in their strategy towards restructuring and agents of change. Enginco used shock therapy tactics to bring about fast change in management culture and workplace organization and relied on external agents to introduce blueprints for a range of production and managerial practices. Vehico's style was much more flexible and adaptive emphasizing, coaching, persuasion and personal contact and drew more heavily on local knowledge. While the firm drew selectively on tacit and embedded knowledge, particularly that of middle managers, the ultimate aim was for workers to gradually absorb the values and ethos of the company.

Although Vehico's approach seems more human centred and purported to build on the tacit knowledge in the locality, an argument can be made that the two firms simply took different routes to the same destination. Instability and turbulence in the market in which Enginco operated meant that it was operating on a shorter time scale and change and restructuring had to be effected quickly. Vehico's longer term commitment to production in the region meant that the pace of change could be slower, nevertheless the firm still exhibited an overriding commitment to instilling a preordained set of attitudes and routines. While the rhetoric was one of accessing tacit knowledge, the reality was one of 'unobtrusive control'.

The case studies suggest that a more nuanced approach is required if we are to understand how firms transfer new knowledge and managerial practices across national boundaries, particularly in peripheral region that are regarded as having a deficit in appropriate knowledge. There is no one 'best way' or blueprint and the dimensions of cultural embeddedness are reflected in the individual corporate strategies adopted. Although ultimately the two firms in question achieved similar outcomes, in the short term at least, Vehico's gradualist and pragmatic approach offers greater potential for regional development by having deeper roots in the economy and by building on local social assets. By contrast Enginco's approach assumed that specific and often codified knowledge had to be implanted in a short time-frame and this translated into shallow and much more fragile roots in the region.

Acknowledgement

I acknowledge the financial support of the University of Hertfordshire.

Note

¹ Both case study firms have been given pseudonyms to protect anonymity.

References

Birkinshaw, J. and Hood, N. (1998), 'Multinational Subsidiary Evolution: Capability and Charter Change in Foreign-owned Subsidiary Companies', *Academy of Management Review*, vol. 23, pp. 773-795.

Cooke, P. and Morgan, K. (1998), *The Associational Economy*, Oxford University Press, Oxford.

Dobosiewicz, Z. (1992), *Foreign Investment in Eastern Europe*, Routledge, London.

Dunning, J.H. (1993), 'Prospects for Foreign Investment in Eastern Europe', in P. Artisien-Maksimenko, M., Rojec, and M. Svetlicic (eds), *Foreign Investment in Central Eastern Europe*, Macmillan, London.

Grabher, G. (ed) (1993), *The Embedded Firm: on the Socio-economics of Industrial Networks*, Routledge, London.

Grabher, G. and Stark, D. (1997), 'Organizing Diversity: Evolutionary Theory Network Analysis and Postsocialism', *Regional Studies*, vol. 31, pp. 533-44.

Hardy, J. (1998), 'Cathedrals in the Desert', *Regional Studies*, vol. 32, pp. 639-652.

Hodgson, G.M. (1988), *Economics and Institutions*, Polity Press, Cambridge.

Kristensen, P.H. (1997), 'National Systems of Governance and Managerial Prerogatives in the Evolution of Work Systems: England, Germany and Denmark Compared', in R. Whitley and P.H. Kristensen (eds), *Governance at Work: The Social Regulation of Economic Relations*, Oxford University Press, Oxford, pp. 3-46.

Lipton, D. and Sachs, J. (1990), 'Creating a Market Economy in Poland', *Brooking Papers on Economic Activity*, 1, Brookings Institution, Washington, pp. 75-147.

Marginson P, Edwards P.K., Armstrong P. and Purcell J. (1995), 'Strategy, Structure and Control in the Changing Corporation: A Survey Based Investigation', *Human Resource Management Journal*, vol. 5.

Mueller, F. and Loveridge R. (1997), 'Institutional, Sectoral and Corporate Dynamics in the Creation of Global Supply Chains', in R. Whitley and P.H. Kristensen (eds), *Governance at Work: The Social Regulation of Economic Relations*, Oxford: Oxford University Press, pp. 139-157.

Nonaka, I. and Takeuchi, H. (1995), *The Knowledge Creating Company*, Oxford University Press, Oxford.

Roth, K. and Morrison, A.J. (1992), 'Implementing Global Strategy: Characteristics of Global Subsidiary Mandates', *Journal of International Management Studies*, vol. 23, pp. 715-735.

Schoenberger, E. (1994), 'Corporate Strategy and Corporate Strategists', *Progress in Human Geography*, vol. 22, pp. 435-451.

Smith, C. and Elger, T. (1997), 'International Competition, Inward Investment and The Restructuring of European Work and Industrial Relations', *European Journal of Industrial Relations*, vol. 3, pp. 279-304.

Smith, C. and Thompson, P. (eds) (1992), *Labour in Transition: The Labour Process in Eastern Europe and China*, Routledge, London.

Swaan, W. (1996), 'Behavioural Constraints and the Creation of Markets in Post-socialist Economies', in B. Dallago and L. Mittone (eds), *Economics, Institutions, Markets and Competition*, Edward Elgar, Aldershot.

Thrift, N. (1998), 'The Rise of Soft Capitalism', in A. Herod, G. O'Tuathail and S.M. Roberts (eds), *Unruly World: Globalization, Governance and Geography*, Routledge, London.

Zukin, S. and DiMaggio, P. (1990), 'Introduction', in S. Zukin and P. DiMaggio (eds), *Structures of Capital and Social Organization*, Cambridge University Press, Cambridge.

16 Paths of Industrial Transformation in Poland and the Role of Knowledge-based Industries

Tadeusz Stryjakiewicz

Introduction

The transformation of the socio-economic system taking place in the post-Communist countries of Central and Eastern Europe, including Poland, has produced profound changes in the spatial organization and functioning of industry. In turn, these changes have implied new relationships among manufacturing firms, new channels for the transfer of knowledge, and new environmental imperatives. This chapter analyzes the different paths of industrial transformation in Poland during the last ten years following the beginning of the process of the market orientation of the economy. The chapter focuses on the role of innovative and knowledge-based industries, particularly from the perspective of the problem of 'catch-up' that firms in these industries face in Poland.

The first part of the paper briefly outlines the general state of innovation potential within the Polish economy at the present time. The second part classifies types of firm-based adjustments that define the processes underlying the transformation of Polish industry. This discussion is illustrated by two case studies: the Optimus computer firm and the ABB group, which offer differing implications for the future of the Polish economy.

Innovative Potential of Industry and its Spatial Variation

One of the most important indicators of the adaptation of industry to the new economic system and its ability to face global competition is the level of innovativeness of enterprises. The ability to generate and assimilate broadly understood innovations (both technological and organizational) determines sustainable development, which is especially emphasized in the concept of regional/local innovative milieus (or systems). Kuklinski (1997) provides a detailed analysis of such regional innovation systems in Poland.

Given that measures of innovativeness are not readily available, in this section, the analysis of the innovative potential of industry is addressed from several perspectives, most notably: (a) the technical condition of enterprises; (b) outlays for the R&D sector; (c) effects of industrial modernization and innovative activity; and (d) the creation and operation of an innovative milieu. The spatial framework for this analysis is the old administrative system which was in force until 31 December 1998 (Stryjakiewicz, 1999). This system brings out spatial variability of phenomena and processes more clearly than the new one.

The technical condition of Polish enterprises hardly makes them competitive at the global scale. Although investment outlays have increased since 1993, the depreciation of fixed assets is still very high (53.9 percent as of 31 December 1998). Worse still, the assets most affected are machinery and technical equipment, that is, those assets that make production modern. Industries with the most modern equipment in Poland are those that are not innovation-carriers in the world economy, notably the tobacco, publishing, printing, and furniture industries. All other industries show a depreciation of machines and equipment of more than 50 percent. Especially alarming is the 75 percent wear and tear of machinery in the capital goods sector, that is industries that build machines and equipment for the rest of the economy, because this weakness produces a vicious circle of technological backwardness that can only be broken through imports.

Regions with relatively modern machines and equipment embrace two types of voivodeship (province). The first comprises highly industrialized voivodeships that have recently attracted large-scale investments, including by MNCs (Bielsko-Biala, Piotrków Trybunalski). The second comprises voivodeships with low levels of industrialization, but a high proportion of privately-owned businesses, especially small and medium-sized firms

(Koszalin, Leszno, Nowy Sacz). The renewal of the productive assets in both those groups of regions is undoubtedly an effect of the transformation towards a market economy, including bankruptcies of many technologically obsolete state enterprises and the growth of entrepreneurship, generated exogenously (the first group) or endogenously (the other group).

In turn, the level of equipment of enterprises with modern production-controlling facilities, such as automatic and computer-controlled production lines, industrial robots and manipulators, and computers controlling and regulating technological processes, is by far the highest in the biggest metropolitan areas (Warsaw, Katowice, Gdansk, Cracow, Poznan) and in the Bielsko-Biala region. In the latter region, the beneficial effects of the FIAT investments can be clearly seen. This observation also corroborates the view that at present multinational automobile corporations are among the most important carriers of innovation at the international scale (see, for example, Bertram and Schamp, 1989; Schamp, 1997). On the other hand, it should be observed that in Poland it is the food industry that displays a relatively high level of computerization. In 1998 it accounted for 19 percent of all computer-controlled technological lines installed in manufacturing, and for 19 percent of local area networks (LANs).

An analysis of per capita investments in research and development (R&D) reveals that such investments in Poland are among the lowest in Europe in relation to national income, though higher than in Greece or Portugal (Figure 16.1 and Table 16.1). Moreover, the share of R&D performed by industrial enterprises is very modest, unlike in most West European countries, where it varies between 55 percent and 75 percent. Within Poland, the main R&D centres are concentrated in the biggest metropolitan areas, especially Warsaw, Cracow and Wroclaw, while 'avoiding' such industrialized voivodeships as Katowice (the Upper Silesian Industrial District), Legnica (the Copper Industrial District), or Plock (the largest Polish refining-petrochemical combine). The spatial distribution of patents granted within Poland in relation to industrial employment is almost identical to the distribution of R&D effort (Stryjakiewicz, 1999). The monopolistic position and state ownership of many big enterprises located in the latter group of voivodeships has not proven conducive to more innovation-oriented strategies, despite the major change in the economic system.

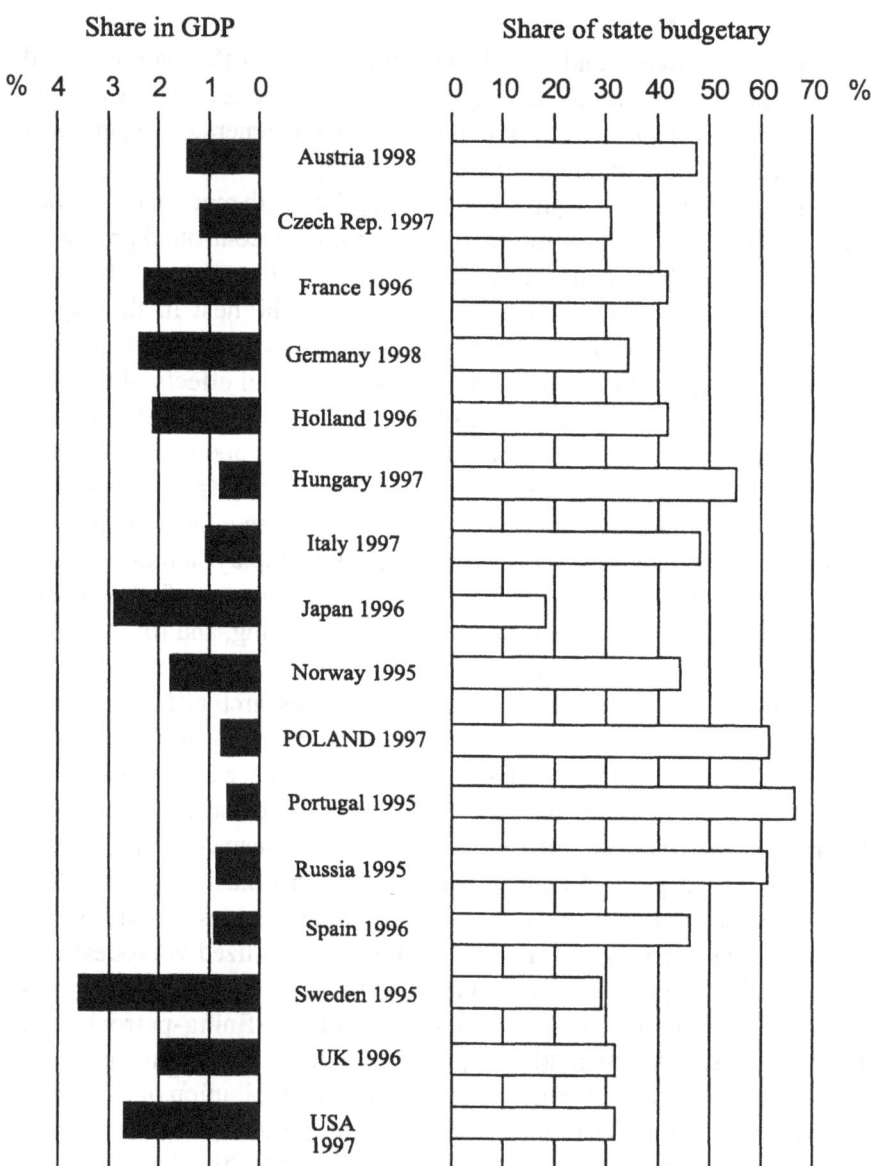

Figure 16.1 Outlays for research and development in selected countries

Source: Main Science and Technology Indicators, OECD, 1998/2; Science and Technology Main Indicators and Basic Statistics in the Russian Federation, OECD, 1998; Maly Rocznik Statystyczny. Warszawa: GUS, 1999.

Table 16.1 Per capita research and development outlays,* 1998

Country	Per capita outlays from state budget (US$)	Per capita outlays of enterprises (US$)
USA	245.5	456.6
Japan	127.7	501.4
OECD average	135.9	258.0
EU average	135.8	188.7
Poland	32.5	18.8

*as measured by the purchasing power parity (PPP).

Source: Pawlowski (1999: 6)

Stanislawski and Uminski (1997) show that there is a growing technological gap among the Polish regions which is much wider than the economic gap. It is an alarming development because, according to a report of the European Commission (1994), the problem of weak regions stems from the fact that they are able neither to generate and develop their own innovative activity, nor adapt innovations developed elsewhere. If innovative foreign firms are located in such areas, they become, to use Grabher's (1992, 1994) expression, "cathedrals in the desert", because the technological gap precludes the formation of local cooperation networks.

It is hard to make an unequivocal assessment of the impact of the transformation on the modernization and innovativeness of Polish industry as measured by the effects produced so far. For example, in the years 1990-1998 the share of car manufacture in total industry output grew (by over two percent, mainly due to foreign investment). However, this growth was offset by a similar decrease in the share of production by the machinery and equipment industries that are key suppliers to the economy as a whole.

Moreover, the biggest increase (3 percent) in the share of industrial output was recorded by food and beverage manufacture, a traditional industry. The proportion of high-tech products (as defined according to the OECD classification) started to climb in the second half of the 1990s, after an initial drop, to reach 11 percent of the total output in 1998 (as against 8.5 percent in 1990).

Between 1990 and 1998 a 'dynamic index' (Id) measured the average rate of growth of production for all industries at 148. Among the industries showing the highest rate of growth of production in this period, as indicated by an Id > 200, are the following: manufacture of radio, TV and communication equipment and apparatuses; car manufacture; manufacture of medical, precision and optical instruments, clocks, and watches; and manufacture of office machines and computers. On the other hand, the dynamic index was equally high, if not higher, in such traditional industries as the manufacture of furniture, cellulose and paper, hardware, and rubber and plastic goods.

In the global context, the low proportion of high-tech products in Poland's foreign trade, as well as the diminishing number of Polish innovations patented abroad, is a worrying feature. Stanislawski and Uminski (1997) point out that the important reasons for this situation relate to the little interest of enterprises in R&D, and to a lack of protection of Polish technologies and inventions in foreign markets. In turn, in the home market, the number of patents granted to foreign inventions has increased while the number of domestic patents has declined (Table 16.2).

Table 16.2 Patents granted in Poland in the years 1980-1998

Inventions	1980	1990	1995	1997	1998
Domestic	5,736	3,242	1,619	1,179	1,174
Foreign	1,962	405	989	1,151	1,242

Source: Maly Rocznik Statystyczny, Warszawa: GUS, 1988, p. 225; Rocznik Statystyczny Rzeczypospolitej Polskiej, Warszawa: GUS, 1999, p. 333.

The described state of innovativeness of Polish industry shows that there is an urgent need to develop and modernize its innovative milieu. During the command-economy period, like the other countries of the Communist bloc, Poland was an almost complete 'desert' in this respect. In the 1990s, so-called innovation and enterprise centres started to appear, thanks to the European Union's assistance funds (the PHARE programme), as well as other sources of funds. Their growth in terms of numbers may seem impressive, especially in the initial stage of the transformation. However, it is not matched by growth in terms of quality. The study by Baranski (1994) shows that only 40 percent of the activities of those centres are connected with industry (as against 60 percent devoted to services), and that industrial enterprises at large are not interested in starting and supporting these kind of units. Only 28 percent of the centres examined by Baranski named industrial plants as initiators or co-initiators of their establishment; in as many as 71 percent of cases the initiative was that of the local authorities. Taking into consideration the modest financial means at the disposal of local authorities, it is hardly surprising that many newly created centres are innovation units by name only, or that they suspend their operation after a short time of existence, for example, when the assistance funds have run out. It would be desirable if the institutional and financial support of the state for R&D, so far rather weak, could be strengthened.

Another important element of the innovative milieu of industry is scientific and engineering education. The transformation to a market economy has produced both advantageous and adverse effects in this context. Thus, between the academic years 1990/1991-1998/1999 the number of university students increased more than threefold (from 404,000 to 1,274,000). There was a similarly mushrooming growth of private schools, which accounted for 26 percent of total enrolment (331,500 students) in the academic year 1998/1999. On the other hand, the number of academic teachers grew by a mere 15 percent over the period analyzed (from 64,500 to 74,100). There was also a brain drain from the science and R&D sectors, both abroad and to other economic sectors at home (for details, see Jalowiecki et al 1994). This trend has affected the quality of teaching and research. Moreover, the drain primarily involves specialists most wanted in a knowledge economy, for example, computer scientists. (On the brain drain issue, see chapter 8 by Fromhold-Eisebeth).

So far, with a few exceptions discussed in the next section, the transformation of the economic and institutional systems has failed to improve relationships between scientific and R&D units (listed in Table 16.3) and industry in any significant way. Table 16.3 shows that, as in the command system, the units are hardly a source of innovation diffusion for industrial firms. The firms largely pursue internalizing strategies: 80 percent of information that they consider basic for innovation is derived from internal sources. Especially alarming is the fact that in the group of the biggest enterprises in the private sector (which participate substantially in the financing of R&D in advanced economies) the proportion is as high as 96 percent. They are the only firms that do not cooperate with units of the Polish Academy of Sciences in any way, and the percentage of those that have contacts with higher schools is the lowest among all types of businesses (four percent). These figures indicate that in developing new industrial networks, scientific and R&D units are a link of minor importance, which is a detrimental development from the point of view of the adaptation of Polish industry to the dominant global trends and challenges of the knowledge economy.

What is needed to reverse the situation is not only the introduction of innovation-oriented strategies in enterprises, but also a change in the mentality of scientific circles and in the organization of science and R&D in Poland. Such an imperative in turn requires a thorough understanding of the technological capabilities of firms in Poland and the various ways they have adjusted to the transformation to a market economy.

Pathways of Industrial Adjustment to the New Economic System

The present stage of transition of most post-Communist economies is what Grabher and Stark (1998) call compartmentalization. It consists in the separation, but also coexistence (owing, among other reasons, to selective state intervention), of effective enterprises, sectors and regions that are adjusting well to the market system, and those that still more or less follow the rules of the old system in their performance. Thus it is difficult to construct an integrated theory or conceptual framework explaining the transition.

Table 16.3 Sources of information considered by enterprises to be basic for innovations, by firm size and ownership sector

| | Internal Sources[1] | E X T E R N A L   S O U R C E S | | | | | |
| | | Scientific and R&D Units | | | | | |
		Market[2]	PAS Units[4]	R&D Units	Higher Schools	Generally Accessible Data[3]	Other
TOTAL	80	60	5	17	10	53	13
public sector	80	62	5	22	11	58	13
private sector	79	59	4	13	9	49	13
Small enterprises (6-50 employees)	63	50	6	8	8	37	10
public sector	69	55	3	10	14	55	17
private sector	61	49	7	8	6	31	8
Medium-sized enterprises (51-500 employees)	78	58	4	15	9	51	13
public sector	77	58	5	19	9	54	13
private sector	79	59	4	12	8	49	13
Big enterprises (501-2,000 employees)	85	66	5	24	14	63	14
public sector	84	67	5	25	14	62	14
private sector	86	65	5	21	15	64	14
Giant enterprises (over 2,000 employees)	90	71	7	33	15	61	10
public sector	88	69	9	38	17	65	10
private sector	96	79	-	14	4	46	11

[1] Management, own R&D activity, sales and marketing, etc.
[2] Suppliers of materials and equipment, customers, competitors, consulting firms.
[3] Published patent specifications, conferences, meetings, professional periodicals, fairs and exhibitions.
[4] PAS – the Polish Academy of Sciences.

Source: Raport o stanie nauki i techniki w Polsce (Report on the state of science and technology in Poland). Warszawa: GUS, 1998, p. 18.

An attempt at generalizing the different ways in which industrial enterprises are adjusting to the new economic system underlies a model of the modes of enterprise behaviour based on the regulation approach. This model, originally proposed by Asheim and Heraldsen (1991), was adopted by Smith (1995) to describe the transformation of the Slovak economy. It identifies the economic spaces of firms as constituting (a) modes of integration into supplier networks; (b) integration into market networks; and (c) integration into networks of competition or cooperation with other firms. In addition, Smith's model brings to prominence intra-firm relations (labour and wages, organization of production, technological orientation) as well as relations between the enterprise and the state, and between the enterprise and the financial system.

Smith (1995, pp. 768-70) distinguishes three modes of enterprise regulation: globalized, de-industrialized, and mercantilist. The last emphasizes the role of big trading companies in regulating industrial firms in Slovakia. In Poland, the connection between merchant and industrial capital is not so strongly pronounced; other forms of regulation and adjustment are more important. In this section, Smith's model is modified to distinguish four basic ways of enterprise adaptation (adjustment) to the new economic conditions in Poland, namely, globalized, home-market oriented, de-industrialized, and paternalistic (Figure 16.2). Each of these categories will be briefly explained.

First, globalized adaptation (Figure 16.2a) is the most desirable from a developmental perspective, but still not common. Enterprises in this group have achieved the ability to compete at the global scale usually with the help of a foreign strategic investor, through changes in management and marketing systems and the introduction of new technologies and flexible forms of labour organization. Their system of spatial links has changed considerably since the period of the command economy, both in terms of material supplies and sales. However, their regional spin-off, or spillover, effects have been rather modest so far, which may confirm Grabher's (1992) fears about new 'cathedrals in the desert' developing (cf. also Hamilton, 1995; Hardy, 1998). This concern also refers to the R&D sector, as has been mentioned above.

Second, home-market oriented adaptation (Figure 16.2b) is the most widespread type at the present stage of the transformation process. It is based on revitalizing all those resources of local entrepreneurship (in particular, of small and medium-sized firms) that were dormant under the previous command-economy system, and on a successful, flexible penetration of home-market niches in the conditions of a big, unmet domestic demand. This

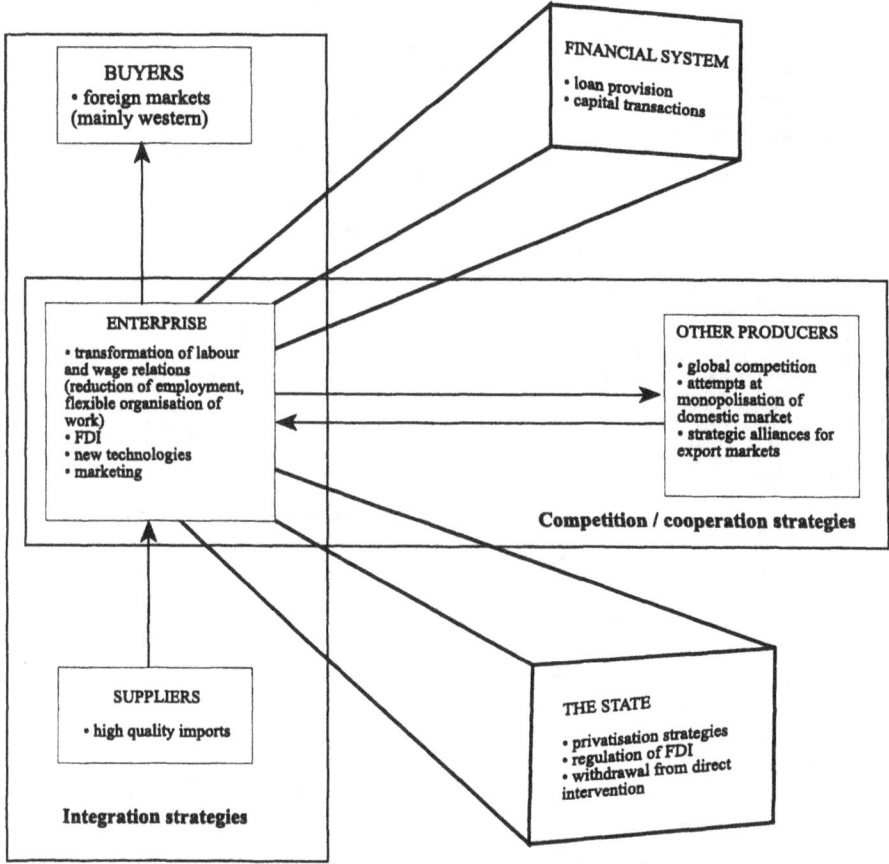

Figure 16.2a Forms of adaptation of industrial enterprises to the new economic system: Globalized type

Source: based on Asheim, Heraldsen (1991) and Smith (1995) with the author's modifications and additions.

pathway of adjustment has been taken, on the one hand, by former state enterprises (e.g. in the food and clothing industries) that have promptly switched to new lines of production and changed their organizational structure, whatever the form of their privatization. On the other hand, this pathway has been pursued by the multitude of newly established private businesses. This process is accompanied by the formation of new supply and distribution networks, often connected with the informal sector (e.g. frontier bazaars). In the opinion of many economists (e.g, Chmiel 1997, p. 202), this form of

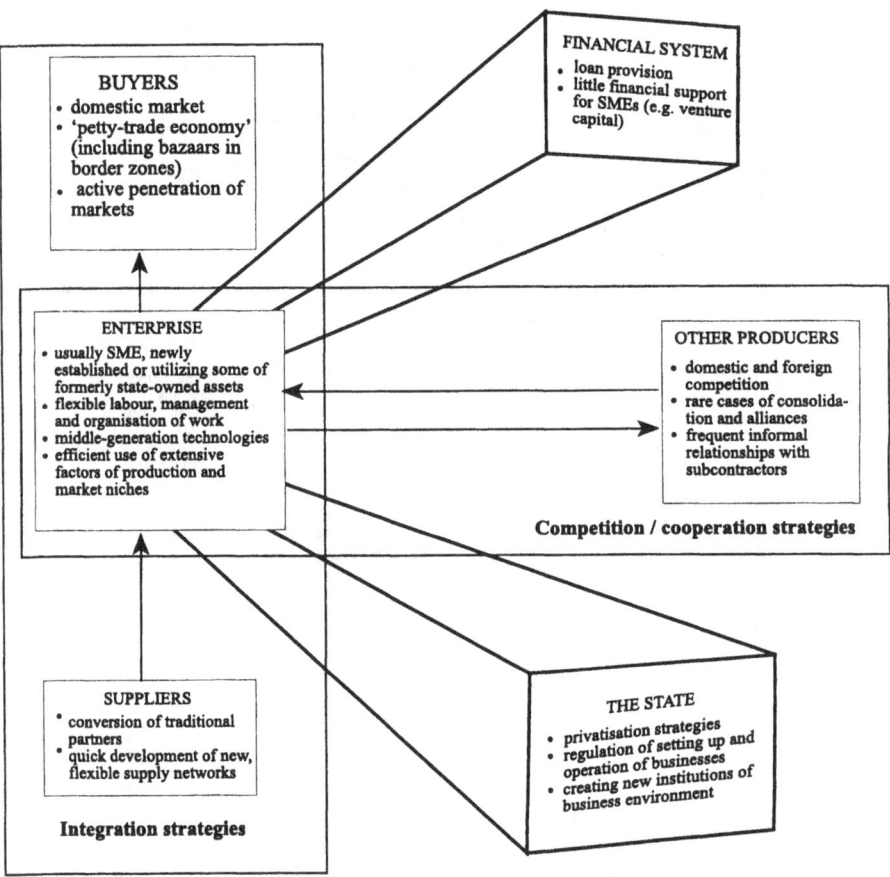

Figure 16.2b Forms of adaptation of industrial enterprises to the new economic system: Home-market oriented type

Source: based on Asheim, Heraldsen (1991) and Smith (1995) with the author's modifications and additions.

adaptation helped to overcome the recession of the initial period of transformation and initiate economic growth after 1992. With time, some firms of this group have entered the path of globalized adaptation, for instance by securing specialized subcontracts from multinational corporations.

Third, de-industrialized adaptation (Figure 16.2c) is an example of unsuccessful adjustment and leads to the closure of a firm or to a significant decline in its employment and output. In Poland it refers especially to enterprises that were oriented towards the Soviet market, technologically backward and with heavy debts or poor management. One more point has to be made here, however: their bargaining power has been too weak for them to secure open or hidden subsidies, cancellation of debts, or further credit, that is, measures which would have enabled them to improve their performance. This is also the case of the next type of adjustment.

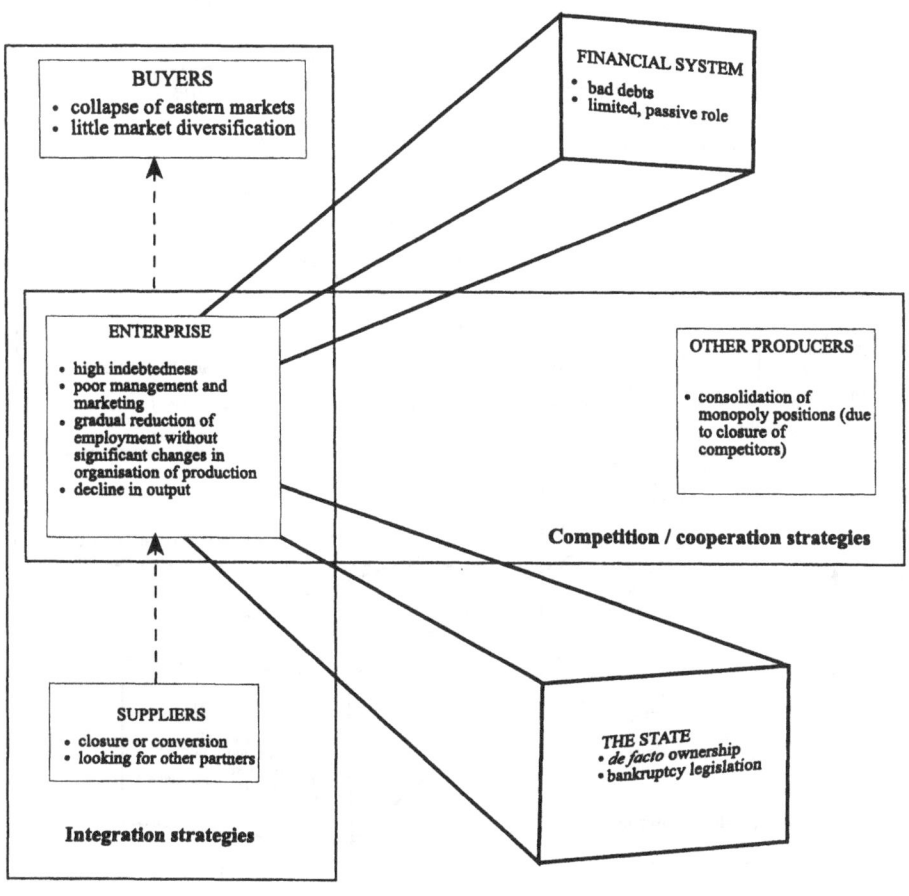

Figure 16.2c Forms of adaptation of industrial enterprises to the new economic system: De-industrialized type

Source: based on Asheim, Heraldsen (1991) and Smith (1995) with the author's modifications and additions.

Fourth, paternalistic adaptation (Figure 16.2d) is in many ways a continuation of firm behaviour typical of the command economy. Low competitiveness and efficiency are compensated for by government protectionism resulting from actions of strong sectoral pressure groups (for example, in mining). In these cases, the state acts towards industrial enterprises as a 'good father'; hence relations of this kind can be called paternalistic. Apart from paternalistic relationships between government and

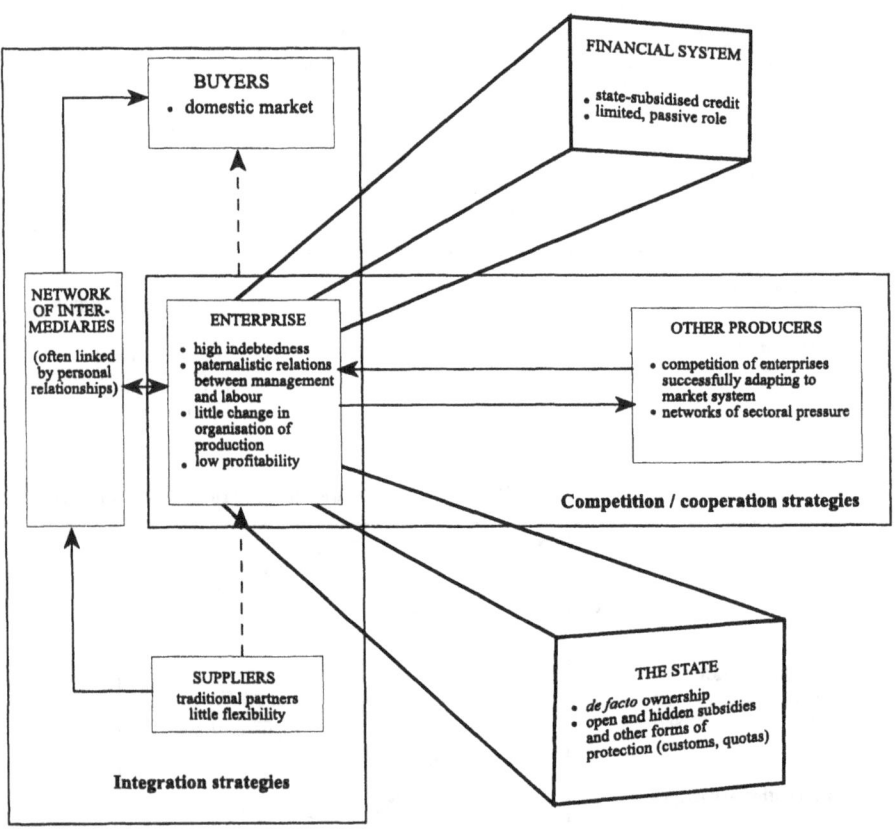

Figure 16.2d Forms of adaptation of industrial enterprises to the new economic system: Paternalistic type

Source: based on Asheim, Heraldsen (1991) and Smith (1995) with the author's modifications and additions.

enterprises, one can also speak about paternalism within enterprises as well as within the supply and distribution networks.

In reality there is a mixture of the above types of enterprise adjustment. With the advance of transformation, however, as the role of market and global forces increases, so does the role of the mode of globalized regulation. A key question therefore is whether this mode of regulation can help Poland 'catch-up' with the leading edge standards of the knowledge economy.

Catching-up with the Knowledge Economy

In view of the contradictory trends occurring in the process of industrial transformation in Poland that were outlined earlier, and the growing impact of global trends, the question of whether and how a peripheral economy can catch-up with the knowledge economy comes to the fore. This issue is not easy to discuss in a country where, after many years of stagnation under a command economy and a recession caused by the initial shock of the transition period, any quantitative increase in industrial output is considered a success. In addition, state and local government policies rely mainly on short-term strategies devised to patch current holes in the budgets while the notions of a 'knowledge economy' or a 'knowledge industry' have not yet found their way in this policy language. Hence, speaking about 'clusters of knowledge industries', 'metropolitan knowledge systems', or 'inter-firm knowledge networks' so frequently encountered when discussing advanced economies, is rhetorical in the Polish case.

It does not mean, however, that in the transformation of the Polish economy no attempts are being made to catch-up with the knowledge economy (cf. Kuklinski and Orlowski, 2000). In the absence of a long-term strategy of promotion and support for knowledge-based industries on the part of the state (despite a multitude of reports and declarations on the subject), two broad development paths are available: (a) an endogenous ('grass-roots') path that spontaneously utilizes local human capital; and (b) an exogenous path that is connected with the activities of transnational corporations. These paths will be now illustrated by two case studies.

The Optimus Computer Firm

Since Poland has no well-developed innovative and knowledge-based industries like aerospace, biotechnology, or an advanced IT sector, nor great financial resources, highly qualified human capital is one of its few competitive advantages that can be used, for example, in the computer industry, especially in designing software. The cost of educating a computer scientist in Poland is about four times lower than in western countries. What is more, in Poland emphasis within the syllabus is placed not only on application informatics, namely operating networks, managing systems, and the implementation of concrete programs, as is done in many other countries, but also on the mathematical foundations of information technology, its logical structure, programming and algorithm construction.

When private economic entities were able to conduct unrestricted economic activity in the 1990s, the above-mentioned competitive advantage in labour skills was soon fully exploited by numerous small computer firms established either near major mathematical scientific centres or technical universities (see, for example, Muent, 2000) or in new greenfield locations. A big role in their operation is played by informal relationships. Many of their owners and employees (often contractual) are also workers of state higher schools in which bureaucratic rules and relics of old networks of relations hamper initiative, independence and flexibility. Sometimes, small, hobby-based ventures develop into larger firms (with new greenfield locations) in association with new inter-firm knowledge networks.

An example of this development path is the Optimus Holding Company. Its founder, Roman Kluska, is a graduate of the Faculty of Cybernetics and Informatics of the Economic Academy in Cracow. In 1988 he started his business by assembling computers in the attic of his parents' house. The economic transformation of Poland allowed him to establish a small computer company 'Optimus' in Nowy Sacz in which he employed young, dynamic people, some of them his old friends. By 1993 Optimus was already the biggest computer firm in Poland. The company was one of the first to be listed on the Warsaw Stock-Exchange, thereby securing means for further development. Within a few years its staff was more than 700 strong and the firm had started to diversify. Today, thanks to skilful mergers, co-operation with well-known strategic partners (such as Intel, Microsoft and Lockheed Martin), and substantial investment, Optimus is a holding company engaged in the production, assembling and distribution of computers, software, multimedia, and the Internet. Regarding the latter, for example, it is the owner of the most popular Polish Internet portals, Onet.pl. Its founder is one of the richest people in Poland and is regarded as the Polish equivalent of Bill Gates.

On 31 December 1999 the Optimus group comprised 38 companies, including 30 subsidiary and seven associated ones. Its distribution network included more than 1,000 dealers. The capital-linked Optimus companies are located in almost all the regions of Poland, and also in Slovakia (Optimus Kosice), Lithuania (Ogmios Sistemos), Ukraine, and France (Printmark). The group's sales, in which service packets such as computer system integration and Internet technologies play an ever increasing role, are largely domestic market oriented (89 percent in 1999). Effective market demand among both individual and institutional customers has been rapid (according to the Optimus 2000 issuing prospectus, at present about 57 percent of firms in Poland have their own home pages, and 82 percent have access to the Internet). The export destinations of Optimus products and services include Ukraine, Lithuania, Slovakia, the USA, Germany, and the Netherlands.

Geographically, the regional context of the location of Optimus in Nowy Sacz is worth noting. Its location effect does not manifest itself, however, in the development of a cluster of local suppliers. It would be hard to find those anyway, because the firm's headquarter office is situated in a typically agricultural and tourist region. Hence, sub-assembled components, as well as software inputs, are supplied by such well-known world corporations as Procomp, Microstar, Hyundai, Samsung Electronics, Fujitsu, Philips, Hewlett-Packard, and Microsoft. As far as computer assembly is concerned, the only local location advantage is the availability of cheap labour in a region with a high unemployment rate, a factor that may have been a major consideration in the initial location decision.

On the other hand, this decision has produced another effect, namely, the attraction and development of highly qualified human capital. The town has been chosen as the seat of one of the best, most dynamically growing private international higher schools in Poland, with a strong informatics faculty, notably the Higher Business School, National Louis University.[1] Many of its students do their practices at Optimus, and the best can find employment in the firm. Employees can further improve their knowledge through contacts with international corporations, that in turn benefits Optimus. For example, such combined sources of know-how has been influential in the development of the firm's new strategy of reducing the role of the assembly and distribution of computer hardware while focusing on software development, the implementation and integration of computer systems, and Internet services (e-commerce, e-finance). A separate company with an R&D centre does the designing of the firm's own licensed technological solutions, like computer-driven cash registers. It is too early to make a full assessment of the firm's regional impact, especially since its line of activity is still very fluid and keeps diversifying; it seems to be evolving towards a financial holding company specializing in electronic banking. Still, it is worth noting that, rather accidentally, a completely new entity has appeared in a traditional region and altered its economic performance. As a result, Nowy Sacz voivodeship has become a leader in terms of the proportion of new and modernized products in its industrial output, and it has been included in several knowledge-based national and global business networks.

Asea Brown Boveri (ABB)

The other path leading Polish industry to the knowledge economy is joining global-local networks through transnational corporations. Its best example is the Swedish-Swiss group Asea Brown Boveri (ABB). The group was one of the first global firms that decided to invest in Poland as early as the first year of transformation, in conditions of high risk. In the 1990s it developed a network embracing not only manufacturing, construction of industrial objects, and maintenance and repair services, but also scientific research, the development of new technologies, and staff training (see Stryjakiewicz, 2000). In 1997 ABB employed 7,738 workers in Poland, or 3.3 percent of its global employment. At the end of 1996 the firm, which earmarks about 8 percent of its total income for R&D, made its most weighty investment decision in Poland, concerning the establishment of a Research Centre in Cracow. So far, this centre is the sole ABB unit of this kind in East-Central Europe, and the ninth in the world. The Cracow Centre, which is supposed to employ some 70 Polish scientists, is intended to perform two kinds of tasks: to help Polish establishments of ABB to adapt technologies designed abroad, and to participate in international research programmes.

According to its director, Dr Franz Schmaderer, the choice of Cracow as the location of the Centre was determined by several factors. These factors relate to: an advantageous structure of the city's higher schools, which made close cooperation possible; an airport with international connections and good infrastructure; and low rents for premises and flats in comparison with, for example, Warsaw. The Centre's first projects deal with the technologies of flue gas treatment (with a special focus on power plants), the problems of diagnosing and monitoring high- and medium-voltage apparatuses, and computer simulation of casting processes. In support of these projects, the ABB Research Centre signed an agreement with the Academy of Mining and Metallurgy in Cracow to conduct a joint research on the utilization of the final product of flue gas desulphurization. A similar R&D cooperation agreement has been signed with the Flow Machines Institute of the Polish Academy of Sciences in Gdansk.

Another manifestation of ABB's treatment of Poland, not only as an 'emerging market', but as an 'emerging knowledge economy' is the location of the Personnel Development Centre at Falenty near Warsaw (in 1995). Its creation is a consequence of one of ABB's long-term strategies which assumes that "the technological and economic aspects do not ensure success in business by themselves; equally important is the creation of a satisfying working milieu favourable to creativity and promoting a worker's development" (ABB, nd., p. 1). Hence, the Centre has set itself the following goals: a) the training of Polish ABB workers of all ranks and in a variety of fields: from the English language through the know-how relevant to the production line of the particular establishments, to the principles of organization and management, marketing, and human resource management; and b) the integration of the personnel of all the ABB establishments in Poland.

The lecturers of the Centre are usually instructors and consultants with some experience of work both in Poland and abroad. It tries to be an intermediary in the transmission of knowledge between foreign experts and the Polish firms. It also performs advisory functions, and is the venue of conferences and meetings with customers. It should be emphasized that the location of such a big personnel training centre in Poland (and similar ones in Prague and Moscow) differentiates ABB from most other multinational companies. The latter, apart from occasional short courses, generally prefer to train their employees in the West and invest in the development of training centres there.

In 1997, Cracow Technological Park was established and, given the impetus already provided by ABB's Research Centre, Cracow is likely to become one of the major clusters of knowledge-based industries in Poland. Making use of the local scientific potential, the Delphi Automotive Systems, a leading world producer and supplier of car parts and sub-assemblies, has also built its Technological Centre in Cracow (the fourth in Europe after those in Paris, Luxembourg and Wuppertal) with a target employment of 350 people. In turn, the American-based Motorola is building a software centre in the city (designed mainly for telecommunications systems) with a target employment of about 500 top specialists, as well as a semiconductor factory.

Conclusions

Despite many difficulties and limitations, foundations have been laid in the 1990s for a market transformation of industry in Poland. The main actors in this transformation have gradually adapted to the (global) conditions of competition. As a result, since 1992 Polish industrial output has grown faster than in the other countries of East-Central Europe. However, this increase has so far depended on the use of simple reserves rather than on any long-term strategy of ensuring competitive advantage in the sense of Porter (1990). Its crucial element is the ability to generate and diffuse innovations. In the future, this ability will probably figure more prominently in determining the role of both individual enterprises and regions in the transformation of industry, and in the creation of its new spatial organization. Even now the importance of traditional industrial districts based on raw materials as well as cheap, unskilled and inflexible labour is clearly on the decline.

Unfortunately, despite the relative success of the process of transition, the Polish economy can hardly be called a knowledge economy yet. At most, we can only talk of enclaves or islands of such an economy. There is still much uncertainty regarding future industrial development in Poland. Nevertheless, it is a future that is connected with progress in Europe, and the already expanding degree of integration will be further stimulated if Poland should gain membership to the European Union, and its socio-economic system.

Acknowledgements

I am grateful for the travel assistance provided by the Institute of Socio-Economic Geography and Space Economy, Adam Mickiewicz University of

Poznan that allowed me to attend the Dongguan meetings of the IGU, and to Marai Kawinska for translation help.

Notes

[1] A similar international Polish-Japanese Higher School of Computer Techniques has been established in Warsaw.

References

ABB (Asea Brown Boveri) (nd), *Training and Development*, Internet, ABB in Poland.

Asheim, B. and Heraldsen, T. (1991), 'Methodological and Theoretical Problems in Economic Geography', *Norsk Geogr. Tidsskr.* vol. 45, pp. 189-200.

Baranski, P. (1994), Rola Inkubatorów Przedsiebiorczosci oraz Centrów Innowacji w Rozwoju Przedsiebiorczosci Lokalnej i Regionalnej. (The role of enterprise incubators and innovation centres in the development of local and regional entrepreneurship). MA thesis written in the Institute of Socio-Economic Geography and Space Economy, Adam Mickiewicz University, Poznan (typescript).

Bertram, H. and Schamp, E.W. (1989), 'Räumliche Wirkungen neuer Produktionskonzepte in der Automobilindustrie', *Geographische Rundschau*, vol. 41, pp. 284-290.

Chmiel, J. (1997), *Male i Srednie Przedsiebiorstwa a Rozwój Regionów*, (Small and medium-sized businesses and the development of regions), ZBSE GUS i PAN, Warszawa.

European Commission (1994), *Competitiveness and Cohesion: Trends in the Regions*.

Grabher, G. (1992), 'Eastern Conquista: the "Truncated" Industrialisation of East European Regions by Large West European Corporations', in H. Ernste and V. Meier (eds), *Regional Development and Contemporary Industrial Response: Extending Flexible Specialisation*, Belhaven Press, London, pp. 219-232.

Grabher, G. (1994), 'The Disembedded Regional Economy: The Transformation of East German Industrial Complexes into Western Enclaves', in A. Amin and N. Thrift (eds), *Globalization, Institutions and Regional Development in Europe*, Oxford University Press, Oxford, pp. 177-195.

Grabher, G. and Stark, D. (1998), 'Organising Diversity: Evolutionary Theory, Network Analysis and Post-socialism', in J. Pickles and A. Smith (eds), *Theorising Transition: The Political Economy of Post-communist Transformations*, Routledge, London, pp. 54-75.

Hamilton, F.E.I. (1995), 'Re-evaluating Space: Locational Change and Adjustment in Central and Eastern Europe', *Geographische Zeitschrift*, vol. 2, pp. 67-86.

Hardy, J. (1998), 'Cathedrals in the Desert? Transnationals, Corporate Strategy and Locality in Wroclaw', *Regional Studies*, vol. 32, pp. 639-652.

Jalowiecki, B. and Hryniewicz, J. and Mync A. (1994), 'The Brain Drain from Science and Universities in Poland, 1992-1993', *Regional and Local Studies*, vol. 11, University of Warsaw, Warsaw.

Kuklinski, A. (1997), 'Regional Innovation Systems in Poland', in A. Kuklinski (ed), *European Space, Baltic Space, Polish Space*, Part II. Hannover-Warsaw: ARL, EUROREG, pp. 326-346.

Kuklinski, A. and Orlowski, W.M. (eds) (2000), *The Knowledge-base of Economy. The Global Challenges of the 21st Century*, Science and Government Series 4. Warszawa: KBN.

Muent, H. (2000), 'Localised Co-operation Between Small Firms in the Socio-economic Transition Process in Poland. Two Cases from the Poznan Region', in J. Parysek and T. Stryjakiewicz (eds), *Polish Economy in Transition – Spatial Perspectives*, Poznan: Bogucki Wydawnictwo Naukowe, pp. 73-88.

Pawlowski, K. (1999), 'Za Duzo Badaczy, za malo Sukcesow' (Too many researchers, too little success), *Rzeczpospolita*, vol. 288, p. 6.

Schamp, E.W. (1997), 'Räumliche Konzentration, ökonomische Kompetenz und regionale Entwicklung. Das Beispiel der oberfränkischen Autozulieferindustrie', *Erdkunde*, vol. 51, pp. 230-243.

Smith, A. (1995), 'Regulation Theory, Strategies of Enterprise Integration and the Political Economy of Regional Economic Restructuring in Central and Eastern Europe: The Case of Slovakia', *Regional Studies*, vol. 29, pp. 761-772.

Stanislawski, G. and Uminski S. (1997), 'Aspects of Technological Integration within the EU', in P. C. Mäller-Graff and A. Stepniak (eds), *Poland and the European Union – Between Association and Membership*, Baden-Baden: Nomos, pp. 169-176.

Stryjakiewicz, T. (1999), *Adaptacja Przestrzenna Przemyslu w Polsce w Warunkach Transformacji* (The spatial adaptation of industry in Poland in the conditions of transformation), Poznan, Wydawnictwo Naukowe UAM.

Stryjakiewicz, T. (2000), 'Implications of Globalization for Regions and Localities in an Economy in Transition: The Case of Poland', in J. Parysek and T. Stryjakiewicz (eds), *Polish Economy in Transition – Spatial Perspectives*, Poznan: Bogucki Wydawnictwo Naukowe, pp. 7-28.

17 Knowledge-based Regions in the Global Periphery: The Case of South Africa

Chris Rogerson

Introduction

In terms of the new global knowledge economy, Africa often is seen as largely excluded from the information technology revolution (Castells, 1998, p. 92). Castells (1996, p. 136) observes that "most of Africa ceased to exist as an economically viable entity in the informational/global economy" and Mabogunje (2000, p.169) writes of the Africa 'dis-connect'. Nevertheless, there are signs that this bleak picture of marginalization may be changing, at least in some parts of the global periphery. Currently, it is recorded that Africa is experiencing a transformation in knowledge access and outpacing the global average for the growth in number of Internet host systems, albeit from a low base (James, 1999). Looking at the continent as a whole French-speaking African countries enjoy a far higher Internet profile than non-French speaking countries and the northern and southern parts of Africa are far ahead of west and east Africa with the situation in central Africa the most undeveloped (James, 1999).

For Castells (1996, 1998), South Africa offers the most hopeful prospects for future development of knowledge-based activities across Africa because of its strong technological linkages in the global economy. These linkages from South Africa might enable "the incorporation of Africa into global capitalism under new more favourable conditions via the South African connection" (Castells, 1998, p. 122). One striking index of South Africa's potential in knowledge-based activities is the listing on the London Stock Exchange during 2000 of the largest information technology company, Dimension Data, which was founded in 1983 in Johannesburg and still remains a South African-based enterprise (Bidoli, 2000). Overall, in terms of indicators of Internet development, South Africa is rapidly advancing with an estimated 370,000 dial-up accounts by 1998 and more than 70 points of presence both in metropolitan areas and small centres (James, 1999). The special role and significance of South Africa's knowledge-based regions for economic development throughout Africa is underscored by such perspective.

The aim of this chapter is to examine emerging spatial trends in knowledge-based activities in South Africa as a case study of knowledge-based economic development in the global periphery. It is shown that in South Africa, the creation of knowledge-based activities reinforces the economic expansion of existing geographical agglomerations of economic activities in the country rather than fosters the emergence of distinctive 'new' industrial spaces or clusters. A particular focus of the chapter is the development of competition for knowledge-based activities between two of the most economically important provinces in South Africa, namely Gauteng, which contains the Johannesburg-Pretoria region, and the Western Cape province, which is centred on metropolitan Cape Town. In both these areas, targeted regional initiatives have been launched to further strengthen their local competitiveness for the attraction of knowledge-based manufacturing and services activities.

The discussion is presented in three parts. First, the chapter's methodology, definitions and research sources are provided. Second, the key elements in the geography of knowledge-based activities are discerned and investigated. Finally, an examination is undertaken of the two competing regional initiatives in South Africa which are targeted at attracting and/or retaining knowledge-based activities.

Definitions, Methods and Sources

The definition of knowledge-based or information-based economic activities is slippery and often contested. In the mainstream literature, the key elements of knowledge-based activities include the following: information technology (IT) services, high technology manufacturing, research and development

(R&D), and even the head office activities of large enterprises. For other observers, an analysis of 'knowledge-based' activities is most appropriately undertaken at a firm or enterprise level of analysis rather than at a sectoral level. In the context of countries in the global periphery, it can be argued that the vast majority of knowledge-based enterprises are those falling into the category of high technology manufacturing or information technology service activities. In addition, R&D and corporate head office functions might be potentially classed as innovative bases for 'smart regions'.

For the purposes of this investigation, however, the key research focus is upon the location of high technology manufacturing and information technology processing activities which taken together provide the basis for identifying South Africa's competitive knowledge-based or smart regions. Head office activities were excluded from the analysis as their location patterns, massively focused in metropolitan Johannesburg, already have been extensively investigated in South Africa (Rogerson, 1996). The location patterns of R&D activities were excluded because of severe data shortcomings. The definition of high technology manufacturing follows that most commonly applied in international research and was used in a previous investigation of high technology manufacturing in South Africa (cf Rogerson, 1997, 1998). International debates and research on high technology activities generally are focused on applying the Butchart (1987) classification which defines high technology activities on the basis of International Standard Industrial Classifications. More specifically, nine types of production activities are classified as forming the basis of a high technology manufacturing economy. These are: synthetic resins and plastics (ISIC 351300); pharmaceuticals (ISIC 352200); office, computing and accounting machinery (ISIC 382500); other machinery (ISIC 382990); electrical industrial machinery (ISIC 383100); radio, television and communications equipment (ISIC 383200); aircraft (ISIC 385500); scientific instruments (ISIC 386100); and, photographic and optical goods (ISIC 386200).

In terms of source material for this investigation of knowledge-based or high technology manufacturing, the research is based on unpublished enterprise data extracted from the University of South Africa Bureau of Market Research Industrial Registers, which provides a national data base of industrial establishments in relation to formal production. For each individual establishment, information is provided as regards type of manufacturing activity (in terms of six digit ISIC code), location of factory, and size of establishment (differentiated in terms of 22 different employment size groups).

The register is continually updated, seeks to aim for a full coverage of officially registered manufacturing establishments and applies a rigorous definition of manufacturing and sub-sectors of manufacturing in terms of the International Standard Industrial Classification. Overall, the industrial register provides a unique establishment-based data source for analysing and examining detailed structural and spatial patterns of manufacturing change in South Africa, not least of high technology manufacturing. Information extracted from the 1999 BMR Industrial Register was used as the basis for defining the location of high technology production or 'knowledge-based' production activities.

Difficulties in the shortage of official data were found in the analysis of the information technology service economy in South Africa. No official information exists to allow the development of a picture of location trends in this important and growing sector of the South African economy (James, 1999). Indeed, aside from anecdotal accounts or newspaper reports, the information technology service sector has not been the focus of detailed sectoral or spatial analysis until recently. Only one limited academic investigation has been undertaken on location issues surrounding the information technology service economy in South Africa (Hodge and Driver, 1999). The most important new source of information derives from the initial outputs of the South African Information Technology Industry Strategy Project (SAITIS). This project is a three year investigation launched by the Canadian International Development Research Centre and South Africa's Department of Trade and Industry seeking to produce a detailed strategy for South Africa's information-technology economy (James, 1999).

In constructing the location patterns of the information technology sector, this analysis utilizes the Matrix Marketing data base on information technology enterprises, which is acknowledged to be the most comprehensive current national existing source of baseline information on the South African information technology service economy (James, 1999). For each establishment, the data base provides information on the nature of activities undertaken by the enterprise, employment size, and location. This data base thus allowed the construction of the first picture of the size and spatial fabric of the information technology service economy of South Africa. Overall, the 1999 Matrix Marketing data base provides a listing of some 1300 information technology-linked enterprises which are spread across 17 different types of service activities.

The BM Register and Matrix Marketing data base together furnish the basis for the analysis which is presented below concerning the competitive position of different regions of South Africa for knowledge-based activities. The listings of individual enterprises in these two sources also provided the sample base for structured interviews which were undertaken in 1999 with a sample of 79 knowledge-based enterprises to examine the key factors behind the location investment decisions made by knowledge-based activities in South Africa. Of these interviews, 39 were undertaken with high technology manufacturing enterprise and 40 were with information-service enterprises. Key themes in the interviews were the performance of enterprises, their location determinants and the relative advantages and disadvantages of Gauteng versus the Western Cape for the operation of knowledge-based activities (Rogerson, 2001).

Knowledge-based Activities in South Africa

In this section, the empirical material derived from the BMR Industrial Register and from the Matrix Marketing data base is interpreted for discerning the basic geography of South Africa's knowledge-based activities. Three sub-sections of discussion are provided which deal in turn with: (a) the picture of high technology manufacturing and its components, (b) the patterns of the information technology service economy, and (c) the overall geographical pattern of knowledge-based or smart economic activities in South Africa.

The Location of High Technology Manufacturing

The key findings concerning the structure and location of high technology manufacturing activities essentially confirm those of previous investigations conducted in South Africa (cf Rogerson, 1997, 1998a; CSIR, 1998; Hodge, 1998; Hodge and Driver, 1999). In total, the BMR Register for 1999 contained 1,460 high technology manufacturing establishments spread across the nine different segments of production. It is estimated that these high technology manufacturers provide a total of 146,576 job opportunities. The largest segment of high technology production, both in terms of numbers of enterprises and employees, is accounted for by electrical and industrial machinery and other machinery segments. Together, these two sub-sectors of high technology manufacturing account for 65 percent of total enterprises and 56 percent of total employment. In terms of employment numbers in high technology production two other sub-sectors are of note, namely pharmaceuticals and radio, television and communications equipment which together encompass a further 17 percent of enterprises and 34 percent of employment. Overall, therefore, the leading four segments of South Africa's high technology production economy account for 82 percent of total enterprises and 90 percent of total employment opportunities.

The spatial patterns of high technology manufacturing disclose a pattern of intense clustering of enterprises and employment opportunities in Gauteng, which is South Africa's leading industrial region (see Rogerson, 2000a). Figures 17.1 and 17.2 show that Gauteng contains approximately 107,000 job opportunities in high technology manufacturing or 73 percent of the national total. By contrast, the Western Cape contains only

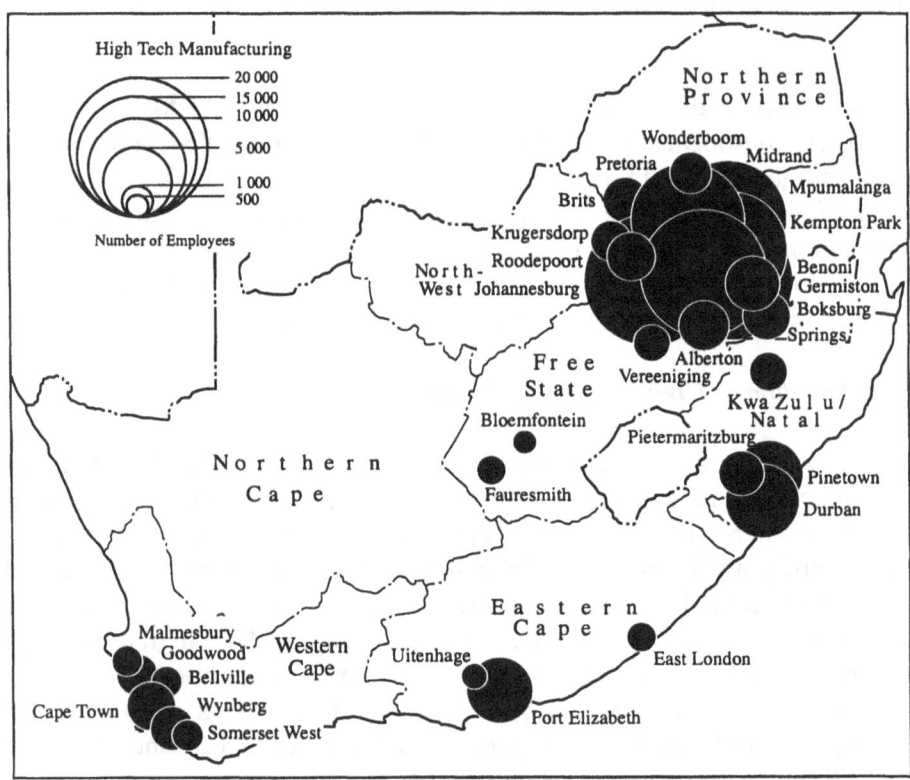

Figure 17.1 Spatial pattern of employment in high-technology manufacturing in South Africa, 1999

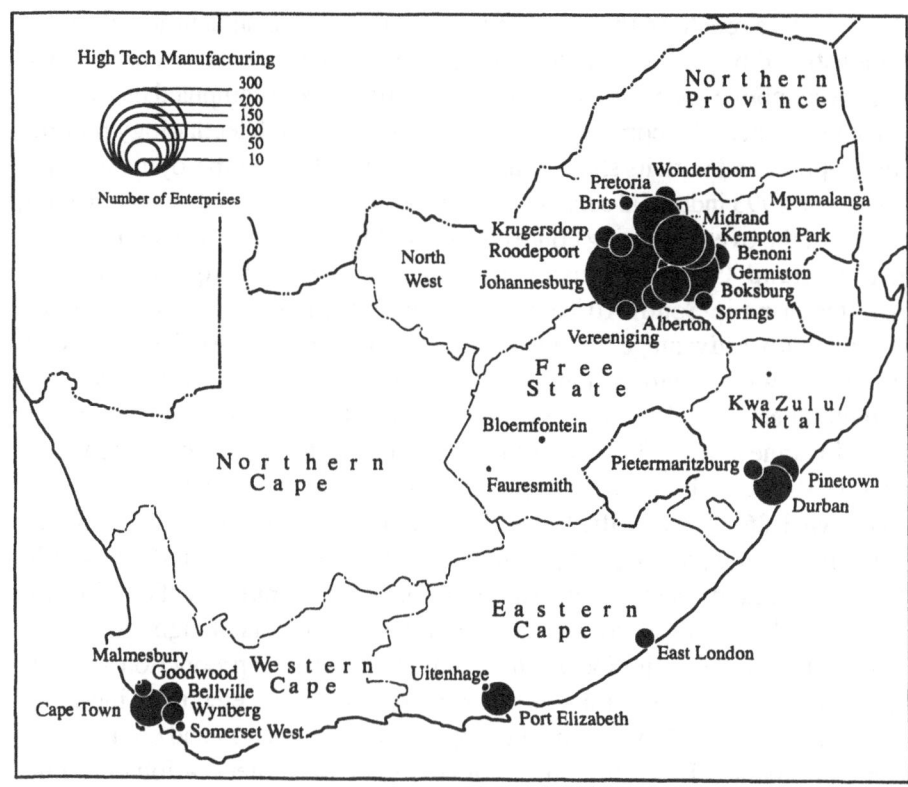

Figure 17.2   Spatial pattern of enterprises in high-technology manufacturing in South Africa, 1999

approximately 10,000 high technology manufacturing jobs or 6.8 percent of the national total. The dominance of Gauteng is shown by the fact that, as indexed and ranked in terms of employment, the leading six individual magisterial districts are all situated there. In absolute numbers of employment the leading ten ranked magisterial districts were as follows: Johannesburg 17,789 (12.1 percent), Germiston 16,399 (11.2 percent), Boksburg 14,790 (10.1 percent), Randburg 13,959 (9.5 percent), Kempton Park 13,720 (9.4 percent), Pretoria 13,275 (9.1 percent), Durban 5,808 (four percent), Pinetown 4,938 (3.4 percent), Port Elizabeth 4,564 (3.1 percent) and Benoni 3,115 (2.1 percent).

The strength of the Gauteng high technology manufacturing cluster is demonstrated by the calculation of simple location quotients as a measure of relative economic concentration. Location quotients were computed on a magisterial district basis by comparing each area's share of high technology production employment with its share of national manufacturing employment (using as base the 1993 industrial census). What emerges from the analysis is a striking picture of the overwhelming concentration of high technology manufacturing in the inland Gauteng cluster and the weakness of high technology manufacturing in South Africa's coastal centres. In Gauteng, strong signals of comparative advantage for high technology production activities are evident in the following location quotient values: Randburg 5.94, Boksburg 5.32, Kempton Park 2.94, Pretoria 2.46, Germiston 2.43, Benoni 2.10, Roodepoort 1.60, Johannesburg 1.33, Alberton 1.20 and Springs 1.06. These signals of relative strength should be compared to the findings for coastal locations: Pinetown 1.26, Port Elizabeth 0.83, Durban 0.42, Cape Town 0.35 and East London 0.31. Unpacking the picture for high technology manufacturing by its constituent sub-segments once more confirms the dominance of the Gauteng high technology cluster. In terms of the four leading sectors of high-technology manufacturing, the major local concentrations of national production all occur in Gauteng. The leading ranked areas were as follows: electrical industrial machinery – Germiston and Johannesburg; radio, television and communication equipment – Boksburg and Pretoria; other machinery – Johannesburg and Germiston; and pharmaceuticals – Kempton Park and Randburg.

With respect to numbers of establishments (Figure 17.2), little change appears in the national picture of knowledge-based manufacturing in South Africa. In total of 1,460 high technology manufacturers, 62 percent of the national share, is situated in Gauteng. The ten leading ranked magisterial districts in terms of enterprise numbers are Johannesburg 281 (19.2 percent), Germiston 161 (11 percent), Randburg 114 (7.8 percent), Pretoria 102 (seven percent), Kempton Park 82 (5.6 percent), Durban 68 (4.7 percent), Cape Town 63 (4.3 percent), Boksburg 61 (4.2 percent), Pinetown 43 (2.9 percent) and Port Elizabeth (2.8 percent). Once again, the dominance of Gauteng is reinforced as it contains six of the leading ten magisterial districts as defined by numbers of high technology manufacturing enterprises.

The location findings from the analysis of the BM Industrial Register data were complemented by the interview material. It is significant that many of the high technology enterprises that were contacted for interviews had ceased manufacturing and instead functioned increasingly either as loci for outsourcing production or for merely importing goods. This result confirms the erosion and restructuring which is taking place within the sector of high technology industry (for pharmaceuticals, see Sellars, 1998), in common with that for South African manufacturing as a whole (Rogerson, 2000a). The stagnant trend of high technology manufacturing is evident in terms of the recent employment trajectories of the sample of high technology manufacturers. Of the sample of enterprises, only 30 percent (12 cases) reported increased levels of employment between 1994-1999, 20 (eight cases) indicated similar levels of employment whereas 50 percent of enterprises observed that employment levels actually had decreased. In several cases, this downsizing in employment between 1994-1999 was of substantial proportions; from 250 to 30 employees in the case of one medical equipment producer and from 493 to 350 employees in another similar manufacturer, both South African branches of multinational corporations. In total, 31 enterprises provided an estimate of projected employment levels in the next five year period. For the period 1999-2004 the employment horizon in high technology manufacturing appears, at best, to be stagnant; only seven firms indicated an anticipated increase in their employment levels, five enterprises expected a decline and the majority (18 cases) expected similar levels of employment to 1999 levels.

With respect to location choice of high technology manufacturers, the key themes are those of access to markets, skilled manpower and infrastructure (see Rogerson, 1997, 1998a). Market issues clearly are important influences in determining the location choice of enterprises. For two-thirds of the sample of high technology enterprises, the major advantages of Gauteng relate to market opportunities linked to availability of suppliers and/or subcontractors. Enterprises also stressed the centrality of the Gauteng location for distribution to the South African market as well as its proximity to the markets in the rest of sub-Saharan Africa. Two other factors were observed as location strengths of Gauteng, namely the availability of skilled workers essential for high technology production, and the quality of infrastructural facilities. In particular, the importance of the Johannesburg International Airport and the quality of the highway system were noted. The list of disadvantages of Gauteng as a business location was headed by the issues of crime and security. Nonetheless, in a number of interviews the caveat was given that while crime was a deterrent to businesses operating in Gauteng, it was also recognized as a national problem, affecting business operations throughout South Africa.

The Location of the Information Technology Service Economy

The information technology service economy of South Africa contains a mixture of strong domestic enterprises, led by Dimension Data and Altech, and the local subsidiaries of a number of large multinational enterprises, such as IBM, Unisys, Microsoft, ICL, Intel, Dell and Compaq. Despite the strength of certain local IT enterprises, one study concluded "that the South African IT [information technology] industry has been effectively created through foreign multinational support and the industry is very dependent on their continuing participation" (James, 1999, p. 68). Recent investigations have drawn attention to the critical influence of the brain drain from South Africa of skilled IT personnel, particularly to the USA, UK, Canada, Australia and New Zealand (Crush et al 2000; see also chapter 8 by Fromheld-Eisebith). Overall, the high demand for South African IT skills in the international market place and corresponding local skills shortages is the most critical factor that threatens the future health of the information service economy in South Africa.

No previous investigation has been undertaken on the location of South Africa's IT service economy as a whole. The findings from the Matrix Marketing data base were of the existence of 1,220 IT service enterprises with an estimated 64,109 employment opportunities. Within the various categories of IT service activities both the largest numbers of enterprises and employment opportunities were in the activities of distributors, systems integrators and specialist retailers of IT equipment and supplies. Taken together, these categories account for 66 percent of enterprises and 76 percent of total employment opportunities within the IT service economy. The largest number of enterprises were systems integrators whereas distributors accounted for the largest individual element of employment.

Analysis of the location of the IT service economy discloses a remarkable degree of spatial clustering or concentration of activity. Figures 17.3 and 17.4 show the geographical patterns of employment and establishments in the IT service economy of South Africa. Of the national total of 64,109 jobs, 78 percent is clustered in Gauteng. The intense spatial agglomeration of the IT service economy is indicated by the fact that 70 percent of national IT employment (nearly 45,000 jobs) is concentrated in the Johannesburg (23,684) and Randburg (21,004) magisterial districts. The third ranked magisterial district in terms of employment opportunities is Cape Town (5,000 jobs) which has a 7.8 percent share of national IT employment.

The only other magisterial district in which there occurs more than 1,000 job opportunities in IT services are Pretoria, Durban, Germiston, and Kempton Park. Overall, the coastal cities fare marginally better in the analysis of the IT service economy than in terms of high technology manufacturing. In particular, Cape Town is much stronger in IT services whereas Durban houses a greater strength in high technology manufacturing. The rank ordering of magisterial districts in terms of the numbers of IT service enterprises shows little variation from the spatial patterns discerned using employment data. Indeed, Gauteng accounts for 68 percent of total enterprises with the largest numbers of enterprises occurring in Johannesburg (404), Randburg (196), Pretoria (114), Cape Town (113) and Durban (54).

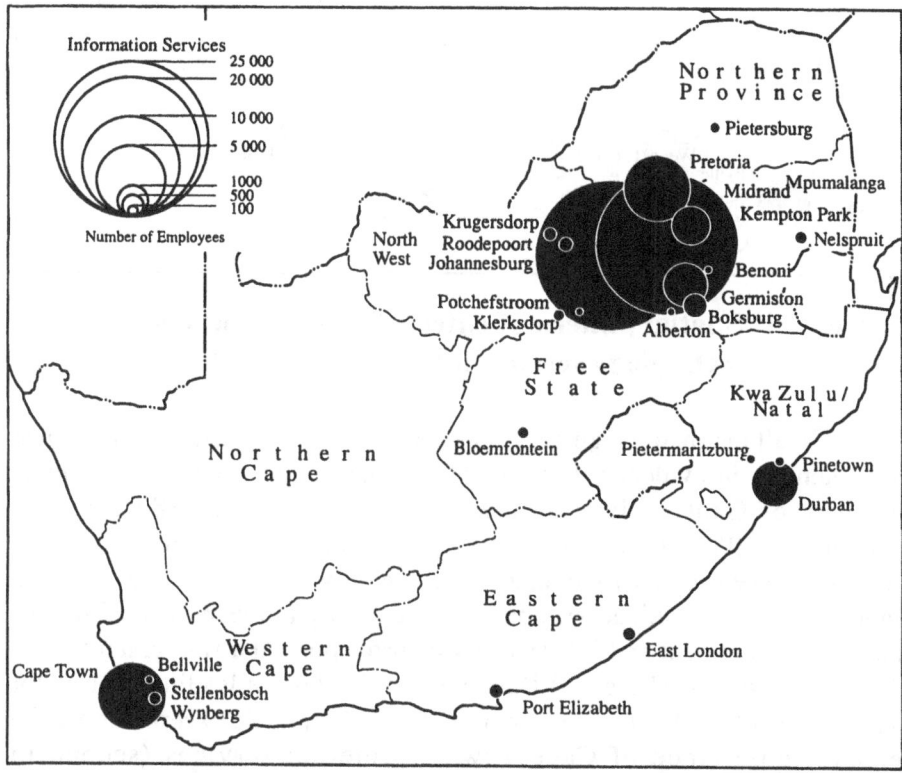

Figure 17.3 Spatial pattern of employment in information technology service activities in South Africa, 1999

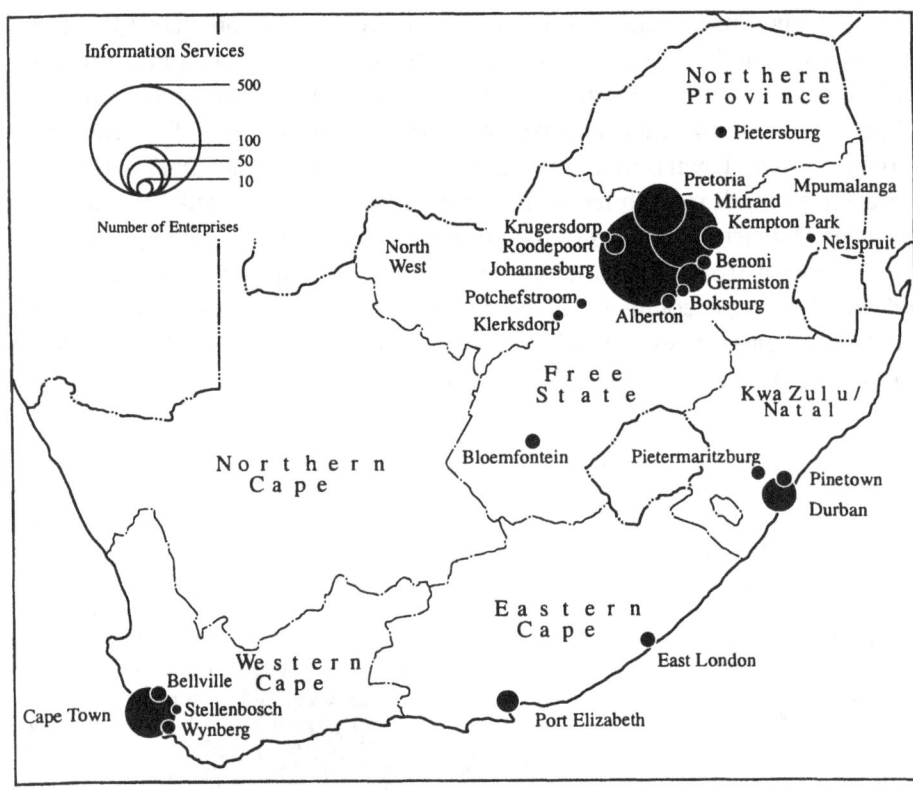

Figure 17.4   Spatial pattern of enterprises in the information technology service in South Africa, 1999

In all the major segments of the IT service economy, the dominance of Gauteng is in evidence. In terms of national employment opportunities the share of Gauteng is as follows; systems integrators (82 percent), distributors (95 percent), netware specialists (91 percent), software development (93 percent) and Internet service providers (68 percent). Outside Gauteng, local specializations appear to exist only in the Western Cape for Internet development and Internet service provision. At a fine-grained level of analysis, Johannesburg and Randburg are by far the leading centres for most aspects of the IT service economy. The only notable exceptions are the relative importance of Cape Town in Internet services (second to Johannesburg) and the leadership of the highly specialized area of Internet development by enterprises based in Pretoria and Cape Town-Stellenbosch.

The spatial organization of the IT service economy appears only marginally different when analyzed in terms of numbers of enterprises. The key difference is that while the dominance of Gauteng is retained, the Western Cape exhibits a stronger level of relative performance. Overall, of the national total of 1 220 establishments, 69 percent are located in Gauteng and a nearly 13 percent share in the Western Cape. The Pretoria-Randburg-Johannesburg cluster centred on Midrand and Johannesburg's northern areas contains 58.5 percent of all national IT enterprises; Johannesburg-Randburg alone accounts for almost half the national total. Cape Town itself accounts for nearly 10 percent of the national share of enterprises, a stronger position than it holds in terms of employment indicators. Finally, outside of Gauteng and the Western Cape, other minor nodes of IT service enterprises are in the Durban-Pietermaritzburg area (seven percent), Port Elizabeth (two percent), East London (one percent) and Bloemfontein (one percent).

As compared to the relatively stagnant picture of employment and enterprise expansion that characterized high technology manufacturing, the recent economic performance and outlook for the information service economy was much brighter. In Gauteng, one recent estimate suggests that the province's IT sector is currently expanding at 17 percent per annum (Dorfling, 1999, p. 2). In the interview research, enterprise growth performance was examined in terms of issues of employment, space usage, investment in new equipment and profits. Signs of positive growth performance were evident throughout the sample of enterprises across both Gauteng and Western Cape. Overall, between 1994-1999 80 percent of IT enterprises (32 cases) had expanded their workforces, 10 percent were at the same level of employment and 10 percent were in employment decline. Optimism about future employment growth is indicated by the findings that in the period 1999-2004, 85 percent of IT service enterprises anticipated further expansions in their employment, the remainder expected to retain present employment levels; only one case was found of an enterprise that anticipated an employment decrease between 1999-2004. Increased levels of employment are leading to an increased need for business space: between 1994-1999 68 percent of enterprises (27 cases) increased their office space – only one case of downsizing was found.

Projections over the next five year period, 1999-2004, suggest that 75 percent of enterprises anticipate further expansion in their need for office space. Increases in employment were linked also to increases in new investment in capital equipment. In total, 95 percent of the sample of South

African IT firms indicated a growth in new expenditures on capital equipment for the period 1999-2004 and at least 75 percent were expecting a further growth in new equipment expenditures for 1999-2004. Finally, the profit performance of IT enterprises was rated as extremely healthy and bright. In total 95 percent of firms recorded increases in their profits over the period 1994-1999 and all were expecting further increases in the period 1999-2004. Although the differences between the enterprise performances of the two groups of Gauteng and Western Cape IT enterprises was marginal, there was an observable trend for a more optimistic future business outlook on the part of Western Cape enterprises.

In terms of the factors that influenced the decisions to locate IT businesses, the core themes relate to market considerations and the location of the informative-intensive users. Among the Gauteng sample, the theme was constantly stressed that South Africa's major IT market was found in Gauteng (see James, 1999). It is estimated that 72 percent of all national IT expenditure occurs in the province (Dorfling, 1999). The location choice for 95 percent of Gauteng's IT enterprises were influenced by the province's status as IT and industrial hub of the country. Other issues that were raised related to availability of skilled personnel as a secondary factor for business location in Gauteng (James, 1999). The disadvantages of Gauteng as a business location drew the familiar responses of crime, traffic congestion and the 'crowdedness'. In the Western Cape, issues of access to local markets again were of core importance in affecting location decisions. Secondary location factors were those cited as the quality of life in the Western Cape and residential preferences of personnel. The disadvantages of the Western Cape related overwhelmingly to the narrowness of the local market. Over half of the Western Cape IT enterprises identified this constraint of the Western Cape as a business location. Interviewee responses were that the "Major IT market is Gauteng", "Market is Limited", "Would Make more Money in Gauteng", "Clients are not as IT oriented as Johannesburg", and several complaints about costs of flights and couriers. A second layer of complaints by Western Cape IT enterprises of local disadvantages, relating to issues of crime and security, is also of note.

Overall, Gauteng was rated as the best potential business location by the national sample of 40 IT service enterprises. The only perceived advantages of the Western Cape as expressed by Gauteng enterprises related to the region's pleasant residential environment. This factor, significantly, is often reinforced by the fact that the Western Cape was the original place of residence and or business of the owners of IT service enterprises in Gauteng.

The Geography of South Africa's Knowledge-Based Economy

Taken together, the two investigations of the high technology manufacturing economy and the IT service economy provide a clear picture of the essential outlines of South Africa's knowledge-based economy. Overall, the two knowledge-based sectors of manufacturing and services provide at least 200,000 job opportunities, two-thirds of which are in manufacturing. Of the total employment opportunities in knowledge-based activities, approximately 77 percent are situated in Gauteng.

The location patterns of these two major segments of South Africa's knowledge based economy reveal both certain common patterns and differences. The most striking finding is the massive dominance of Gauteng in both the activities of high technology manufacturing and IT services. Overall, the IT service economy is even more strongly anchored in Gauteng than is the high technology manufacturing. With the exception of certain specific aspects of IT services in the Cape Town area, the broad picture is of the essential weakness of the coastal cities for knowledge-based activities. Even weaker still is the performance of South Africa's group of secondary cities and small towns for virtually all forms of knowledge-based activities. Minor exceptions are the small group of high technology production activities situated in Brits (Gauteng) and of specialized computer services in Stellenbosch (Western Cape). In relative terms, the most striking feature of South Africa's knowledge based economy is the extraordinary performance of the Randburg district, which contains the largest section of the Midrand high-technology cluster (cf Hodge, 1998).

In terms of the combined absolute numbers of employment in knowledge-based activities, the findings underscore the strength of the Gauteng 'smart agglomeration'. Of the ten leading ranked magisterial districts, seven are found in Gauteng. In terms of employment, the major centres are as follows: Johannesburg 41,473 (19.7 percent), Randburg 34,963 (16.6 percent), Germiston 26,520 (12.6 percent), Pretoria 17,860 (8.5 percent), Boksburg 15,404 (7.3 percent), Kempton Park 15,319 (7.3 percent), Durban 8,051 (3.8 percent), Cape Town 7,535 (3.6 percent), Port Elizabeth 3,132 (1.5 percent) and Alberton 2,691 (1.3 percent). In terms of numbers of knowledge-based enterprise, the rank ordering is as follows, Johannesburg

328   Knowledge, Industry and Environment

685 (25.8 percent), Randburg 310 (11.7 percent), Pretoria 216 (8.1 percent), Germiston, 197 (7.4 percent), Cape Town 176 (6.6 percent), Durban 122 (4.6 percent), Kempton Park 105 (3.9 percent), Boksburg 68 (2.6 percent), Port Elizabeth 65 (2.4 percent), and Pinetown 55 (2.1 percent).

Finally, it is significant to note that in at least three areas of South Africa, employment numbers in the IT service economy currently exceed those of the (much larger nationally) high technology production economy. The three areas are Johannesburg, Randburg and Cape Town. These centres clearly represent the leading geographical edge of the knowledge-based service economy of South Africa.

In sum, the findings on the spatial fabric of knowledge-based activities in South Africa underline that Gauteng clearly deserves to be recognized as South Africa's 'smart region' with the Western Cape as a secondary region for knowledge-based activities. In this example of the global periphery it is significant that there has been no emergence at the national scale of new economic spaces associated with the growth of knowledge-based enterprises (cf Castells and Hall, 1994; Castells, 1998). Indeed, only at the local scale of analysis can any new economic spaces be discerned in this South African analysis. The development of Midrand, situated between Johannesburg and Pretoria, can be noted as a new decentralized cluster which is associated with knowledge-based activities, more particularly of service-related activities (see Hodge, 1998; Rogerson, 2000a). Not surprisingly, therefore, in the South African press increasingly Midrand is referred to as the 'Silicon Rand'.

Regional Competition

The potential importance of knowledge-based activities for promoting regional and local economic development has not gone unnoticed in post-apartheid South Africa. Against the background of the national government's retreat from top-down spatial planning, there has occurred a surge of sub-national initiatives and programmes for economic development (Bloch, 2000). Impetus for the burgeoning of sub-national development initiatives in South Africa derives variously from the impact of globalization, national programmes for decentralization of functions to local level, and local responses to post-apartheid economic restructuring (Rogerson, 2000b). Among a growing number of sub-national strategies for promoting economic development, the recent initiatives undertaken in Gauteng and the Western Cape surrounding knowledge-based activities warrant some attention. Essentially, these different initiatives reflect the intense regional competition that is developing in South Africa for potential new levers for regional and local growth, not least for knowledge-based activities.

The programme of Spatial Development Initiatives (SDI) in South Africa is characterized by intensive interventions which seek to fast-track investment (both public and private sector) into areas identified as having inherent and under-utilized economic potential. The SDIs represents a strategic investment initiative launched since 1995 in support of South Africa's new macro-economic framework, the Growth, Employment and Redistribution (GEAR) strategy, for building an internationally competitive economy (Rogerson, 1998b). The SDIs seek to foster new investments in strategic, sustainable and competitive economic sectors at locations that will have maximum economic and social development impacts. As of mid-2000, 12 SDIs had been established and were functioning at various stages of implementation. In Gauteng, the planned so-termed 'special economic zones' project represents one of the most recent additions to the portfolio of SDI programmes in South Africa.

The Gauteng Special Economic Zones project is essentially a programme for reinvigorating the base of the provincial manufacturing and service economy (Rogerson, 2000a). In the initial planning framework for the Gauteng SDI, strategic options were developed for a proposed Gauteng Technology Triangle (CSIR, 1998) with emphasis upon promoting existing competitive high technology activities. As the SDI plans have evolved, the vision has consolidated into developing the region into a 'smart hub' (DTI, 1998, p. 1), which would be focused upon knowledge-based economic activities. Within the planning for Gauteng as a 'smart hub' are several projects. The most notable is a planned 'innovation hub' in the form of a development corridor that runs along the highway that links Johannesburg and Pretoria. It is recognized that of critical importance in this innovation corridor are "the regional economy of creative industries, innovation, IT and telecommunications 'connectivity' into global information networks and markets" (DTI, 1998, p. 4). In February 2000 an initiative was established which serves to link a set of business, education and research institutions in the Pretoria area in order to create, nurture and grow technology-led or high-growth knowledge-based businesses. Long-term planning involves developing at a new greenfield site a high technology industrial park with high bandwidth connectivity through fibre optic cables, an incubator and innovation centre, a venture capital company and an investor information and support system (Dorfling, 1999, p. 2). Overall, the projects coming on stream as part of the Gauteng SDI are geared to strengthen the province's competitiveness for high technology manufacturing and information service activities (Rogerson, 2000a, p. 337).

In the Western Cape there have been several regional initiatives designed to boost the province's competitiveness for knowledge-based activities. The local development agency, WESGRO, prioritizes high technology activities as one of the "core clusters" for development in the Western Cape (WESGRO, 1998). Accordingly, there is extensive place marketing of the Western Cape as South Africa's technology hub, using the backdrop of Cape Town's scenic Table Mountain as the basis for asserting the region's high quality of life and environmental attractions. It is argued that the Western Cape "has an industrial development advantage through its 'density of intellectual talent" (WESGRO, 1998, p. 66) and "a good chance of capturing significant segments of high tech related research and production activities" (WESGRO, 1998, p. 49). A key future challenge is to sharpen the province's focus on high-technology activities in competition with Gauteng. More especially, it is asserted that the "challenge lies in creating high tech firms which will build on the intellectual talent in the (province's) tertiary institutions and which at the same time provide growth opportunities in promising niches" (WESGRO, 1998, p. 66).

The Cape Information Technology Initiative (CITI) is seen as potentially a model for regional high technology cluster initiatives in South Africa (James, 1999, p. 52). The CITI is an independent Section 21 company that seeks to catalyze the development of a dynamic cluster of IT industries in the Western Cape. The goal is to establish the Western Cape as the IT 'Gateway to Africa'. The initiative is grounded in the current theory and practice of cluster development and arose out of discussions held in 1997 between the private sector, local Universities and provincial government. It was conceded that while the brain drain of IT professionals from South Africa was serious and growing, regions such as the Western Cape could become a magnet for high technology investors. Among its activities the CITI has sought to address the IT skills shortage by spearheading an inquiry into fast-track immigration, developed plans for an incubator for start-up IT enterprises and increased the level of networking taking place in the regional IT economy as a whole (James, 1999). In addition, during 1999 it was announced that CITI had secured seed funding for developing a world-class Web site, which is seen as pivotal for galvanizing the IT cluster in the Western Cape.

Another high-profile initiative to support the development of the Western Cape as a competitive base for knowledge-based enterprises is the development of South Africa's largest science or technology park in Cape Town. The Capricorn Technology and Industrial Park is marketed as the "Western Cape's technology innovation hub", a "world class location" which provides "an opportunity for companies to locate their manufacturing, office, research and distribution activities on one fully integrated site" (see Rogerson 2001). Despite the hype, however, the successes of this technology park in terms of attracting knowledge-based enterprises have been relatively meagre. Indeed, most attention has been from property developers rather than high technology and IT enterprises; during 1999 it was proclaimed that a religious organization was taking space in the technology park in order to establish a Christian convention centre. Increasingly, the Capricorn park is taking on the mantle of a property initiative rather than a science park per se. Behind its difficulties are its poor choice of location within Cape Town (close to low-income settlements), the weak human resource base in the province as a whole, the lack of many innovative firms and, most important of all, the limited market base in the Western Cape as compared to the several advantages of locating in the Gauteng agglomeration.

The most detailed and recent analysis of the issues that challenge the knowledge-based economy in the Western Cape is by Hodges and Driver (1999). Their study shows that the critical factor determining an enterprise's location in the Western Cape relates to local market opportunities or an entrepreneur's place of residence in the metropolitan Cape Town area. In respect of the key location determinants for high technology activities, the study draws attention to a number of critical issues. It stresses that the much-vaunted 'quality of life' factor in the Western Cape was not a critical determinant in the growth of knowledge-based enterprises. In the first instance, the key location factors for such enterprises relate to issues of markets, availability of inputs into the production process, including the input of skilled personnel, and an excellent infrastructure (Rogerson, 1998a; Hodges and Driver, 1999). Of vital concern to the future of high technology manufacturing is the finding that the majority of Western Cape firms complain of shortages of skilled labour. Although access to skills has traditionally been seen as an advantage of locating in Cape Town, especially for high technology activities, "it seems that this advantage is in danger of being eroded" (Hodges and

Driver, 1999, p. 31). In fact, it was argued that the Cape's quality of life is "proving an insufficient base on which to maintain the region's advantage in the supply of skilled labour" with skilled personnel preferring to work overseas or in Gauteng where there exist greater opportunities and higher salaries than in the Western Cape (Hodges and Driver, 1999, p. 35). Moreover, in terms of regional infrastructure, it was observed that while better communications infrastructure and services could be factors to support existing Western Cape enterprises, "there appears not much in the local infrastructure that actively attracts firms to the region or gives local firms a significant advantage" (Hodges and Driver, 1999, p. 33).

Conclusion

Internationally, it is evident that knowledge-based economic activities are becoming increasingly significant drivers of growth and patterns of spatial development (Castells, 1989, 1996, 1998; World Bank, 1999; Mabogunje, 2000). In North America and Western Europe, there is considerable evidence that such knowledge-based activities represent key drivers in shaping the directions of development of national and regional spatial systems.

South Africa's knowledge-based economic activities are seen as important outliers in the new international informational economy as they offer a potential linkage for Africa as a whole (Mabogunje, 2000). This study suggests that currently knowledge-based activities provide direct employment opportunities to an estimated 200 000 persons with signs that the information service economy is fast-expanding as compared to a stagnation in the high technology manufacturing economy. In terms of location, the research disclosed that both the high technology manufacturing economy and the information-service economy are overwhelmingly clustered in Gauteng. Indeed, the information service economy is presently even more strongly centred in Gauteng than that of high technology manufacturing. The dominance of Gauteng as South Africa's 'smart province' must be explained partly in terms of the depth of agglomeration in the province of the major users of the products and services of the knowledge-based economy. Other key factors, however, are those of its comparative advantage for available skilled personnel, infrastructural base, and location advantage for African markets. Important regional level initiatives are in place to maintain the competitive edge of the Gauteng knowledge economy against aggressive regional competition from Western Cape.

Finally, certain parallels can be drawn with the international literature on the location of knowledge-based activities. What emerges from the South African case study is that the creation of a knowledge-based economy has not resulted in the emergence of notable new industrial spaces or geographical clusters of activity. High technology manufacturing has become anchored on the market, infrastructure and human resource advantages offered by the Gauteng agglomeration. Moreover, the findings on the location and ranking of potential areas for IT activities in South Africa parallel those in the international literature which underline the critical significance of close location access to the major informative-intensive using activities in the economy (cf Castells, 1996, 1998). New initiatives to change the geography of knowledge-based activities, in terms of the building of new technology or science parks have met with little success. Overall, the case of South Africa shows that the advantages of the existing agglomerations have been reinforced in cumulative fashion by the building of knowledge-based activities upon and integrated into the advantages of the long-established cluster of economic activities (particular head offices and manufacturing).

Acknowledgements

Financial support for this research from the National Research Foundation, Pretoria is gratefully acknowledged. For their assistance with this project thanks are extended to Anna Kesper, Danielle Burger and Anthony Prangley. Mrs W. Job prepared the diagrammes which accompany this chapter.

References

Bidoli, M. (2000), 'Dimension Data – It's no Ordinary Company', *Financial Mail* (Johannesburg), vol. 158 (2), June 30, pp. 6-7.
Bloch, R. (2000), 'Subnational Economic Development in Present-day South Africa', *Urban Forum*, vol. 11, pp. 227-271.

Butchart, R.L. (1987), 'A New UK Definition of the High Technology Industries', *Economic Trends*, vol. 400, pp. 82-88.

Castells, M. (1989), *The Informational City: Information Technology, Economic Restructuring and the Urban-Regional Process*, Basil Blackwell, Oxford.

Castells, M. (1996), *The Information Age: Economy, Society and Culture: Volume 1 – The Rise of the Network Society*, Basil Blackwell, Oxford.

Castells, M. (1998), *The Information Age: Economy, Society and Culture: Volume 3 – End of Millenium*, Basil Blackwell, Oxford.

Castells, M. and Hall, P. (1994), *Technopoles of the World: The Making of Twenty-First-Century Industrial Complexes*, Routledge, London.

Crush, J., McDonald, D., Williams, V., Mattes, R., Richmond, W., Rogerson, C.M. and Rogerson, J.M. (2000), *Losing Our Minds: Skills Migration and the South African Brain Drain*, Southern African Migration Project, Migration Policy Series No. 18, Cape Town and Kingston.

CSIR, (1998), Facilitating the Provincial Trade and Industrial Strategy through the Gauteng Spatial Development Initiative, Unpublished report prepared for the Gauteng Provincial Government.

Dorfling, C. (1999), 'Gauteng Gets Ready to Mine IT Highway', *Martin Creamer's Engineering News*, (Johannesburg), April 2-8, pp. 1-2.

DTI (Department of Trade and Industry), (1998), *Special Economic Zones Project: a Spatial Development Initiative in Gauteng*, Unpublished Report, Department of Trade and Industry, Pretoria.

Hodge, J. (1998), 'The Midrand Area: An Emerging High-technology Cluster', *Development Southern Africa*, vol. 15, pp. 851-973.

Hodge, J. and Driver, A. (1999), *Understanding the High Tech Sector in the Cape Metropolitan Region: A Contribution to the Development of a Regional Strategy for High Tech Industry*, Unpublished report Development Policy Research Unit, University of Cape Town.

James, T. (ed) (1999), *SAIT'S Baseline Studies: A Survey of the IT Industry and Related Jobs and Skills in South Africa*, Unpublished report prepared for the International Development Research Centre. (Available at www.saitis.co.za).

Mabogunje, A. (2000), 'Global Urban Futures: An African Perspective', *Urban Forum*, vol. 11, 165-183.

Rogerson, C.M. (1996), 'Dispersion Within Concentration: The Changing Location of Corporate Headquarter Offices in South Africa', *Development Southern Africa*, vol. 13, pp. 567-79.

Rogerson, C.M. (1997), 'Locational Influences on High-technology Clustering: The Role of Infrastructure', *Trade and Industry Monitor*, vol. 4 (December), pp. 18-21.

Rogerson, C.M. (1998a) 'High-technology Clusters and Infrastructure Development: International and South African Experiences', *Development Southern Africa*, vol. 15, pp. 875-905.

Rogerson, C.M. (1998b), 'Restructuring the Post-apartheid Space Economy', *Regional Studies*, vol. 32, pp.187-197.

Rogerson, C.M. (2000a), 'Manufacturing Change in Gauteng 1989-99: Re-examining the State of South Africa's Economic Heartland', *Urban Forum*, vol. 11, pp. 311-340.

Rogerson, C.M. (2000b), 'Local Economic Development in an Era of Globalisation: The Case of South African Cities', *Tijdschrift voor Economische en Sociale Geografie*, vol. 91, pp. 397-411.

Rogerson, C.M. (2001), 'Knowledge-based or Smart Regions in South Africa', *South African Geographical Journal*, vol. 83, in press.

Sellars, C. (1998), 'Restructuring in the South African Pharmaceuticals Industry: A Tentative Review', paper presented at the Trade and Industry Policy Secretariat Annual Forum, Glenburn Lodge, Muldersdrift, 20-22 September.

WESGRO, (1998), The Western Cape Economy on the Way Towards Global Competitiveness with Social Stability: Review of Western Cape's Economic Performance, prepared for the Provincial Legislature, WESGRO, Cape Town.

World Bank, (1999), *World Development Report 1998/99: Knowledge for Development*, Oxford University Press, Washington DC.

18 The Sustainable Renovation of the Industrial Complex in Inner Tokyo: The Case of the Japanese Machinery Industry

Atsuhiko Takeuchi and Hideo Mori

Introduction

The sustainable development of industry should be investigated on the assumption of a symbiotic relationship between industry and the preservation of the natural environment. Industrial development has strongly influenced natural and socio-cultural conditions throughout the globe. However, geographic variations in the formation of regional industrial systems have varying implications for sustainable paths in different areas. This chapter addresses the Japanese case, especially with respect to the urban industrial dynamics occurring within the heart of Tokyo.

In Japan, after World War II, the industrial sector achieved a high rate of growth under a nationwide production system anchored around various metropolitan core regions, most notably Tokyo. In the 1970s, environmental problems were widespread in Japan. Industry was identified as the major culprit of pollution and for this and other reasons, such as high land costs and the need to provide jobs in outlying regions, it became the target of expulsion from urban areas. However, despite the policies of industry expulsion, certain location imperatives underlying nationwide industrial systems did not change. Indeed, a number of studies by economic geographers argued that small and medium sized enterprises (SMEs) that were embedded in metropolitan cores and vital to the health of the national economy, were threatened by the policies of dispersal (Sato, 1981; Seki, 1993; Watanabe, 1998). Since the 1980s, advice offered by these studies has gradually been reflected in policy.

In the 1990s, the steep appreciation of the yen, the extended recession, and the rapid internationalization of Japanese industry through direct foreign investment (DFI), led to the closure of domestic plants and threats to the survival of industry, especially in high cost areas such as Inner Tokyo. In Inner Tokyo, in particular, industry has responded strenuously to these changes and continues to survive around a core of hardware-based technologies that represent a new type of knowledge industry complex. Furthermore, the local government has strongly promoted new policies in support of this industrial complex in order to sustain jobs in ways that also preserve the environment.

The objective of this chapter is to reveal and clarify the dynamics of the sustainable renovation of the SME-dominated industrial complex of Southern Tokyo, specifically Ota-ku (city), that is the technological core of Japanese machinery industry. In this study, sustainability refers to the maintenance of job and community viability, as well as to environmentally friendly or non-polluting activities. The study is based on detailed interviews of 120 plants that were conducted by the authors as well as a land use analysis of Ota-ku (city) in 1998 and 1999. There are three main parts to the chapter. The first part of the chapter outlines the changes in the industrial complex of Southern Tokyo. The second part emphasizes the role of 'second generation' entrepreneurs in renewing the vitality of SMEs in this locality. The last part of the chapter notes how government policy has sought to meet environmental goals. With respect to the study region, it might be noted that Southern Tokyo is the largest of the four zones comprising Inner Tokyo, and that Ota-ku is the main part of Southern Tokyo.

Changes in the Industrial Complex of Southern Tokyo

The Tokyo metropolitan area has long been the largest industrial region in Japan and the biggest core of the national industrial system (Takeuchi, 1994). Traditionally, a heavy concentration of plants has existed in the Ku or inner city area. Within this inner area, the southern part (Southern Tokyo) features a concentration of machinery industry and metal processing activities. The centre of Southern Tokyo is Ota-ku (city) near to Haneda airport. Although there has been a reduction in the number of plants in recent years, Ota-ku still has the largest number of plants in Inner Tokyo.

The Concentration of Industry and The Utilization of Land

Although the number of plants has declined in Ota-ku, their spatial distribution pattern has not changed (Figure 18.1). Moreover, mixed in with this largest concentration of small-scale plants in Inner Tokyo, is also a large concentration of people. Consequently, land use within Ota-ku is characterized by a mixture of manufacturing plants, housing (for offices as well as blue collar workers) and a variety of service activities in support of the local population (Takeuchi, 1974). In the 1970s, the mixture of plants and houses was looked upon unfavourably as a 'confused' pattern. As a result the government sought to evict manufacturing from the Inner area.

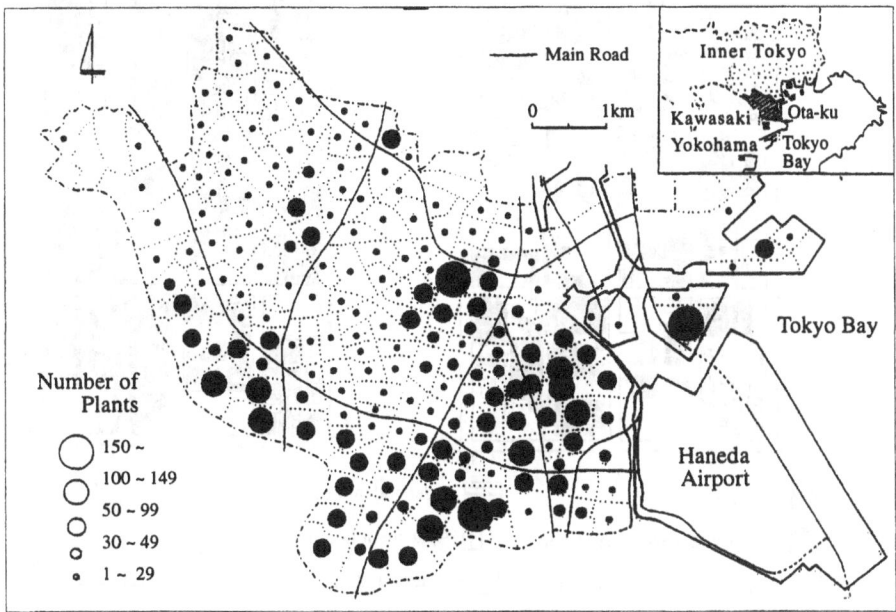

**Figure 18.1 Concentration of industry in Ota-ku:
Distribution of plants, 1993**

In the long run, however, this persistent policy of encouraging plants to withdraw from the area has not been reflected in fundamental changes in the overall pattern of land use. Indeed, a comparison of the land use of a typical block with a high density of the plants in Ota-ku between the late 1960s with 1999, reveals that the percentage of land utilized by manufacturing plants of the total area has actually increased and now exceeds 50 percent (Figure 18.2). This trend might be considered surprising in light of the government's dispersal policy. In addition, during the period of high economic growth in the 1980s, the plants could be sold for high prices and it was feasible for those managers who had acquired a high level of technical know-how to become employees of multinational corporations (MNCs) and receive high salaries.

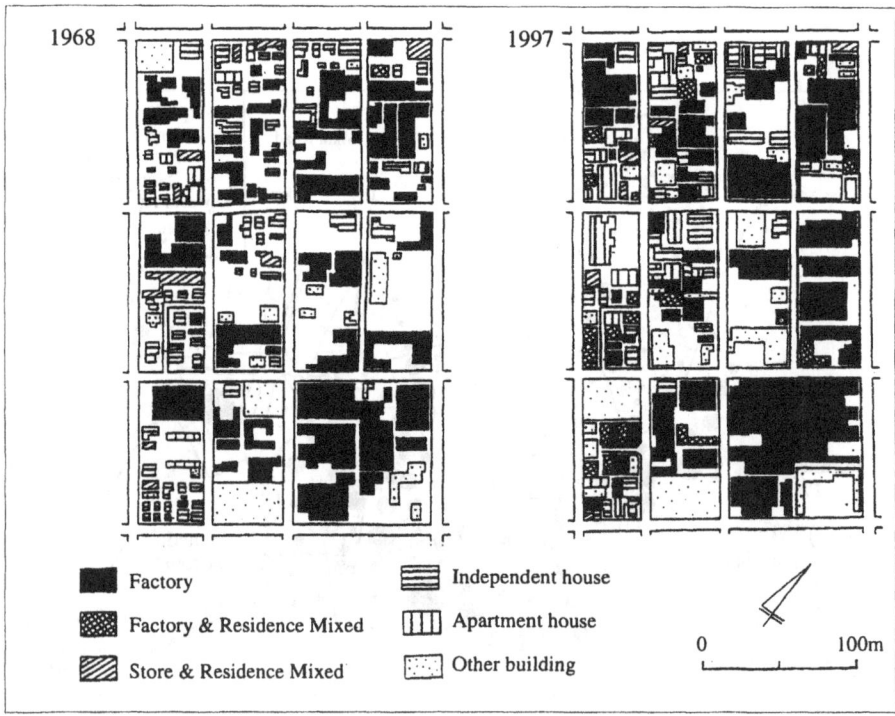

Figure 18.2   Concentration of industry in Ota-ku: Change in land use in Shimomaruko district, 1968-1997

Given such circumstances, an interesting question is why the land use system remains dominated by manufacturing activities. A related question is why did many of the younger generation entrepreneurs participate as managers or engineers in the SMEs of this area. The answers to these questions first need to be briefly placed within the context of the manufacturing structure of Ota-ku.

Composition of Manufacturing Industry

According to the criteria of the standardized industrial classification, an overwhelming majority of manufacturing production in Ota-ku features the machinery and metal processing industries. However, there are no longer many large-scale plants of finished products in Ota-ku and the standardized industrial classification scheme hides the heterogeneity of work organized by SMEs in the area. A detailed examination of each plant located in Ota-ku provides an appreciation of how highly diversified the plants of this area actually are.

In practical terms, a majority of the plants consist of various processors (Table 18.1). Indeed, there is such a wide range of capability that it is hard to think of a processing activity that could not be done in the area. These processors form a common root underlying the machinery industry and are labeled as 'basic industries' (Takeuchi, 1972). Within the industrial mix, there is an extremely wide range of special machines and of parts manufacturers. In terms of the size distribution of plants, the majority of the plants in Ota-ku have less than 30 employees and therefore must be considered 'small', with functional capabilities that are different from medium scale firms often considered leaders within manufacturing systems. Indeed, over half of the small firms are micro plants comprising less than four employees each (Table 18.2). The number of plants with 300 or more employees declined from 33 in 1975 to just 13 in 1995. Consequently, it can be said that the SMEs that sustain the industry of Ota-ku are those in the strata of extremely small-scale processors, and the machine manufacturers based on them. Moreover, the industrial complex of Southern Tokyo that centres on Ota-ku is the technological core of the nationwide system of the machinery industries and for core firms that are now located elsewhere. Therefore, it is important in the analysis of the production systems, not to confound the idea of 'core firms' with that of 'core area'.

Table 18.1 Composition of manufacturing in Ota-ku

Composition of firms, 1998

Casting, forming	99
Pressing	247
Cutting	1,028
Grinding	77
Polishing	50
Canning	235
Surface finishing	180
Assembling	382
Heat treatment	38
Machine elements	11
Prime movers, hydraulic machines	20
Machine tools	61
Molds	126
Automatic machinery	35
Precision machinery, instruments	78
Other machine instruments	245
Electric machinery	486
Other	919
Total	4,317

Source: Data base of industry of Ota-ku.

Formation of the concentration of the machinery industry in Ota-ku started in the latter half of the 1950s. Many large-scale plants of electrical equipment, precision instrument, machine tool and automobile companies were located in Southern Tokyo together with their many subcontractors. In the 1960s, the industry gradually moved into the southwestern outskirts of Tokyo, spatially extending the industrial area. This movement, however, did not cause a decline or weakening of industry in Ota-ku and Southern Tokyo.

Table 18.2 Composition of manufacturing in Ota-ku

Size-structure of the plants (%)

Number of workers	1960	1995
1 ~ 3	43.8	48.7
4 ~ 9	36.5	32.5
10 ~ 19	10.0	10.5
20 ~ 29	4.7	4.4
30 ~ 49	2.4	1.8
50 ~ 99	1.6	1.1
100 ~	1.0	1.0
Total	100.0	100.0

Source: Manufacuring in Ota-ku.

Southern Tokyo's Industrial Complex: Formation and Transformation

From the latter half of the 1960s, national and local governments looked upon manufacturing activities as the main source of pollution and forcibly sought to move manufacturing plants out of Inner Tokyo. This policy was also supported by the desire to promote regional development by the dispersion or decentralization of industrial activity on a nationwide basis. As a result of the migration of many plants, the relative importance of Inner Tokyo as a manufacturing centre declined in Tokyo as a whole. In fact, most economists and urban planners forecasted that manufacturing would vanish from Inner Tokyo. Contrary to these expectations, however, a majority of the SMEs remained in Southern Tokyo. The surviving manufacturers are mainly small in scale and responded to market dynamics by their own efforts. In particular, they raised their technological levels to become highly original manufacturers or processors in support of the increasingly sophisticated demands placed on them by machine manufacturers.

Within the area, many plants strengthened their cooperative relationships to collectively formulate technological complexes (Takeuchi, 1983). The large-scale machinery plants had increasingly located in the surrounding areas of the metropolitan region forming the leading force of industrial systems of the Tokyo metropolitan industrial region. Simultaneously, the industrial complex of Southern Tokyo shouldered an important role as the common technological base of the machinery industry of the entire metropolitan industrial region (Takeuchi, 1980). This trend comprised the first step towards a sustainable path of industrial development in Inner Tokyo. Furthermore, during this time, there was growing participation of micro firms.

Transformation Towards a New Hardware Centre

At the beginning of the 1980s, the SMEs of Inner Tokyo faced severe demands from their customers to reduce the costs and to improve levels of precision in manufacturing processes. Firms respectively acquired mechanical engineering (ME) technology through their own initiatives. Thus, they transformed the existing production system that was based on craftsmanship skills into a new production system based on scientific technology that itself was further refined by 'learning by doing'. In this way, the renewal of manufacturing in Ota-ku occurred and a 'new hardware centre' serving the whole Tokyo region was created (Takeuchi, 1994). That is, a learning system for the innovation of firms in Inner Tokyo was developed, shaped by individual managers and engineers to form a 'working community', imbued with its own unique cultural features. This trend is the second feature in the creation of a sustainable path of industrial development in Southern Tokyo. However, despite the attainment of high technological levels through individual initiative, the SMEs of Inner Tokyo were still mostly positioned at the end of vertical and sub-contractual relationship organized by large enterprises.

Entering the 1990s, the Japanese economy, as a result of a sudden increase in the value of the Yen and the intensity of international competition, was plunged into a recession that became a long-term affair. Under such circumstances, the stagnation of the inner areas of major cities was far more severe than in the outskirts of these cities or in provincial areas. Nevertheless, even in these difficult circumstances, manufacturers in Southern Tokyo sought to renew themselves and their complex through their own efforts, as was the case in second half of 1980s. As a result of these efforts, the nature of the industrial complex of the area is today different from its nature in the 1980s.

New Generation Entrepreneurs and Manufacturing Renewal

In the 1990s, the outstanding institutional feature in the renewal of Inner Tokyo is the participation of a new generation of entrepreneurs and engineers who have replaced the first or founding generation. Most of the first or founding generation started their business in the 1950s and by the 1980s were beginning to retire. At least half of the members of this first generation of entrepreneurs were graduates of elementary schools or junior high schools and they used 'craft skills' to boost the technological expertise of their firms, while accommodating to the changing demands of customers. In comparison, the new generation of entrepreneurs is well educated with considerable investment in formal academic backgrounds. Most of them are in their 30s or early 40s and 70 percent of them are graduates of university or technical colleges. Many had studied ME and other high technologies or new management systems. The two generations are opposed in philosophy and ways of thinking regarding engineering and management. However, through discussions and practical development of new products, the highly developed craft skills of the first generation have been effectively combined with the ME technology of the new generation. In effect, the two generation-based sources of knowledge have been mutually combined, 'crystallizing' in the form of new products.

An example of crystallization is provided by the manufacturer of a bill counting machine in which the vacuum pump is an indispensable component. The heart of the vacuum pump is a drilled-hole that is measured in fractions of a micron. The know-how for the original technology was created by a craftsperson who created his own SME as a pump manufacturing firm. However, the development of computerized technology that makes continuous production feasible depended on the ME expertise of the second generation entrepreneur, the founder's son who is in his mid-thirties. In 1999, this SME has 22 employees and enjoyed a 90 percent share of the world market.

One of the conditions that facilitates the crystallization between crafts skills and modern, computerized ME expertise possible is the close relationships between father and son, the feeling of pride in the behaviour of the first or founding generation, and the desire to make a complementary contribution by the son beyond the innovations of the father. Indeed, such values are part of the culture of the area. The new generation of entrepreneurs has brought with them a new set of human relationships, new sets of qualifications, nurtured by open minds, and has become the motivating power for the formation of new types of inter-firms networks and inter-regional networks as well. Basically, those plants that have failed in Southern Tokyo have occurred because of their inability to cope with technological change. For the surviving plants that have successfully addressed technological challenges, the role of the 'second generation' entrepreneur has been vital in grasping the dynamic shifts and changing trends throughout the industrial regions of Japan.

The Renewal of Manufacturing Plants: A Classification

Although there are many different kinds of plants in Inner Tokyo, it is possible to divide them into the following three categories (Table 18.3). The first category comprises the majority of firms and involves firms that have acquired high precision processing technology based on fractions of a micron. The second category refers to the manufacturers of single models of finished products, parts and moulds. To illustrate this category, mention can be made of a SME that has only seven employees. This firm developed a new high tech precision mold and won an award from the University of Tokyo. There are many similar examples. The third classification covers requests from MNCs for the development, trial manufacture and production of prototypes, and the prescribing of production methods to be employed. The technological sophistication of such firms is high. It is predicted that this type of firm will increase in numbers in the future. Such increases will occur not in MNCs nor NC machines but where standard equipment is mainly employed in manufacturing.

In this manner, the overall concentration of the plants in Ota-ku features the acquisition of high levels of technological expertise combined with craftsmanship skills. This area is imbued with a high level of processing skills and may be labeled as a 'specialized technology district' (Storper, 1992). Of the enterprises that were surveyed, 75 percent are world leaders and in some cases they have a technology that is identified as the only one of its kind. In the case of Southern Tokyo, the 'key firms' that lead innovation are not MNCs but SMEs. Furthermore, their main strength lies, not in the 'chuken firms' or medium to large-sized (but not giant) firms that have recently been the focus of attention (Nakamura, 1990; Hayter, 1997), but the very small firms.

Table 18.3 Categories of renewed enterprises

Category	Example
(A) Ultra high level of processing	• Unique processing • Grinding of non-spherical lens
(B) Particular products	• Specialized mechanics or parts • Precision molds
(C) Development and trial manufacturing	• Trial or prototype production for large companies • Proposal of know how for other firms

Strengthening of the Localized Industrial Complexes

In the case of the manufacturing complex of Southern Tokyo, close inter-firm relations among a wide variety of numerous processors is evident, and reinforced by land use dynamics. Thus, manufacturing plants in Ota-ku are almost all smaller than 50 square meters in area, thereby limiting the extent of in-house processing. It is also risky for these firms to enter new fields. Such diversification automatically necessitates mutual reliance on procurement processors with high levels of technology and skills that are well known to them and located in the same area. At present, there are many plants in Southern Tokyo that are moving into the outskirts of Tokyo. However, they invariably retain planning, development, and trial or pilot production operations in Southern Tokyo. This trend is further conditioned by the fact that they are able to access numerous associate processing plants with high technological levels that are able to share their work. What is important in Southern Tokyo is not the scale of the enterprises but the economies of scope they collectively share.

Thus, in the 1990s, through the crystallization of a merging of two generations, the technological capabilities of Southern Tokyo have been boosted to globally leading edge levels and shares of top quality production. Furthermore, SMEs in the area have developed strong intra-regional networks, exchanging parts, finished products and information. In this highly innovative, technological complex, the 'key plants' are capable of both development and processing. In addition, most of these key plants are small in scale and are literally micro-firms.

In recent years, SMEs have increased their autonomy as individual enterprises, helping to create a more self-sustaining industrial complex. In the machinery industry, the SMEs are transforming themselves from vertically linked suppliers to MNCs to horizontally linked suppliers within the industrial complex of southern Tokyo (Figure 18.3). Such a trend is a prerequisite for

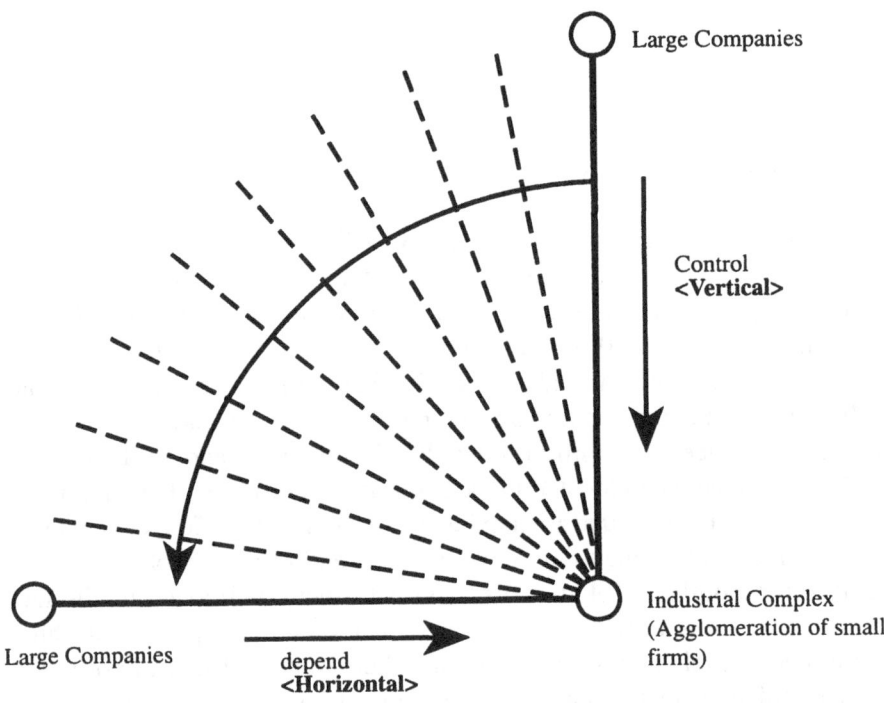

Figure 18.3 Transition from vertical to horizontal inter-firm relationships

the sustainable renovation of industrial complexes in Inner Tokyo. This is also the way for skill-based industrial complexes to develop new types of knowledge, accompanied with relevant hardware. It is impossible to relocate such an autonomous and high level of localized technological complex to provincial areas or overseas. It is extremely difficult to formulate such innovative systems and its milieu in a short time. Under such circumstances, one of the most important problems faced by Japanese industrial policy makers is how to sustain and develop the technological complexes that are composed of SMEs in Southern Tokyo centering Ota-ku. Indeed, this industrial complex is not only vital to Japanese industry but also to the machinery industry of Asia that has developed, at least in part, on the basis of strong technological ties with Japan.

Community, Government, and Environment in the Renovation of Ota-ku's Industrial Complex

The new generation of entrepreneurs in Ota-ku has changed the characteristics of the community (Figure 18.4). Originally, the community was created on the basis of competition and cooperation among craftsmen entrepreneurs, and community culture reflected their attitudes and behaviour. Subsequently, a second generation of entrepreneurs has added new skills, based on the application of ME, to the community and enhanced local and non-local networking among firms. The crystallization of the two generations has also helped shape the evolution of community character. Until the first half of the 1980s, there was a bigger concentration of plants than today. At that time, the first generation was strongly bound by the common experience of organizing their processing as vertically linked suppliers to MNCs. Although they formed a 'productionist culture' that assumed a 'lord of the castle mentality' (Wittaker, 1996), they were immature as an 'industrial community' (Takeuchi, 1973). During the latter half of the 1990s, the number of plants declined in Southern Tokyo and new houses were built on many of the former manufacturing sites. However, an important feature of organization of the area, namely the integration of plants in the homes of management and workers, has not changed.

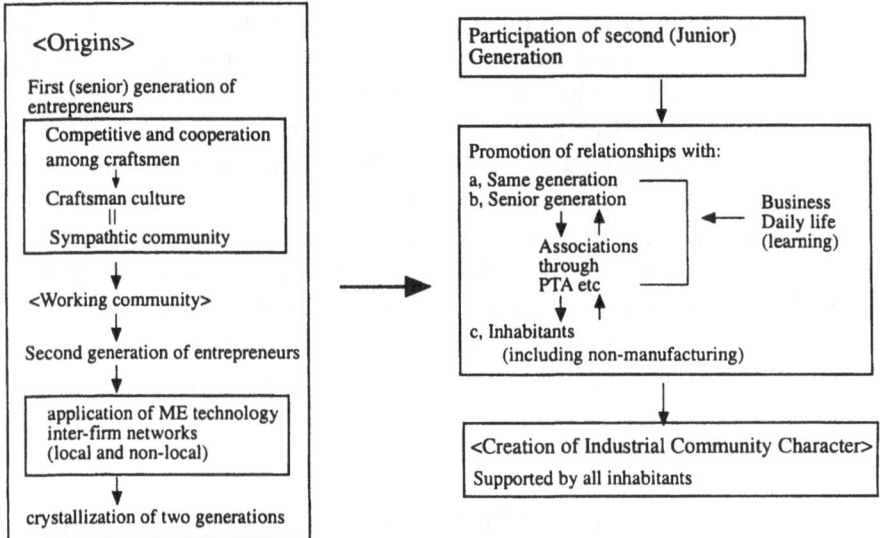

**Figure 18.4 Participation of new generation of entrepreneurs
and evolution of community spirit**

The younger generation that became community leaders in the 1990s was brought up in the area and have been friends since childhood. Therefore, they are not only working together in this area but are continuing their association through daily life patterns. As a natural consequence, they have become fully acquainted with each other's businesses, extending the mutual understandings associated with the first generation. Much local networking, such as mothers engaged in parent teacher association activities, occurs beyond business. Indeed, compared to the times when there were more plants, the density of daily association has become closer and tighter. In Inner Tokyo, there is very little distinction or awareness of the difference in social class among people. The succession of the new generation and the retirement of the previous generation has not brought about social segregation of the kind apparent in the western society (Scott, 1988).

Consequently, many different social groups are fused and joined together, forming a supportive community. The addition of the new generation has become one of the more important aspects in favor of the sustainable renovation of not only the industrial space but also the social space. As a result, the "industrial community" has become increasingly consolidated.

The consolidation of the community cemented by close personal relationships, serves to fortify the technological complex and is an important condition for the embeddedness and sustainable development of industry in the area.

The Role of Government

Until the 1980s, the manufacturing industries in Inner Tokyo were considered to be injurious and harmful, resulting in the introduction of a policy favouring forcible extrusion of plants. However despite such severe pressures, the plants of Inner Tokyo have survived; indeed they are robust. As the result of detailed fieldwork conducted by economic geographers after the 1980s, government became more aware of the embedded nature of SMEs in the area and their importance to the national economy. As a result, government policies were drastically revised.

Particularly in the early 1990s, MITI for the first time correctly became aware of the significance of the concentration of plants in Ota-ku. With the enactment of 'The Law of Temporary Measures Concerning the Activation of Specific Industrial Accumulation', that was based on the idea of geographers, the unjustified severe restrictions upon the location of plants in Ota-ku were alleviated (Takeuchi, 1999). In the 'Revitalization project of the Tokyo Bay area' of the central government, emphasis was also placed on the role fulfilled by the accumulation of the plants in Ota-ku. In Ota-ku, manufacturing is deemed to support, not only the economy, but also the cultural and social aspects of citizens' lives.

The fundamental pillar of the new policy for Ota-ku is to establish a 'real industrial city' that promotes the further expansion of industry through comprehensive action throughout the private sector (Wittaker, 1996). In practical terms, a database is being prepared on the products and technologies of all enterprises in the area, an information exchange system that is funded by investment from both the public and private sectors. Action is also being taken to support a study that advised the development of learning systems for both management and technology, and a system for corporate interchanges. At present, there are five active cooperative groups, comprising firms in dissimilar industries, catalyzed by the government. These policies will help further embed technology-oriented industry in Ota-ku. The government is providing assistance to firms that participate in trade fairs held overseas while also sponsoring national and international trade fairs in Ota-ku, thereby helping to develop ties with MNCs. In terms of physical planning, the construction of industrial parks, industrial apartments and an industry promotion center (PIO) are in progress.

Co-existing with the Environment

Upon first glance, Government policies appear only to give priority to the activation of industrial activities. In practice, they have also contributed to the improvement of the physical environment. Thus, the construction of industrial parks and apartment complexes on reclaimed land are based on a fusion of manufacturing and residential functions. Industrial buildings for high tech industry have also been included in land use plans. For instance, in industrial apartment complexes, plants are often located on the ground floor while the second floor and above are devoted to the residence of managers or workers. Such combinations of manufacturing and residential activities can reduce pollution, especially as regards transportation, and possibly to improve external appearances.

Industrial parks have provided a solution to serious pollution problems in urban areas and have contributed to the creation of pleasant, landscaped operating environments for the manufacturing sector. Therefore, the promotion of manufacturing does not run counter to the interest of the maintenance of a pleasant environment and the realizing of a harmonious land use. Currently in Southern Tokyo, the government recognizes the importance of the area's industry not only to the regional economy but also to the nation as a whole. Ideally, policy should seek to strengthen the industrial community based on a symbiotic relationship between industry, living conditions and environment. Such a symbiosis should be the basic direction in which sustainable renovation in Southern Tokyo is to proceed.

Conclusion

The development of industry and its role in the economy as a whole differs among countries. Especially with respect to SMEs, differences in entrepreneurial performance, related attitudes and government policy can differ greatly among countries and regions. Accordingly, it is impossible to properly assess 'global' developments solely on the basis of western standards or experience. Furthermore, the same observation applies to domestic conditions in Japan because rural and urban areas differ greatly. This is particularly important when considering the significance of SMEs and their complexes in Inner Tokyo and their sustainable development. It is impossible to characterize the industrial complexes of Southern Tokyo according to western concepts of industrial district.

Particularly in the case of Inner Tokyo, after many years of policies of forcible expulsion, there is still no protection provided for industry by the central government. Yet, industry in the area comprises the technological core of the national system of production. The industrial complexes of Southern Tokyo are independent and are endowed with an exceedingly high ability to foster development. As a core area of the machinery production system, it has evolved as an independent technological complex that features powerful horizontal and equivalent relationships. New types of knowledge industry complexes have emerged. The sustained development of these industrial complexes of Southern Tokyo is an extremely important issue, given the role of the area in the national system of machinery industry. The renovation of industrial complex in Southern Tokyo is realized by the promotion of a symbiotic relationship among the regional economy, the community, land use and the natural environment. This relationship is essential for an appropriate regional and industrial policy for Inner Tokyo.

References

Hayter, R. (1997), *Industrial Location: The Factory, The Firm and the Production System*, John Wiley and Sons, Chichester.

Nakamura, H. (1990), *The New Mid-sized Firm Theory*, Toyokeizai, Tokyo (in Japanese).

Ota-ku (1995), *Report of Research on Industrial Structure of Ota-ku*, Ota-ku (in Japanese).

Sato, Y. (1981), *Small Firms in the Metropolitan Area*, Shinhyoron,Tokyo (in Japanese).

Scott, A.J. (1988), *Metropolis – From the Division of Labor to Urban Form*, The University of California Press, Los Angeles.

Seki, M. (1981), *Beyond the Full-set Industrial Structure*, LTCB, Tokyo.

Storper, M. (1992), 'The Limits of Globalization: Technology Districts and International Trade', *Economic Geography*, vol. 68, pp. 60-93.

Takeuchi, A. (1972), 'The Bottom Structure of the Machinery Industry in Japan', *Jinbunchiri* (Human Geography), vol. 24, 4, pp. 36-54 (in Japanese).

Takeuchi, A. (1974), 'The Integrated Industrial Residential Areas of Tokyo', *Geographical Review of Japan*, vol. 19, 2, pp. 40-57 (in Japanese).

Takeuchi, A. (1980), 'Industrial System of Tokyo Metropolitan Area', *Report of Research, Nippon Institute of Technology*, vol. 1, 2, pp. 1-40.

Takeuchi, A. (1983), *Industrial Complex and Industrial Community*, Taimeido, Tokyo (in Japanese).

Takeuchi, A. (1994), 'Location Dynamics of Industry in the Tokyo Metropolitan Region', *Report of Research, Nippon Institute of Technology*, vol. 23, 3/4, pp. 195-220.

Takeuchi, A. (1999), 'Revitalization of Tokyo Bay area and Formation of the New Industrial Complex', *Report of Research, Nippon Institute of Tech nology*, vol. 29, 3, pp. 1-9.

Watanabe, S. (1998), *Social Division of Manufacturing Industry of Japan*, Yuuhikaku, Tokyo (in Japanese).

Wittaker, D.H. (1996), *Small Firms in the Japanese Economy*, University of Cambridge Press, Cambridge.

19 Globalization and the Reorganization of a Metropolitan Knowledge System: The Case of Research and Development in Frankfurt/Rhein-Main, Germany

Eike W. Schamp

Introduction

The increasing economic integration of Europe, if not the world, that is 'globalization', clearly results in rising competition between firms and regions. Unlike the 'old' interregional division of labour, the new rivalry of firms and regions focuses on innovations and access to knowledge resources to gain competitive advantage that is at least temporarily monopolistic. Cities play a crucial role in this process, not only because they spatially concentrate people and firms, such as clients, suppliers, and competitors, but also because they are focal points in knowledge production and diffusion. The theoretical approaches to the organization of knowledge, the growth of an urban economy and the changing role of cities in globalization are still unclear. In the debate about cities as centres of the production, or at least of the diffusion of innovations, urban knowledge spillovers are considered the principal motor for economic growth. It is said that knowledge spillovers are a particular form of external (urban) advantages to firms that improve their innovativeness and competitiveness.

This chapter explores the relationships between innovation, urban growth and the organization of knowledge with specific reference to Frankfurt/

Rhein-Main, Germany. Specifically, this chapter discusses the hypothesis that the city is an indispensable or even the driving part of a regional innovation system, at least in the context of Europe's societies and economies.

Macro-systemic Concepts and Actor-behaviour in Metropolitan Knowledge Systems

In current macro-economic debates on urban economics, however, there are different hypotheses on how knowledge spill-overs work. On the one hand, spill-overs are assumed to occur within industrial sectors that are highly interrelated, whether in the sense of the Marshallian industrial districts (MAR effects) or in the sense of Porterian clusters. On the other hand, major spill-over effects are expected to occur between very different sectors in the city (Jacobs' effects). While these hypotheses have been tested in different sectors, the effectiveness for total urban growth of a city and its region still seems to be rather unknown. However, prosperous city regions show both: a specialization in growing sectors and a multiplicity of clusters at the same time.

According to Crevoisier (1999), this macro-economic debate fails to realize the role of social relationships and their institutional settings in knowledge spill-overs. Therefore, Crevoisier suggests a more micro-analytical approach to understanding the city as a meeting place of actors who, under certain institutional conditions, are able to launch the process of spill-overs, that is, the process of production and diffusion of innovations. In what Crevoisier (1999, p. 68) calls a "meta-place of interaction" different actors interact, local and non-local, public and private, economic and non-economic. Important in this interaction is the continuity of collective learning protected by particular institutional forms.

Recent evolutionary approaches to innovation highlight the importance of national innovation systems (Edqvist, 1997). An innovation system is "a set of organizations, institutions, and linkages for the generation, diffusion, and application of scientific and technological knowledge" (Galli and Teubal, 1997, p. 345). In the national innovation system, public and private research infrastructures, as well as the educational system, are of tremendous importance because they generate and maintain the human capital that is essential for the production of knowledge and the diffusion of innovations. The process of knowledge creation is currently seen as an interactive process between firms, academic establishments and rules set by the political and social system. Within this nexus, Crevoisier's concept can be introduced. It is consistent with current reflections about the communicative character of knowledge creation among the three different sub-systems involved: academia, that is, the world of science in universities and public research laboratories ('basic research'); the business sector; and the government sector, all of which function at local and non-local levels.

While communication and interaction are paramount for the creation of new ideas within the sub-systems, the recently proposed Triple Helix model refers to the "network overlay of communications and expectations" between the three sub-systems involved as basis of innovativeness (Etzkowitz and Leydesdorff, 2000, p. 109). This model assigns universities more focus in the innovation process, than is often the case, and emphasizes the complex web of communications between the sub-systems, the unpredictability of technological trajectories and the difficulties for political actors to 'anchor' knowledge creation to cities and regions.

Many organizations acting within the Triple Helix are spatially concentrated. If it may be assumed that their likelihood to interact decreases with distance from each other, particularly in the case of dealing with non-codified, 'tacit' knowledge, it seems plausible to expect that a system of regional (urban) innovation systems emerges (see, for example, Sternberg, 2000 concerning Germany, and Cooke, 1998). The emerging current debate on regional innovation systems, however, has hardly tackled the question of the role of the city as a territorial and organizational part of regions (Figure 19.1).

There are different arguments to support the central thesis of this chapter, namely that the city is an indispensable or even the driving part of a regional innovation system, at least in the context of Europe's societies and economies. First, it should be recognized that there is a large and heterogeneous number of economic and non-economic actors who cooperate in the production of knowledge and in learning. Heterogeneity is a mandatory base for the production of knowledge. Second, specific possibilities for enabling relationships, communication and social interaction among heterogeneous actors, that are necessary for innovation, occur mainly within value chains. As a result, it is existent or emerging clusters that establish the trajectories of innovations, and this clustering is facilitated in cities. Third, cities are politico-territorial entities in which local political actors make

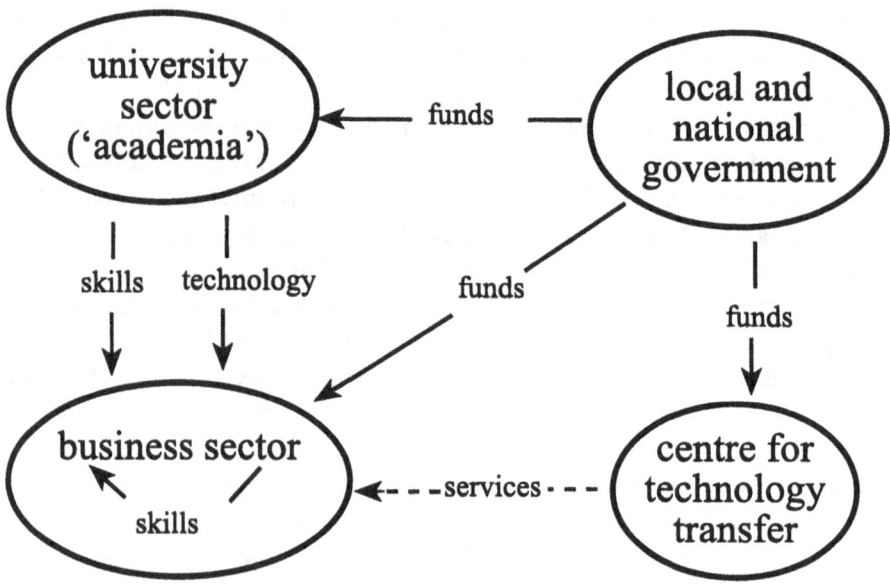

Figure 19.1 A Triple Helix model of innovation system

decisions on the location of public knowledge infrastructures. Typically, successful cities invest in a larger number and heterogeneity of public research and educational establishments that can be part of a regional knowledge system. If urban political actors contribute actively to the formation of regional knowledge systems, cities can host public intermediaries in the innovation process such as centres of technology transfer or technological consultancy.

In general, it is large cities that have the most appropriate characteristics for innovation. The city is more than a container of a large number of economic actors. It offers the institutional infrastructure necessary for the emergence of local knowledge and innovation systems. It seems, however, as if cities no longer can perform this task alone. Läpple (1997), for example, rightly talks about the "dissolution of the city into its region". With suburbanization, many important firms as well as public research laboratories and educational establishments have located outside the traditional boundaries of the city. These trends apply particularly for poly-central metropolitan regions, such as the Rhein-Main region. Thus, the regional knowledge system extends the boundaries of the central city, sometimes considerably, with two main consequences. First, the concept of a regional knowledge system has to take into consideration that there is an interplay between the metropolis and its region. Second, attempts for a political formation of regional knowledge systems can only be based on the development of an understanding of what the metropolitan region might be. The institutionalization of a metropolitan region therefore becomes a constitutive part of a regional knowledge and innovation system.

Metropolitan regions and regional knowledge systems thus form a potentially necessary but sometimes conflicting entity. Regional innovation systems require a metropolitan region as an advantageous location for its actors, and, at the same time, are presumably far less coherent than national innovation systems. In other words, it may be assumed that regional innovation systems as part of metropolitan regions are largely fragmented. Fragmentation develops along two lines. Thus, regional innovation systems are characterized, first, by a high degree of openness and of non-local determination. National regulation and industrial structures may have more importance on its trajectory than regional efforts. Second, a regional innovation system of a metropolitan region may include various clusters of economic activities that may differ according to the phase in their technological trajectory and their internal governance structures. In the context of the present chapter, a major hypothesis of the case study of the metropolitan region of Frankfurt/Rhein-Main is that in a long process of economic restructuring, Fordist structures and sectors of production are replaced only slowly by post-Fordist ones. Third, the regional knowledge system as a whole, including all relevant clusters is only taken into account by political actors of local administrative units which aim at safeguarding the economic power and prosperity of the region.

While the concept of a national innovation system almost unavoidably deals with macro structures and relationships, the Triple Helix approach can be used on the regional scale as a heuristic concept by which the communication and collaboration of actors in sustaining and changing knowledge systems becomes clearer. In a knowledge system, several dimensions matter simultaneously. These are: first, how firms improve both their internal resources for innovations by efforts in internal research and development (R&D) and their external resources by cooperation with other organizations; second, how the public education system, public research laboratories and intermediary organizations of technology transfer offer specific resources; and, finally, how political agents manage to establish an enabling institutional framework in general and specific assistance in particular.

In this chapter, I use Porter's (1990) concept of cluster around leading economic sectors and exemplify this concept with a study of public and private R&D that forms a particular Triple Helix system in a particular city region, namely Frankfurt/Rhein-Main. Empirically, the chapter describes the structure of the regional knowledge system as well as its changes, in recent years, and, then, discusses the consequences for the role the city region plays as an economic node in the world system.

This task is not easy. First, concepts such as globalization, knowledge system and their meaning for urban growth are rather fuzzy, and they are even more fuzzy when put into relation to each other. Second, the role of Frankfurt and its city region as a globalizing city within the context of an integrating Europe is still rather unclear. Third, political actors do not agree on a common concept of the city region of Frankfurt/Rhein-Main. While the chapter addresses these latter points rather marginally, it starts by outlining the peculiarities of the city region of Frankfurt/Rhein-Main. The main focus is the presentation of the current structures of the local knowledge system and its dynamics. I intend to show how the transformation from Fordist production of goods to post-Fordist production of services affects the role of this city region as a node in the European city-system.[1]

The Arena of Knowledge Creation in the Metropolitan Region Frankfurt/ Rhein-Main

The metropolitan region Frankfurt/Rhein-Main provides a special arena for the actors participating in the regional knowledge system. Given its slightly more than four million inhabitants, it is a rather small metropolitan region in the European context. The city of Frankfurt itself does not have more than 660,000 inhabitants. However, it has enormous economic power, comparable to regions like Ile de France.

Frankfurt's surrounding metropolitan region shows a poly-central, diversified urban pattern that includes some other large cities such as Wiesbaden or Darmstadt. Although Frankfurt (as a financial centre), Wiesbaden (as state capital and spa) and Darmstadt (as a 'science city' with its technical university) suggest a metropolitan region which is fully oriented towards services, this region used to be one of the major industrial regions in Germany.

With nearly half a million employees, the industrial sectors still provide approximately 27 percent of the total employment of the region. Important clusters include chemical and pharmaceutical industries, car manufacturing with a wide variety of suppliers, and, in former times, electrotechnical and engineering industries. These sectors grew rapidly during the golden age of Fordism. Despite the reduction of employment by approximately 25 percent in the last 25 years, manufacturing industries are the second largest employer in the metropolitan region. Business services in its proper sense prevail among other sectors, followed by financial services. These are activities that are spatially concentrated directly in the core of large cities. In the city of Frankfurt, for example, manufacturing activities provided only 15 percent of total employment in 1998. It seems that those economic activities that participate in the regional innovation system are increasingly located at the urban fringe (Figure 19.2).

Nonetheless, the region Frankfurt/Rhein-Main is one of the most important regions in Europe in terms of innovativeness. In 1994, a report from the EU Commission identified 11 leading innovation islands in Europe, all of them metropolitan regions, including Frankfurt/Rhein-Main. In fact, both input and output factors of R&D indicate a powerful regional innovation system. The regional pattern of R&D employees is spread unevenly in Germany. At the top of the list are the Southern German metropolitan regions of Munich and Stuttgart with 36,000 and 34,000 employees in industrial R&D in 1993, followed by Frankfurt/Rhein-Main at the third rank with 30,000 employees (Specht, 1999). In absolute figures of R&D expenditures, the most important sectors in Germany are the electrical engineering industries, car manufacturing and the chemical industry. However, the share of these sectors measured in expenditures and employees has dramatically decreased in recent years. Among the 'losers' are the electrical engineering and chemical industries, in particular. At the top of the winning sectors is car manufacturing (Wissenschaftsstatistik, 2000). The spatial concentration of these sectors in metropolitan regions is reflected by the spatial distribution of their R&D. This is the reason why the geographical pattern of R&D can be largely explained by both a sector effect and a size effect.

It comes with no surprise that registration of patents is also unevenly distributed in spatial terms. As in the case of R&D, Frankfurt/Rhein-Main takes up the third position in Germany, after Munich and Stuttgart (Greif and Schmiedl, 1998).[2] Frankfurt/Rhein-Main is the leading region in patents of

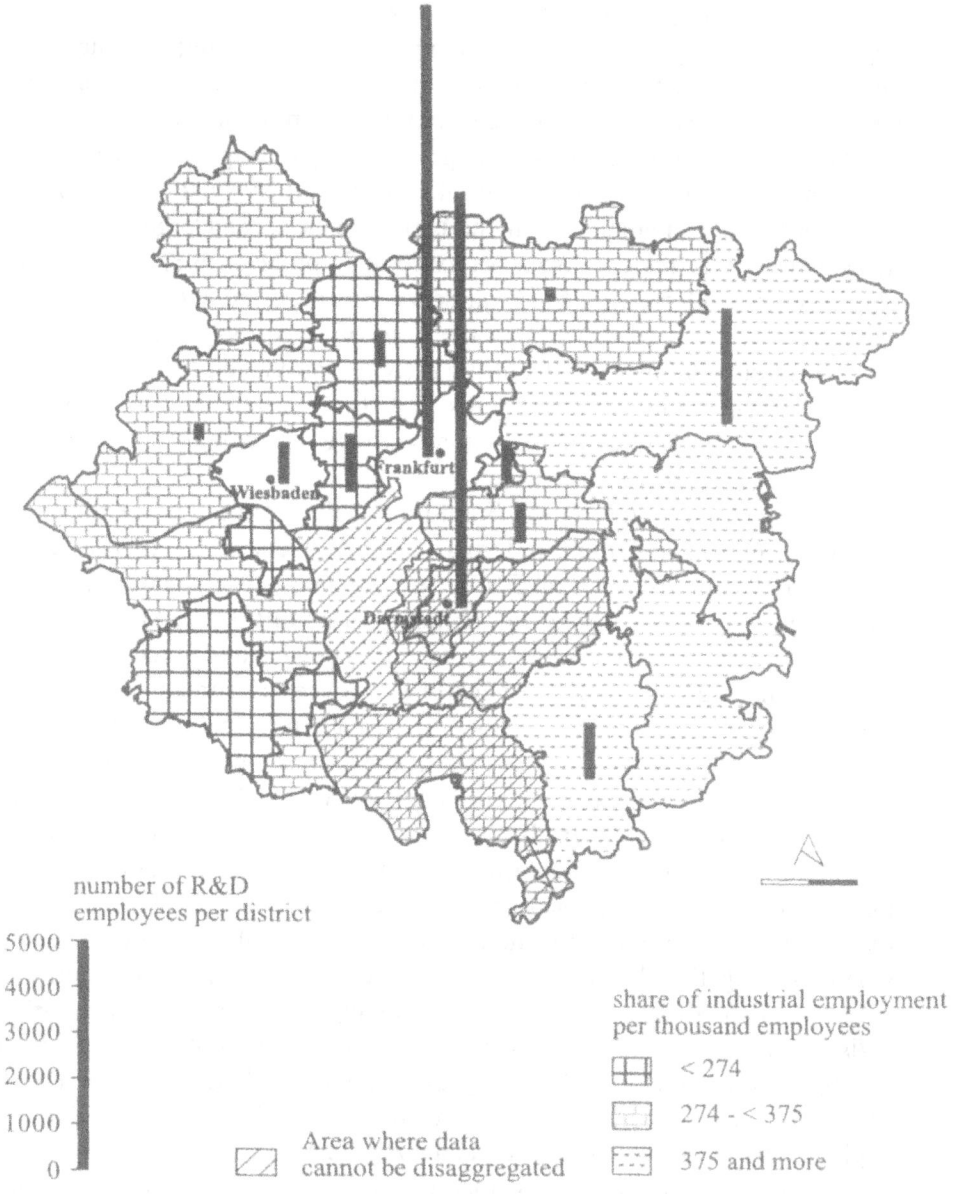

number of R&D
employees per district

5000
4000
3000
2000
1000
0

Area where data
cannot be disaggregated

share of industrial employment
per thousand employees

< 274

274 - < 375

375 and more

**Figure 19.2 Research and development employment in the
districts of metropolitan region,
Frankfurt/Rhein-Main**

inorganic chemistry and occupies a third rank in patents of gauging and testing. A description of a regional innovation system has to take into account the actors who are the driving forces for the patent registration. According to Greif and Schmiedl, there are three different types of these actors: first, enterprises, second, scientific establishments, and third, independent inventors. In the metropolitan region of Frankfurt, registration of patents is dominated by enterprises. There is a sectoral bias as there are less independent inventors in chemistry than in other technologies such as electronics. This information already intimates that the regional knowledge system of Frankfurt/ Rhein-Main may mainly comprise firms, not academia or public research laboratories. Hence, it is possible that the Triple Helix is incomplete.

An Incoherent Knowledge System in Manufacturing

The regional innovation system of Frankfurt/Rhein-Main is thus widely based around its traditionally dominating industrial clusters, first in chemical industries, second in car manufacturing, and third, in electrical engineering (Table 19.1). These sectors that emerged during Fordism dominate the R&D expenditures in Germany, as well. Today, they only possess a medium level of technology. It may be claimed that the Fordist organization of R&D in these sectors, namely around a linear model of innovation, was never capable of creating a coherent regional innovation system. In any case, the ongoing restructuring of these sectors has an enormous impact on the shape of the regional innovation system. By way of context, the following section describes the decrease of a Fordist sector in the region, namely the chemical industries, and the restructuring of another sector into a neo-Fordist organization, namely car manufacturing. The focus is on the economic actors pursuing different strategies according to their resources and their global markets. The way in which the regional industrial clusters change currently results also in a specific way of restructuring of the regional R&D system. Taken together, these trends will completely alter the character of the regional knowledge system in Frankfurt/Rhein-Main.

Table 19.1 Research and development (R&D) employment in manufacturing sectors in the region Frankfurt/Rhein-Main, 1993

Sector	Regional R&D Employment	R & D Intensity[1]	Total Employment in Germany	As % of Total Employment
Chemical Ind.	12,757	13.1	53,234	24
Electrical Engineering	5,960	8.3	80,021	7
Car Manufacturing	5,838	9.0	50,024	12
Mechanical Engineering	2,688	3.9	34,942	8
TOTAL	29,245	5.7	276,813	11

[1]Intensity is ratio of R&D employment to total employment in industry

Source: Specht 1999, p.146.

The major sector in the metropolitan region of Frankfurt/Rhein-Main is the chemical industry. Approximately 44 percent of all employees in regional industrial R&D were active in this sector in 1993 (Specht, 1999, p. 146). Furthermore, the chemical industry in the region has been more R&D intensive than the average for Germany (13.1 percent compared to 8.7 percent, see Specht, 1999). This sector is characterized by the dominance of some very large firms, with Hoechst AG the largest, and some medium sized firms which possess R&D laboratories of their own. According to a recent study, the sector's production is still organized in a Fordist way (Bathelt, 1997).

Chemical firms were among the first in Germany to internationalize their R&D by relocating to foreign countries. In this region, easily the largest chemical enterprise is Hoechst AG. Being a global player for so long, the company began early to de-territorialize its R&D system. While it started considerable research cooperation with American organizations in the 1980s, it shifted R&D expenditures to foreign countries by more than 50 percent during the 1990s. In explaining his reorganization strategies to the public, the chairman of the board of Hoechst, Dormann, stated in an interview in 1998, that he saw a lack of knowledge and knowledge resources in the region for the further growth of the firm. In fact, although there have been some research cooperation between firms in the chemical industry in their long history, there was no coherence of a sustainable regional innovation system. Regional universities and colleges were not integrated into the research process. The small number of public R&D laboratories did research on different areas of technology.

In recent years, in which the company had undergone radical restructuring, the number of R&D employees in the Frankfurt laboratories decreased enormously. The firm as a production unit has been dismantled. Some activities have been merged with European and American firms. The company itself changed into a holding company which has given up traditional activities in favour of the creation of a 'life sciences' company (Becker and Schumm, 2001). The headquarters of the new company, whose new name is Aventis, has been relocated to Strasbourg, France. The main production site in Frankfurt changed into an industrial park with different chemical industries. The central R&D laboratory has been closed down. With a staff that has been reduced to 1,200 persons, it is currently only a small part of the global R&D organization of Aventis. Total employment in the chemical industry has decreased tremendously at those locations where Hoechst Company dominated. Within six years, employment in Frankfurt was cut in half to 20,000 persons in 1998. The strength of the regional innovation system in chemistry, demonstrated in Table 19.1, no longer exists. Even the central industrial research laboratory in basic research has disappeared.

However, foreign, mainly American, companies have established new research laboratories in the sector during the 1990s. In fact, since World War II the region Frankfurt/Rhein-Main has been the most successful in attracting foreign capital in Germany. Companies such as DuPont, Procter and Gamble or Japan's Takeda have invested in greenfield sites or taken over German companies in the region. Their R&D laboratories, however, scarcely have a relationship to the regional knowledge system. They are oriented towards their companies' global R&D systems, and are responsible mainly for observing technological developments and for adjustments of products to the German market. Hence, their task is market-oriented (Specht, 1999). As a consequence, the traditional regional knowledge system, based more on technical research in the business sector, will be undermined. Market knowledge becomes more important than basic technical knowledge.

In contrast to the chemical industry, restructuring of the R&D system in the second Fordist cluster, namely of car manufacturing and its suppliers, that has survived has been successful. Since the crisis at the beginning of the 1990s, car manufacturing has been radically re-organized in Germany (Schamp, 1995). As part of this re-organization, concepts such as lean management and the pyramid-shaped organization of suppliers have been adopted in the development of new models. Furthermore, increasing competition has compelled firms to shorten product cycles and to improve the quality of the models. Hence, new, often electronic components for cars have been required. As a consequence, car manufacturers both intensified their own R&D and increasingly bought R&D services from the first tier suppliers and technical consultancy firms. Thus, R&D expenditures in the sector increased by two thirds form 1990 to 1997. In 1997, the car industry spent 28 percent of R&D expenditures of total manufacturing in Germany and was by far the leading sector. The large and still increasing share of R&D that is out-sourced in the car industry, however, differs from behaviour in other sectors. External R&D expenditures in the car sector rose to account for 62 percent of all external R&D expenditures in German manufacturing in 1997 (Wissenschaftsstatistik, 2000).

In the new process of simultaneous engineering in cooperation with external partners, it becomes necessary to bundle R&D competencies both in time and space. As a result, the large R&D centres of car manufacturers have become the focal points for new engineering offices and suppliers' R&D laboratories. Next to Munich and Stuttgart, Frankfurt/Rhein-Main is a leading location for car production, due to Opel, a GM subsidiary, and even more so due to the supplier industry (Lompe and Müller, 1991). The International Technology and Development Centre (ITEZ) of Opel has been assigned as the global R&D centre for all GM models outside of the Americas. This responsibility was cancelled early in 2000, due to the reorganization of GM resulting from its cooperation with Fiat. Nevertheless, employment in the centre has increased by 4,000 to over 8,000, completed by some other subsidiaries in software production and motor development that employed 1,500 and 800 persons, respectively. Many suppliers became just as important as Opel in the local innovation system, even though their headquarters are not within the region. A case in point is the American company Meritor that has concentrated 90 percent of its global development of sun roofs and roof modules in Frankfurt. The Frankfurt laboratory controls smaller R&D laboratories in the USA, Japan and India. Further car suppliers, both from

Germany and abroad, invested in new R&D laboratories during the 1990s in Frankfurt/Rhein-Main. Additionally, rapidly growing technical consultancy firms have opened up large branches in the region, often near the R&D laboratory of Opel (Rentmeister, 2001).

The local knowledge system in car manufacturing has been even more strengthened by new R&D laboratories of Japanese companies such as Honda, Mazda, Mitsubishi, Subaru and Isuzu. These companies had established small offices for market research by the end of the 1980s, often in close neighbourhood to their earlier distribution centres for Germany. While these offices were responsible for tests and inquiries in market and technological trends, that is for activities that are near to the market (Park and Schlunze, 1991), they received more responsibilities in model development in the 1990s. This trend was part of a new product strategy that took the European market more seriously.

As a consequence, a dense net of R&D laboratories in car manufacturing has emerged that is of international importance. This net is nearly completely independent from academic institutions and public R&D laboratories in the region. Because the driving force originates from restructuring within the sector, one could claim that car model development is an island in the regional knowledge and innovation system. This island is still increasing but receiving most impulses from an international, if not global network. The governance of this flexible R&D system is still concentrated in the hands of car manufacturers, the powerful clients of suppliers and technical consultancy firms. This R&D system can be characterized as 'neo-Fordist' as it is lacking the typical decentralized organization of post-Fordist structures.

As a first conclusion, it can be stated that the trajectories of the chemical and car sectors that are so important for the region Frankfurt/Rhein-Main have brought about the following characteristics:

- the regional knowledge system remains largely incoherent;
- the regional knowledge system is restricted mainly to R&D within and among the industrial firms and receives little support from academic research, public consultancy and technology transfer;
- the possibilities of local governance of the metropolitan knowledge system have been reduced by cuts in the chemical industry and by the international integration of foreign firms of the car sector;
- in parts of these manufacturing sectors the strategic responsibility of R&D laboratories is changing from technology to market competence.

New Technological Clusters

In the face of these processes, which were considered negatively for the region, local authorities ordered a study in the early 1990s to identify emerging new technologies in the region and to suggest a regional strategy in technological development (UVF, 1993). The authors found only one technological competence centre in the region, specifically in the city of Darmstadt, which is located beyond where political actors consider the boundaries of the metropolitan region to be. Darmstadt and its neighbourhood focus on information and telecommunications (IT) technologies, a knowledge base formed by a technical university, a technical college, public research laboratories in information technology and data processing, the main research laboratory of the now privatized German Telekom, some large software companies (such as Software AG) and, more recently, Internet firms (such as T-Online). A host of new spin-out and spin-off firms form a major German centre in software development, second only to Munich (Sternberg and Tamásy, 1999). Furthermore, recent local and combined initiatives from academia, local politics and local business, including banks, foster the creation of technological firms. There is still no detailed survey of these activities; but in many respects it can be assumed that a German competence centre in Internet services and multimedia business has emerged in and around Darmstadt. Such a centre would not be the only one, as the multimedia sector currently receives much support in cities such as Munich, Cologne or Berlin.

While a science based island of technology has partly emerged spontaneously at the fringe of the region Frankfurt/Rhein-Main, political actors of the city of Frankfurt have come to the conclusion from the aforementioned study that it is necessary to create a new knowledge cluster. They believe that the biotechnology sector has potential in this regard and, therefore, should be promoted. The national innovation system in Germany, however, is not favourable to biotechnology. It is biased on technology fields with low growth expectations, such as biotechnology services and platform technologies (Giesecke, 2000). Nonetheless, public authorities in the region have taken initiatives to promote the sector. This initiative, however, seemed to be a rather defensive strategy, as the following example after the loss of Hoechst plants makes clear.

Thus, the state government of Hesse established a so-called Future Capital Company, together with the Hoechst Company, in March 1999. During the next decade, this company is expected to provide business start-ups in the biotechnology sector with risk capital. The University of Frankfurt established a new institute that combines biochemistry with information sciences. The public Max Planck Research Institute of Biophysics (which, by the way, hosts a Nobel prize winner) will be relocated to a new biotechnology campus. An International Max Planck Research School for highly qualified PhD students will be established, and, last but not least, the city council will invest in a biotechnology park. Although the region has lost in the BioRegio contest initiated by the German Federal Government some years ago[3], political actors of both state and city now cooperate in order to create a new knowledge cluster in biotechnology.

It is odd that, according to recent studies, while the business sector shows considerable weaknesses, the two other axes of the Triple Helix knowledge system underlying biotechnology, namely 'academia' and 'public support', are much in evidence. Only a few young biotechnology firms with innovative products stand against the dominating large companies that, until now, have formed the incoherent cluster of the chemical industry. The majority of other start-ups are service firms of biotechnology. These are supplemented by a great number of distribution subsidiaries of large American companies. This business structure does not stimulate cooperation between firms – which had been one of the reasons why the region Frankfurt/Rhein-Main had been unsuccessful in the BioRegio contest.

Can Service Clusters be Initiators in Regional Knowledge Systems? The Case of Finance.

Research on regional biotechnology has shed some light on an interesting detail: there is an exceptionally large number of consultancy and investment firms for biotechnology in Frankfurt. Recently, investment firms for Internet start-ups have completed this trend. So, is the true cluster and knowledge system located in finance?

Innovation statistics such as that published by the OECD and innovation research generally start from a technological concept of innovations and assume that innovations happen exclusively in manufacturing. It has become clear, however, that a lot of knowledge-based services contribute to the emergence of new technological innovations; in the first instance engineering services, in software production and other fields of technology. Unfortunately, it is still not possible to measure their contribution to a regional knowledge system (Strambach, 1997).

A more radical question is whether or not the services sector by itself could become an innovative motor in regional development. In the region Frankfurt/Rhein-Main, this sector at issue is finance. In general the financial system is understood as a supporting part of national and regional innovation systems. It is rather unusual to understand new services in finance as an output of a non-technical, service system of innovation. There is still no procedure to measure either inputs (for example, in R&D expenditures) or outputs (for example, in patents) in this sector in the same way as it is done for technological innovation in industries. Bankers claim that they do R&D of their own. This may be particularly the case in production sites of financial services, that is, the large financial centres where there is a high rate of financial innovations. Because there are no patents given on non-technological innovations, however, there is still no way to measure this kind of research.

Yet, there are signs that a complete Triple Helix of a knowledge system in finance is emerging in Frankfurt. The increasing role of Frankfurt as a financial centre in Europe is well known (Sassen, 1999). The business sector in finance is characterized by strong spatial concentration in the centre of Frankfurt, by a multiplicity of firms and domination of informal over formal relationships among firms (Grote, Harrschar-Ehrnborg, and Lo, 2000). This superstructure has largely emerged without the base of a complex local knowledge system in the Triple Helix sense.

Today, both firms and academic institutions are making efforts to establish such a knowledge system, thereby securing competitiveness, especially with regard to other financial centres in Europe such as London or Paris. A few such efforts may be mentioned renewed activities of the private bank academy that is financed mainly by the banks; at the public university, the faculty of economics specializes in teaching and research on finance; and the university has established a PhD college for finance and cooperates with the private sector in the Centre for Financial Studies. The results of such efforts are still unknown. Neither the amount of input nor its effect on the innovativeness of the financial sector are clear. The whole development seems to be too early to assess, given the fact that it only started with the first steps of deregulation of financial markets in Germany in the 1990s. It is even less clear what the emergence of this knowledge cluster will mean for the metropolitan region, as the total financial sector contributes by only 69,000 persons or 15 percent to the regional labour market.

A Portfolio Approach to Regional Knowledge Systems in Metropolitan Regions

Clearly, the metropolitan knowledge system is characterized by an enormous fragmentation and heterogeneity on at least two levels. First, two out of three dimensions of the Triple Helix are almost non-existent in all sectors. In general, academic training and research and a specific policy are lacking. The only exception seems to be the telecommunications cluster in and around the city of Darmstadt. In particular, those regional knowledge systems that are based on Fordist industries are mainly restricted to the business sector. Efforts made by political actors and academic institutions are seen in emerging knowledge systems, but they are still very weak. Many observers only concede a small prospect for success. Second, the still ongoing deindustrialization of the region seems to undermine its capacities to innovate. What then is the regional knowledge system and where will it go?

For a preliminary response to this question, a regional portfolio approach can be used to combine the different fragments of industrial clusters and different fields of technology. Such a portfolio is, first, a procedure for describing the current state and denoting possible trajectories of a regional knowledge system. This portfolio for the region Frankfurt/Rhein-Main can be constructed along two axes, namely an axis describing types of organization of R&D in the business sectors and another one identifying the sectors involved. On the one axis, Fordist, neo-Fordist and presumably post-Fordist organization of sectoral R&D are differentiated. On the other axis, traditional and emerging sectors are distinguished. The still major sectors in terms of employment are set in bold (Figure 19.3).

Obviously, the portfolio of Frankfurt/Rhein-Main is mainly characterized by a Fordist sub-system of the knowledge system in chemical industries. However, as mentioned, this sub-system is currently being undermined by the restructuring of the sector. In contrast, the sub-system in car manufacturing is increasing. But this is quasi-hierarchically governed by a focal core enterprise, that is the car manufacturer, and hence the characterization of this sub-system as neo-Fordist. Sub-systems which are

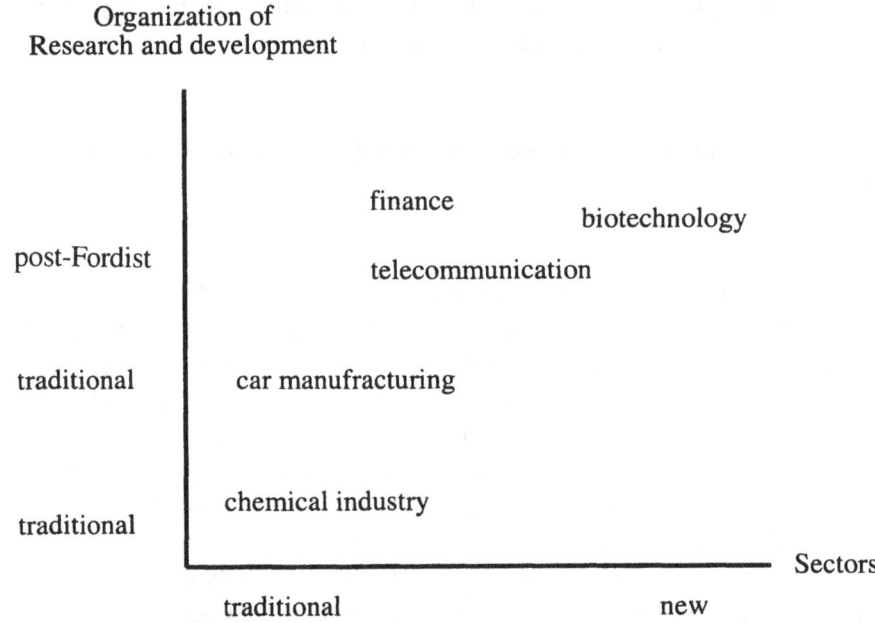

Organization of
Research and development

post-Fordist

finance

biotechnology

telecommunication

traditional

car manufracturing

traditional

chemical industry

Sectors

traditional

new

**Figure 19.3 A portfolio of the regional knowledge system of
Frankfurt/Rhein-Main**

based on cooperation between firms, academic institutions and political actors
may be judged sceptically, however, as they have not yet shown their contribution
to regional development.

This portfolio can be interpreted differently concerning the long-
term trajectory of the region. First, past success prevents the rapid production
of new knowledge. A lock-in of still Fordist and neo-Fordist sectors may
establish barriers for the transition to new knowledge in Frankfurt/Rhein-
Main. Second, the regional knowledge system in Frankfurt/Rhein-Main has
slightly abandoned its traditional trajectory and lost important fields in
technology without assurances of gaining access to new fields of knowledge.

Third, it can be claimed that the character of the regional knowledge
system – and by implication the total regional trajectory – is changing from
a technology-driven to a market-driven system. Both changes in the character
of R&D in the chemical industries and in parts of car manufacturing, as well
as the nature of the strengthening of the financial sector, support the conclusion

that Frankfurt/Rhein-Main is returning to a very old – even pre-industrial – role in the European city system. In particular, it is once again becoming a focal trade and financial centre in Central Europe. This would require a further shift of the regional knowledge system from manufacturing to services.

Notes

[1] This chapter has partly resulted from a broader comparative research on Frankfurt and Tel Aviv metropolitan areas as 'emerging nodes in the world system' (see Felsenstein et al 2002). Financial support by the German-Israeli Foundation (GIF I-003-402.02/95) is acknowledged.

[2] The comparability of the spatial distribution of R&D personnel and patents is restricted. First, patents are not always registered at the place of their invention. As a result, there are two different spatial distributions, one on the location of the declaration and the other on the location of invention (Greif and Schmiedl, 1998). Second, different types of regions are in use: the patent statistics refers to regional labour markets whereas the statistics on R&D employment refer to Physical Planning Regions.

[3] In the BioRegio contest of 1996, the Federal Government tried to give momentum to the development of the biotechnology sector by spending 50 million German marks in each of the three winning regions. One of the prerequisites was an existing network between academia, political actors and the business sector (see Dohse, 2000).

References

Bathelt, H. (1997), *Chemiestandort Deutschland*, Berlin, sigma.

Becker, St. and Schumm, W. (2001), 'Begrenzte Vernetzung, Restruckturierung von Grossunternehmen der Chemischen und Pharmazeutischen Industries in der Region Rhein-Main', in J. Esser and E. W. Schamp (eds), *Metropolitane Region in der Vernetzung – der Fall Frankfurt/Rhein-Main*, Frankfurt/Main, Campus, pp. 105-30.

Cooke, P. (1998), 'Introduction. Origins of the concept', in H.J. Braczyk, P. Cooke and M.Heidenreich (eds), *Regional Innovation Systems*, London, UCL Press, pp. 2-25.

Crevoisier, O. (1999), 'Innovation and the City', in E.J. Malecki and P. Oinas (eds), *Making Connections. Technological Learning and Regional Economic Change*, Aldershot, Ashgate, pp. 61-77.

Dohse, D. (2000), 'Technology Policy and The Regions – The Case of the BioRegio Contest', *Research Policy*, vol. 29, pp. 1111-1133.

Edqvist, C. (ed.) (1997), *Systems of Innovations. Technologies, Institutions and Organizations*, London, Pinter.

Etzkowitz, H. and Leydesdorff, L. (2000), 'The Dynamics of Innovation: from National Systems and "Mode 2" to a Triple Helix of University-industry-government Relations', *Research Policy*, vol. 29, pp. 109-123.

Felsenstein, D., Schamp, E.W. and Shachar, A. (eds) (2002), *Emerging Nodes in the Global Economy. The Tel Aviv and Frankfurt Metropolitan Regions Compared*, Kluwer, (forthcoming).

Galli, R. and Teubal, M. (1997), 'Paradigmatic Shifts in National Innovation Systems', in C. Edqvist (ed) (1997), *Systems of Innovations. Technologies, Institutions and Organizations*, London, Pinter, pp. 342-370.

Giesecke, S. (2000), 'The Contrasting Roles of Government in the Developing of Biotechnology Industry in the US and Germany', *Research Policy*, vol. 29, pp. 205-223.

Greif, S. and Schmiedl, D. (1998), *Patentatlas Deutschland. Die Räumliche Struktur der Erfindungstätigkeit*, München, Deutsches Patentamt.

Grote, M., Harrschar-Ehrnborg, S. and Lo, V. (1999), *Technologies and Proximities: Frankfurt's New Role in the European Financial Centre System*, Goethe University, Working Papers, Finance and Accountancy 46.

Läpple, D. (1996), 'Städte im Umbruch. Zu den Auswirkungen des gegenwärtigen Strukturwandels auf die städtischen Ökonomien. Das Beispiel Hamburg', in ARL (ed), *Agglomerationsräume in Deutschland. Ansichten, Einsichten, Aussichten*, Hannover, Akademie für Raumordnung, pp. 192-217.

Lompe, K., Müller, Th., Rehfeld,D. and Blöcker, A. (1991), *Regionale Bedeutung und Perspektiven der Automobilindustrie. Die Beispiele*

Südostniedersachsen und Südhessen, Düsseldorf, Hans-Boeckler-Stiftung.

Park, S.J. and Schlunze, R. (1991), 'Forschungs- und Entwicklungsmanagement Japanischer Konzerne in der Bundesrepublik Deutschland', in S.J. Park (ed), *Japanisches Management in der Bundesrepublik*, Frankfurt/M., Campus, pp. 147-182.

Porter, M. (1990), *The Competitive Advantage of Nations*, New York, Free Press.

Rentmeister, B. (2001), 'Vernetzung Wissensintensiver Dienstleister in der Produktentwicklung der Autoindustrie', in J. Esser and E. W. Schamp (eds), *Metropolitane Region in der Vernetzung – der Fall Frankfurt/Rhein-Main*, Frankfurt/Main, Campus, pp. 154-180.

Sassen, S. (1999), 'Global Financial Centers', *Foreign Affairs*, vol. 78, pp. 75-87.

Schamp, E.W. (1995), 'The German Automobile Production System Going European', in R. Hudson and E.W. Schamp (eds), *Towards a New Map of Automobile Manufacturing in Europe? New Production Concepts and Spatial Restructuring*, Berlin, Springer, pp. 93-116.

Specht, J. (1999), *Industrielle Forschung und Entwicklung: Standortstrategien und Standortvernetzungen*, Wirtschaftsgeographie 14. Münster, Lit.

Sternberg, R. (2000), 'University-Industry Relationships in Germany and Their Regional Consequences', in Z.J. Acs (ed), *Regional Innovation, Knowledge and Global Change*, London, Pinter, pp. 89-120.

Sternberg, R. and Tamásy, Ch. (1999), 'München as Germany's No.1 High Technology Region: Empirical Evidence, Theoretical Explanations and the Role of Small Firms/Large Firm Relationships', *Regional Studies*, vol. 33, pp. 367-377.

Strambach, S. (1997), Wissensintensive Unternehmensorientierte Dienstleistungen – ihre Bedeutung für die Innovations – und Wettbewerbsfähigkeit Deutschlands, *Vierteljahrshefte zur Wirtschaftsforschung*, vol. 66, pp. 230-242.

UVF (Umlandverband Frankfurt) (1993), *Technologieprofil der Region Rhein-Main*, Frankfurt/Main, Umlandverband.

Wissenschaftsstatistik (ed) (2000), *Forschung und Entwicklung in der Wirtschaft 1997-1999*, Essen, Wissenschaftsstatistik GmbH.

20 Enhancing Competencies, Networking and Institutions for the Knowledge Economy: Singapore's Garment Industry

Leo van Grunsven

Introduction

A pervasive idea in the Asian industrial restructuring literature is the assumption of a (rapid) decline of 'mature' industries. However, such developments are often contradicted in practice. An illustration is provided by the experience of one of the most mature industries in the Asian Newly Industrialized Economies (NIEs), the garment industry. While the industry has indeed undergone significant changes over the past decade(s), the majority of (production) firms are not set on a course of 'extinction'. Rather, many production firms have succeeded in sustaining and reconfiguring their connections to global networks (Gereffi, 1999, 2000). A learning-based enhancement of their knowledge base, with augmented competencies and capabilities, is the key to this adjustment (Ernst et al 1998; Gereffi, 1999). The industry has been further marked by diversification and by the entry of international firms controlling large parts of the global commodity chain and international production networks. As such, the role of East Asian NIEs as regional trading and sourcing hubs has been growing.

From a global commodity chain perspective, the evolution of Singapore's garment industry has both paralleled and differed from the general Asian NIE development pattern. With the introduction of knowledge-based economy policy initiatives in Singapore, the national environment of the industry (and for all mature industries) is about to change drastically again in the years ahead. As resources are diverted to new knowledge-driven industries and activities, in addition to the increasingly higher costs with which the industry has been confronted over the past decade, there are serious implications for the garment industry in the city state, even relative to the other Asian NIEs. Official national perspectives envisage mature industries becoming more irrelevant than already is the case .

In this chapter, I argue, contrary to official thinking, that the government's new knowledge-based economy strategic policy initiatives can provide a significant catalyst for positive change within Singapore's garment industry. These initiatives also have detrimental effects. Nevertheless, it is postulated that the opportunities for sustaining a viable garment industry stem from a congruency of overall visions and policies pursued (even if these are primarily directed towards other sectors and branches of the economy) with what should be accomplished as a next step in the development path of the industry.

The chapter is organized into three sections. First, the evolution of the garment industry in East Asia is sketched. Second, the dynamics of Singapore's garment industry is outlined and, finally, the opportunities available to individual firms associated with initiatives related to the knowledge-based economy are briefly indicated. Space precludes detailing the nature of Singapore's strategy for a knowledge-based economy (see Ministry of Trade and Industry, 1998).

The East-Asian Garment Industry: Upgrading in Global Apparel Trade and Production Networks

Asian countries, mostly the NIEs, were incorporated in the globalization of the garment industry at an early stage. From the 1970s onwards, with USA and Europe-based buyers as drivers of the internationalization of the commodity chain, contract manufacturing of apparel has fuelled their export drive (Bonachich et al 1994; Dicken, 1998). Although by the early 1980s a truly Global Apparel Commodity Chain (GACC) had emerged, at this time some 80 percent of total USA apparel imports were sourced from Asia (primarily from Hong Kong, Taiwan, South Korea and China). The key to success of (South) East Asian firms in developing a strong position in the GACC was to move from mere assembly manufacturing to original equipment manufacturers (OEM). Subsequently, firms in the East Asian NIEs began to produce and market their own brands (OBM) (Gereffi, 1999).

Significant changes can be observed from the mid-1980s. The overall share of Asia in USA imports of clothing started to decline to 59 percent in 1996 (Gereffi, 1997). This trend reflects a change in the geographical landscape of production and distribution. The decline came mostly from the big three centres (Hong Kong, Taiwan and Singapore), while China gained substantially. Countries in the Southeast Asian region also experienced growth in their share of USA imports (with the exception of Singapore). Overall, imports into the USA and European Union (EU) markets from Asia came from an increasing number of countries after the 1990s. In addition, there has been a shift of production to Latin America (Mexico) and the Caribbean. Despite the decline, Asia's overall role in the global apparel trade remains significant due to the position of China and the Southeast Asian late-industrializers (for Indonesia, see Dicken and Hassler, 2000).

A number of inter-linking forces can be identified behind the more recent shifts. First, changes in markets, market differentiation and segmentation, and consequent changes in buyer (sourcing) strategies, have favoured certain production locations over others (at least for certain products and segments) because of proximity and lead-time considerations. Second, the capabilities and competencies of Latin American producers have been extended, rendering the region a more serious competitor. The liberalization of trade and regional trading agreements between the USA and Latin America, most notably Mexico (under NAFTA) and the Caribbean Basin countries, have reinforced this trend, confirming their role as intervening opportunities to the North American market. Third, the international trade regime and regulations in the form of non-tariff barriers, especially quota restrictions organized mainly under the Multi Fibre Agreement (MFA), have increased as more product categories and countries have been included. This trend had the unanticipated effects of stimulating industry upgrading in developing countries, and dispersing industry over a larger area. The abolition of the MFA, to be achieved by 2005, may have the opposite effect.

Fourth, the dynamics associated with the changing competitiveness of Asian countries and firm level adjustments and specialization in these countries, has led to the development of an intra-regional division of labour (Clark and Kim, 1995; Chiu et al 1997; Dicken and Hassler, 2000). As the comparative advantages of the Asian NIEs have changed, their function as production locations has surely been affected. However, a designation of this industry as a 'sunset' industry, if with some basis, is misleading in many respects (see also Bernard and Ravenhill, 1995). It would be more accurate to speak of sunset activities or even segments. While firm closures and diversification into other activities have occurred, the industry has shown remarkable resilience and a range of adjustments (Simon, 1995; Taplin and Winterton, 1997). In part, firms have succeeded in shifting the basis of competitive advantage to higher end activities in the chain (Gereffi, 1999, 2000). In a number of cases (e.g. Hong Kong), infusion of higher end commodity chain segments from outside has contributed to this change of orientation of the industry. Certainly, there has not been an en masse exit of NIE firms from the industry.

Roughly two types of adjustment strategies have been applied, often simultaneously. First, retention and wage depressing strategies focus on shifting the burden of labour cost or shortages elsewhere in order for firms to still compete. This strategy involves relocation of production, international subcontracting or the use of foreign labour, producing intricate regional networks of production and distribution that comprise many layers or tiers in many different countries. The outcome in most cases is 'triangle manufacturing', that is, orders from buyers are shifted to offshore factories in lower wage Asian countries and the finished goods are exported directly from third countries to the buyer country (Gereffi, 1997, 1999). The 'old NIE manufacturers' have become orchestrators of these networks, providing inputs in the form of entrepreneurial and technical skills, components (Schmitz and Knorringa, 1999), production planning and co-ordination, whereas the new manufacturers work mostly on a Cut-Make-Trim basis (CMT).

Second, pro-active strategies aim at upgrading and changing the competitive advantage of firms. Here, a distinction can be made between upgrading within production and upgrading beyond production. The former refers to producing better (higher quality) and faster (shorter lead times, increased flexibility), involving investment in the latest technology to improve processes and products. The latter refers to moving into other activities and positions in the apparel chain (e.g. towards OBM, design, marketing, retailing and distribution), through which specialization in certain segments in the chain has ensued (Schmitz and Knorringa, 1999). This has offered opportunities for trajectories that move firms beyond 'lock-in' dependent positions in the commodity chain (see Gereffi, 1999).

The trajectories of East Asian apparel firms point towards continuity in the global chain. These shifts derive from a larger range of competencies and capabilities, acquired through learning in the processes of developing multi-layered regional sourcing networks, playing critical co-ordinating roles in the full-package production process, and venturing beyond these roles or functions (Ernst et al 1998; Gereffi, 1999). This observation is enhanced by a 'consolidation of networks' strategy adopted by buyers in response to trends in consumer markets (especially in quality driven segments). Having become more mindful of their operations in developing countries, buyers now prefer to continue to work with core producers on the basis of their track record, production practices and standards and organizational capabilities. Investments in the networks and the transaction costs sustained in reconfiguring the sourcing base, has also impacted the attitude of buyers towards contractors. Industry dynamics are further fuelled by upgrading associated with the entry of firms performing functions in the core of the commodity chain. The large number of overseas buyers, agents and other intermediaries, that have set up regional offices in Hong Kong have significantly shaped the structure of its apparel industry and its function in the GACC.

The Singapore Apparel Industry: An Overview

The apparel industry contributed significantly to Singapore's industrialization in the 1960s and 1970s, especially in terms of labour absorption. Foreign-based apparel firms were attracted to Singapore as a production location because labour was low cost and disciplined (von Alten, 1995). The industry grew rapidly in this period, from 162 establishments employing an estimated 10,000 workers in 1970, to 374 establishments employing around 27,000 employees in 1980 (Münch and van Wijk, 2000). The industry was largely domestically owned and export oriented, with the USA and Europe as its main export markets. Production was initially mostly limited to contract manufacturing on a Cut-Make-Trim (CMT) basis. Most inputs were directly supplied by the buyers and not sourced locally. Marketing and distribution were likewise in the hands of the buyers. There was a segment of small and micro firms, focused on the domestic market, or working as subcontractors for the bigger producers.

In the late seventies, environmental changes, consequent government policies and increased international regulation (in particular in the main markets, the USA and the EU) exerted pressure on the industry, initiating a first wave of relocation to lower cost locations in the region, mainly Malaysia and Indonesia. However, relocations exerted little structural impact on the industry, in terms of number of establishments and employees. In terms of positioning in the apparel commodity chain, producers managed to develop into full package suppliers, sourcing affiliated inputs and expanding production networks regionally to fulfil buyers' orders.

The early 1990s marked a turning point. At this time, wage increases, labour shortages, a downturn in the industries' main markets and a quota bidding system, introduced by the government in the late 1980s, started to take their toll among the (smaller) producers. Consequently, the producer segment of the industry declined (in terms of local production and exports). A second wave of production relocation abroad occurred. Subcontracting networks started to extend abroad as well. In tandem with the 1997/1998 economic crisis, this trend negatively affected smaller firms in the industry that functioned as subcontractors, or were completely domestically oriented.

Official statistics (Census of Industrial Production, SEDB), indicate that in 1997 some 170 production establishments remained, employing some 8,200 workers (SEDB, 1999). Domestic exports had dropped from a peak of S$1,793 million in 1990 to S$620 million in 1997. These figures suggest that the industry has drastically declined since the early 1990s; indeed the industry has been labelled a sunset industry. However, the actual dynamics of the industry prove otherwise. First, from an inventory of a variety of sources, a different trend can be identified in the production segment of the industry. The above-mentioned 170 production establishments all employ more than 10 workers. Another 62 establishments, still producing garments in Singapore, employ less than 10 workers (and are therefore excluded from the Census). In total, in November 1998, 231 production establishments employing an estimated 10,680 workers were counted (Münch and van Wijk, 2000). Another 138 establishments were confirmed to be still operating in the industry, but no longer with (affiliated) production facilities in Singapore (53 had fully relocated production overseas, while the other 85 fully contracted out production). These establishments employed in total 2,050 workers. A third group involving some 125 establishments was identified, a quarter of which had closed down or withdrawn from the industry, while the current status of the remainder could not be ascertained. Obviously, the methodology employed in the Industrial Census has led to an underestimation of the size of the production segment of the industry.

Second, diversification has taken place in the industry as a whole. On the sourcing, marketing and distribution side of the industry, developments through both local and foreign investments and activities can be identified. International (branded) companies, retailers and buying agents, both international and local, have set up regional buying and sourcing offices. International branded companies use the city-state as a regional marketing and distribution centre. Local brand and designer companies have been set up and are starting to gain some recognition in the rest of the region. Overall, some 40 international buyers, agents, and distributors have operations in Singapore.

Third, individual firms' strategies on the production side also involve intricate regional production networks including affiliated branch plants and/ or subcontractors in lower-cost countries in the region. The functional change and development of activities in the Singapore establishment, accompanying production strategies, and consequent shifts in the positioning in the global commodity chain, remain hidden in the official statistics.

These dynamics are now examined more closely, aided by the outcomes of research conducted on the Singapore apparel industry between 1998 and 1999.

Production Characteristics of Garment Producers

A survey conducted among both producers and buyers provides up-to-date empirical data on the Singapore garment industry. Some 15 percent (57) of the establishments categorized as either having production facilities in Singapore, producing in overseas establishments only, or producing only through subcontractors, were interviewed. The survey (somewhat biased towards large producers) covered products, markets, buyers, suppliers, brand manufacturing, subcontracting, changes over the past ten years (in terms of company size, products, markets, business and institutional environment), corporate responses to these changes, and finally the competitive position and future plans. Another set of interviews were conducted with 26 buying/ sourcing offices and agents (further referred to as buyers) about the position of Singapore in their global sourcing networks, the expected changes in the global industry and Asia as a sourcing region that might affect Singapore's positioning within the commodity chain.

Ownership of the establishments is predominantly domestic, and most (almost 80 percent) employ less than 100 workers. The average number of non-production workers was 19, while the average for production workers was 68, a ratio of 1 to 3.6. This is remarkably low for such a labour-intensive industry. The reason will become clear when foreign direct investment (FDI) and subcontracting patterns are considered. Most companies (approximately 90 percent) exported at least part of their production, and more than half exported more than 75 percent of output. The main export market was the USA, to which 65 percent of respondents exported about 70 percent of output. The share of output going to the EU was quite low. A smaller share of the respondents exported to Asia, but the share of output going to this market was higher on average. Local sales went mainly to department stores (many foreign-owned) or affiliated shops and boutiques (local brand producers).

The main products manufactured were knitted shirts and polo shirts (the only two quota categories completely filled were the so-called 'hot items'). Knitted apparel is considered to be the strong point of the Singapore/ Malaysian region. Many of the producers indicated they produced a whole range of products. Complex and skill intensive products are hardly produced. As these require many different steps to make and specific labour skills, the restrictions in the availability of skilled labour are especially felt here. The majority of establishments are anchored in international sourcing networks, play a pivotal role in regional production networks, and continue to produce to the orders of buyers. Some 70 percent indicated that they sell at least 50 percent of output to buyers; 50 percent are totally dependent on such sales. The big branded companies, such as Polo Ralph Lauren, Gap, Nike, and Tommy Hilfiger, are the most prominent, followed by (local) department stores, and to a lesser extent mass merchandisers, retailers and catalogue companies. Manufacturers usually have more than one buyer and relationships are mostly long-term. Buyers supply producers with designs and product specifications and audit factories to ensure quality levels, labour standards and agreements on subcontracting are upheld. Procurement of inputs is left to the producers, although buyers do work with nominated suppliers, implying that the choice of suppliers has already been made.

Although only seven percent of manufacturers exclusively produced/ marketed their own brand(s), more than half carried these, which were sold through department stores or in affiliated retail outlets. Almost a third of the establishments indicated that at least half of their output were affiliated brands; 14 percent produced almost exclusively (more than 90 percent of output) for own brands. OBM appears to have seriously taken off since the 1990s. Own brands were locally or regionally marketed.

In the past 15 years the multi-establishment company has become more prominent: two-thirds of the surveyed firms are part of a larger company with overseas manufacturing plants. About one-third (31 percent) was part of a company with at least three overseas establishments. The status of the Singapore establishment within such large companies was mostly that of an independent headquarter/unit or a parent company. More than half of the establishments were engaged in investments overseas. On average more than 40 percent of production of the entire company took place overseas. Another 27 percent is attributed to local or foreign subcontractors. Indeed, more than one-third of the establishments no longer performed any production activities and 44 percent indicated a reduction in production capacity in the Singapore establishment over the past ten years.

Roughly 35 percent of the establishments with offshore production sites established the first overseas facilities in the 1980s. In the early 1990s many more companies started or expanded overseas investments to three, four or even more establishments. Most investments involved manufacturing facilities, with the exception of those in Hong Kong, which were all sales and marketing offices. The early investments were sited in Malaysia and Indonesia. Later investments went to China, and more recently Cambodia. Serious labour shortage and lack of quota in Malaysia stimulated relocation further afield.

Almost all establishments (88 percent) made use of subcontractors for parts of production, and 40 percent indicated that they have increased subcontracting over the past ten years. Some now even rely totally on subcontracting. The main reasons for contracting out work were: limited capacity in affiliated establishment (e.g. in peak season or times of big orders), lower costs, increase in flexibility and use of specialized skills. The first three involved contracting out of entire garments, including or excluding cutting. In the latter case it involved embroidery, printing, washing and finishing. Most producers (60 percent) used local subcontractors, or subcontractors in neighbouring Malaysia and Indonesia. In total, 58 percent of the producers contracting out work made use of overseas subcontractors. In addition, overseas branches/subsidiaries subcontracted work locally. In general, subcontracting networks 'stay closer to home' than FDI. As the local textiles industry is virtually extinct, most fabrics and other raw material inputs were imported (mostly from nominated suppliers).

Competitive Adjustment and Repositioning in the GACC: Perspective at the Level of the Firm

The overall profile of producers presented above reveal some of the defensive or offensive strategies adopted at the firm level in response to competitive pressures. Labour strategies employed to address the shortage and cost of labour include: higher wages and better fringe benefits; employment of foreign labour; increase overtime work; and international relocation and/or subcontracting of production, allowing reduction in production capacity in affiliated Singapore establishments. Subcontracting of entire production involved mostly local brand companies which focused more on the design and marketing of the products.

There have also been recent shifts in the focus of strategies, namely from competition based on costs towards the development of competitive and market positioning. Several tactics have been featured. Thus, investments have been made in new technologies, mainly involving the larger manufacturers because of capital requirements and the nature of these technologies. Technological innovations have been applied mostly in pre-assembly stages and the organization of production on the work-floor. The latter have involved computer aided design, computerized marking, grading and cutting, conveyor belt or hanger systems for transportation between production units, and attachment technologies to machines, allowing more efficient use of sewing machines and quick change-over between different products. Industrial engineering has sped up the production process and increased efficiency and productivity. New technologies have reduced the need for labour, and allowed upgrading product quality and production capabilities. In addition, there have been improvements of product quality (mainly by stepping up quality control and raising quality standards), aimed at upgrading of products and shifting towards higher value added market segments as well as towards increased flexibility.

A shift in the buyer or client base towards branded companies and higher end retail chains and department stores has also occurred. Indeed large discounters are virtually absent as buyers in Singapore and producers are no longer competing in these highly priced competitive segments of the market. Related to this strategy, the production of own brands, which is a form of upgrading since it often entails a repositioning of the firm within the commodity chain, especially if firms also diversify into marketing and retailing (as some have done), is another trend.

These strategies suggest a pro-active industry that has moved into high value added, high quality products and the provision of fast, flexible services to buyers and clients. However, from some buyers' viewpoints, flexibility and lead times could be improved (many attributed the latter problem to the lack of local inputs). Moreover, the quality of products manufactured by Singapore producers was not substantially higher than other manufacturers in the region.

The most common firm-level development trajectories include upgrading within production (moving into higher end, higher quality products; yet not specialized products or activities) and moving into co-ordination and intermediary export roles through relocation of production. Branding is a strategy adopted by many companies to spread risk and become less dependent on buyers. However, it is premature to speak of a clear shift towards OBM, or a clear upgrading beyond production. With the exception of a minority of local brand manufacturers with their own retail outlets, many still see contract manufacturing as their main business. Figures 20.1 and 20.2 summarize the configuration of production and distribution networks of Singapore OEM and OBM producers.

Although the local institutional environment has been ambivalent towards the industry at best, the role of government can by no means be ignored. Government responses to critical international competitiveness issues have induced firms to adjust. At the same time, because of the promotion of the city-state as a regional shopping centre, many international brands and fashion houses have set up retail outlets and boutiques in Singapore while many firms use the city-state as a centre for regional distribution and retailing of their products. The strong retailing function has stimulated local designers to set up their own boutiques and retail outlets where local producers can introduce and market their own brands. The government and other institutions are trying to create and stimulate a business system focusing on non-production and downstream activities.

Its trading, sourcing and retailing functions have also prompted the idea of the city-state as a possible regional fashion centre. This is viewed by local institutions as a way to fundamentally reposition Singapore in the apparel industry and place it on the world map of fashion, as will be discussed later. The government has provided incentives to buyers setting up regional trading and sourcing offices. To this end, the government also has continuously upgraded its supporting and infrastructural services (including telecommunications).

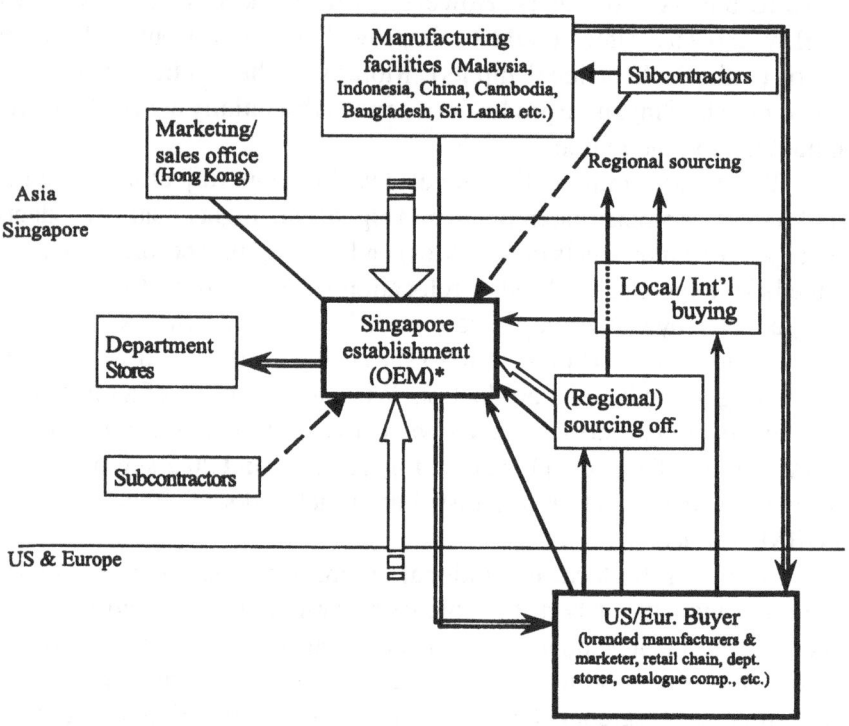

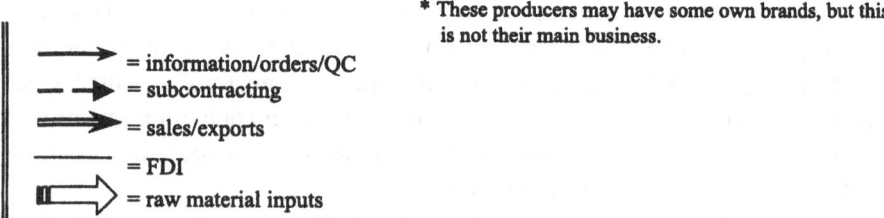

Figure 20.1 Configuration of production and distribution networks of OEM producers

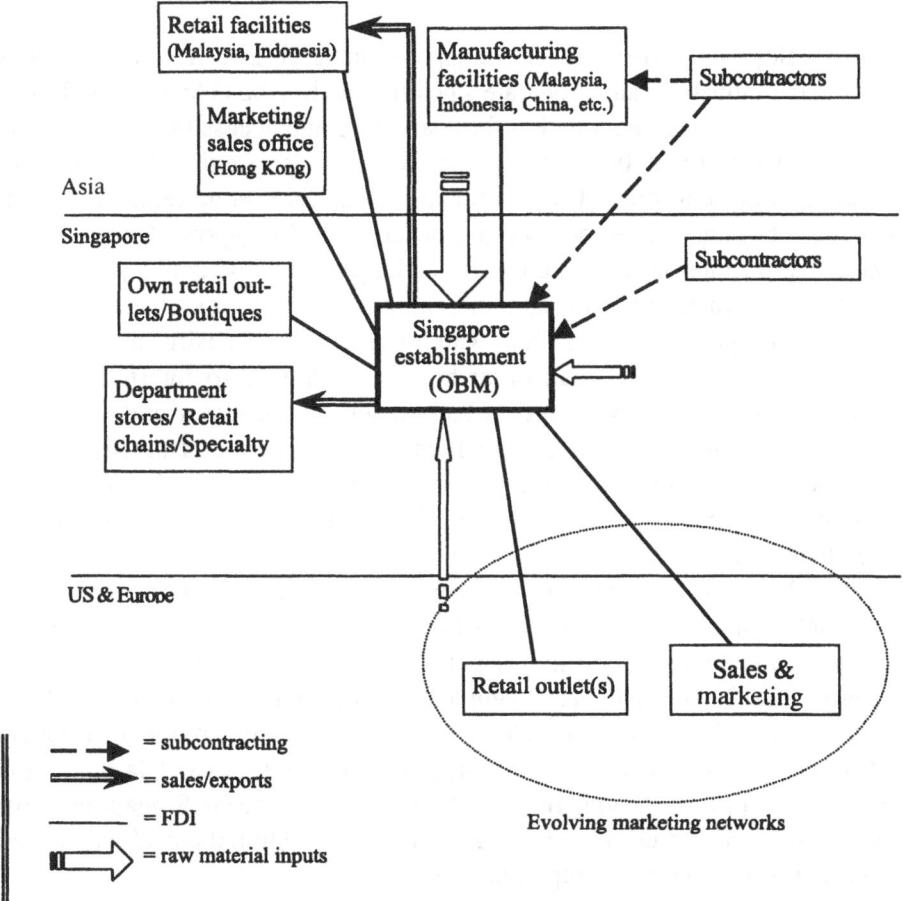

Figure 20.2 Configuration of production and distribution networks of OBM producers

International Buyers and Distributors: Another Role of Singapore in the GACC

The development trajectories of Singapore's apparel firms have meant some shift in the industry's and city-state's position within the GACC. In addition, Singapore's geographic location, infrastructure and accessibility, as well as its human resources, bode well for its function as a regional trading and sourcing hub. A number of international buyers have set up regional branch offices – buying houses or sourcing offices – in Singapore. Currently, an estimated 40 international companies and agents use the country as a base for their regional sourcing and merchandising activities. These include branded companies, such as Polo Ralph Lauren, retail chains such as May Department Stores, and mail order houses, such as Otto International, as well as international buying agents, such as Jebsen and Jessen and Swire and Maclain. In addition, there are local buying agents functioning as intermediaries between buyers and producers. One buyer, Polo Ralph Lauren, has made its Singapore office the global centre for all sourcing and merchandising activities outside the USA.

Most of the offices were set up in the 1980s and 1990s, and gave a substantial impulse to a shift of position in the commodity chain with more connections to non-production activities. In this respect, the role of international buyers has been complemented by local producers that have moved along the value chain. However, in recent years a number of buyers (Tommy Hilfiger, Land's End, Phillips and van Heusen, Liz Claiborne, the Venator Group) have pulled out or reduced their presence in Singapore. Most buyers still see and use Hong Kong as their main regional base for Asia and consider Singapore a sub-regional hub.

The Industry in an Economy Gearing up for the Knowledge-based Era

Notwithstanding increasingly adverse environmental conditions, Singapore's apparel industry has shown marked resilience. It is widely believed (also by circles in the industry itself) that the impact of the current knowledge-based economy initiatives and the new priorities will be such that the industry cannot be sustained. It is indeed not realistic to expect a change of attitude of the institutional environment towards the producer segment of the industry, and the industry at large. Yet, a viewpoint which stresses the unsustainability of the industry, and production firms in particular, runs counter to the interests of the stakeholders. This view is not particularly helpful, and, above all, is unnecessary. The knowledge-based economy initiatives may instead be conceived as constituting a vehicle to enable the industry (and the production segment in particular) to maintain and transform its position so as to continue to be relevant in rapidly changing structures. The key seems to be the opportunity to leverage the knowledge-based economy for enhanced competencies, through networking, to achieve operational and strategic shift. There are several features vital to the maintenance of a viable industry in Singapore. These include the position of production in Singapore, the continued viability of production in a regional hub and spoke system, OEM or OBM in regionalized production arrangements, the emergence of Singapore as a fashion centre, and recognition of the role of buyers as a component of the industry.

Labour shortages and wage increases, combined with rising costs of land and facilities, mean that Singapore producers will continue to face rising costs of production at home. Shortages of labour with specialized skills are a particular concern. The share of foreign workers in the industry (an estimated 40 percent of the workforce) conveys the importance of this problem. Yet, foreign workers bring their own problems and costs with them, such as obtaining permits, government imposed levies, workers overstaying their permits, and facilities to be provided. Current supply-side labour market developments stimulated by the knowledge-based economy initiatives will only work against the labour shortages facing the industry.

Producers, buyers and institutions seem to agree that for this reason there is little or no future for production in Singapore. While producers continue to be 'based' in Singapore, production will increasingly have to be organized through production networks abroad, in and outside the region (apart from some highly automated processes locally). This direction is already discernible in the plans for further subcontracting and investments abroad.

As to the viability of regional production arrangements, it may be noted that competitiveness in production is becoming more dependent on regional conditions. In terms of products or services there is hardly any specific competitive advantage, clearly distinguishing Singapore-based producers from competitors elsewhere in the region. While costs of production in many other countries in the region are substantially lower, recent trends suggest that quality levels are levelling out. As Singapore producers are competing for the same buyers and often in the same product categories as their competitors elsewhere in the region, the question is whether they continue to rely on contract manufacturing and if so, whether they will still be able to service clients better. Concurrently, the regional production system is firmly linked to buyer networks. Many companies indeed intend to keep producing to the order of buyers, and relatively few companies (35 percent) had plans to introduce or further develop their own brand(s). Among the latter, a number are branded companies, already producing exclusively own brand apparel.

In the effort to sustain network positions on the basis of OEM, Singapore producers may benefit from consolidation among buyers. The advantage for 'included producers' arises as the consolidation encompasses specific producers, regardless of location. However, this consolidation per se does not make demands from buyers in terms of higher quality, stricter labour and environmental standards, more flexibility and shorter lead times less issues of concern. Given the production arrangements, competitive advantage will rest more on efficiency and speed of organization in all facets of production, including the regional sourcing of inputs. Enabling advanced technology (Electronic Resource Planning and Quick Response Systems) is already applied. More important, given the newly emerging knowledge-economy, Singapore producers are certainly in a much better position than their rivals in the region to access technological innovations enabling better production organization. The opportunity to tap into a number of the knowledge-economy based initiatives, for example, the expansion of the range of information technology applications, the comparatively early development and application of internet-based procurement systems etc, may give Singapore producers the advantages they need.

Singapore producers may then be able to consolidate and upgrade their position within broader regional networks by transforming establishments at home into coordination centres, where HQ functions are performed. Singapore's role as an intermediary and coordinating centre may be consolidated as a result. An additional factor is the upgrading and improvement of facilities, with many producers planning to invest in new technologies within the next five years. The viability of OEM notwithstanding, the production segment of the industry is moving to a critical juncture as to product and market orientation. The main suggestions in this regard are to upgrade to more rewarding roles by deepening the manufacturing and export roles and to advance 'along the value and production chain', as well as to less dependent positions in the commodity chain. These trends revolve partly around shifting to niche markets and speciality products and partly around shifting further to OBM.

A further move towards OBM can be brought about also by the entry of new firms starting out as branded firms without own production facilities. This has occurred already and will be further stimulated by a growing local/regional fashion industry. This raises the question of the viability of Singapore as a (regional) fashion centre. Some success has been achieved with local brands. Despite the lack of a fashion tradition and local consumers not being overly supportive of local designs and brands, Singapore will be able to manifest itself internationally within Asia as a fashion centre for local designs if and when OBM is given more emphasis. It was agreed by all actors that within South East Asia Singapore has the best potential. Some of the strategies and programmes related to knowledge-based economy initiatives can function as a catalyst, if properly leveraged by the industry.

The range of activities to be carried out by, and capabilities required of, local firms under the scenario outlined above will be more extensive than hitherto. These trends imply a greater emphasis on information-based and knowledge-related activities. Developing international brands and a distinct image requires more direct information links with markets and consumers, higher risk taking and more investments in brand-image creation and development, development of (or use of outside) design capabilities, and so on. Of countries in the Southeast Asia region Singapore will offer the best opportunities for the needed organizational learning. Again, much can be achieved by tapping into a number of the knowledge-based economy initiatives, for example, those related to markets and products, information and communication technology, human resources, and entrepreneurship. There will be no lack of an environment with ample resources and opportunities for further learning.

In terms of location, infrastructure and human resources, Singapore is, in principle, in a good position to further develop its function as the Southeast Asian trading, sourcing and marketing hub. Most buyers indicated that they would retain their business in Singapore for at least the next five years. Singapore's competitive advantages, in terms of infrastructure, combined with a hinterland that is still considered attractive (especially given the potential of Indonesia) are the main reasons given for this view. The sustenance – even growth – of this function in the longer term should be viewed in first instance in the context of factors in the wider regional context, rather than simply the local environment. These factors include the long-term competitiveness and attraction to buyers of South East Asia as a production region and market and the competition from Hong Kong as a location for regional headquarters. It is difficult to make definite predictions. Government could provide incentives to regional producers (sellers) to set up sales offices in Singapore. In any case, from the point of view of infrastructure, facilities and other assets relevant from an operational point of view, the knowledge-based economy initiatives will only lead to further improvements.

Institutions of the Knowledge-based Economy, Networking and Business Attitudes

I have argued that tapping into and leveraging the opportunities provided by the knowledge-based economy initiatives allows firms in the garment industry to enhance competences and capabilities. Leveraging will require engaging in networking with relevant institutions of the knowledge-based economy. In the past networking with institutions has not been a feature of the 'mode of operation' of producers and related associations in this industry. Rather, firms have had negative perceptions of government agencies. Paradoxically, pro-active adjustments and business attitudes marked by conservatism and resistance to change go hand in hand. Upgrading towards other functions within the chain, beyond production, requires substantial investments, risk taking and a very pro-active attitude, with long term vision. These attributes, however, are not among the majority of apparel producers in Singapore. For many firms, business is good as long as there is money coming in. Exceptions are local designer and branded companies, but these companies do not comprise a dominating share of the industry. Further upgrading would therefore require some extent of change in attitude from producers.

Conclusion

While the economic policy and development framework for the garment industry in Singapore in the recent past has not been particularly benevolent, the production segment of its garment industry has displayed remarkable resilience. This has been the result of production firms responding in an offensive way to both internal and external pressures impinging on their international competitiveness. Many have succeeded in sustaining their connection to global networks by reconfiguring the organization of production. Some repositioning in the global commodity chain has been achieved by pursuing new value chain integration strategies. The key to these developments has been the building of additional competencies and capabilities on the basis of knowledge acquisition. As to the industry at large, diversification has occurred on the retail and marketing side as Singapore's role as regional trading and sourcing hub has grown.

The current economic policy framework seeks to transform Singapore into a knowledge-based economy. The future prospects of the garment industry under this strategic regime seems bleak. Nonetheless, I have developed an argument that production in the industry will continue to develop in a number of directions. Knowledge-based economy initiatives provide a positive framework enabling firms to build the necessary enhanced competences and capabilities, and the industry to transform itself to remain viable and relevant in the Singapore context in the decades ahead. While the key is leveraging available opportunities, the onus will be on the companies themselves to take advantage of any opportunities. In this context, the necessity of attitudinal change is vital, specifically for firms to consider themselves as learning organizations and to recognize the potential of the local environment to facilitate further learning in new areas.

References

Alten, F. von, (1995), 'The Role of Government in the Singapore Economy', *European University Studies*, Series 5, Economics and Management, vol. 1695, Frankfurt am Main.

Bernard, M. and Ravenhill, J. (1995), 'Beyond Product Cycles and Flying Geese: Regionalization, Hierarchy and the Industrialization of East Asia', *World Politics*, vol. 47, pp. 171-209.

Bonacich, E., Cheng, L., Chinchilla, N., Hamilton, N. and Ong P. (eds) (1994), *Global Production. The Apparel Industry in the Pacific Rim*, Temple University Press, Philadelphia.

Chiu, S.W.K., Ho, K.C. and Lui, T.L. (1997), *City States in the Global Economy: Industrial Restructuring in Hong Kong and Singapore*, Westview Press, Boulder, Colorado.

Clark, G. L. and Kim, W.B. (eds) (1995), *Asian NIE's and The Global Economy: Industrial Restructuring and Corporate Strategy in the 1990's*, Johns Hopkins University Press, Baltimore.

Dicken, P. (1998*), Global Shift, Transforming the World Economy*, Paul Chapman Publishing Ltd., London.

Dicken, P. and Hassler, M. (2000), 'Organizing the Indonesian Clothing Industry in the Global Economy: The Role of Business Networks', *Environment and Planning A*, vol. 32, pp. 263-280.

Economic Development Board Singapore (SEDB) *Report on the Census of Industrial Production*, various years, SEDB, Singapore.

Ernst, D., Ganiatsos, T. and Mytelka, L. (eds) (1998), *Technological Capabilities and Export Success in Asia*, Routledge, London.

Gereffi, G. (1997), 'Global Shifts, Regional Response: Can North America Meet the Full-Package Challenge?', *Bobbin*, vol. 39, pp. 16-31.

Gereffi, G. (1999), 'International Trade and Industrial Upgrading in the Apparel Commodity Chain', *Journal of International Economics*, vol. 48, pp. 37-70.

Gereffi, G. (2000), 'The Regional Dynamics of Global Trade: Asian, American and European Models of Apparel Sourcing', in M. Vellinga (ed), *The Dialectics of Globalization. Regional Responses to World Economic Processes: Asia, Europe and Latin America in Comparative Perspective*, Westview Press, Boulder Co., pp. 31-62.

Grunsven, L. van and Westen, A. van (2000), 'Global Forces, State Responses and Industrial Development in Singapore and Malaysia', in: M. Vellinga, (ed), *The Dialectics of Globalization. Regional Responses to World Economic Processes: Asia, Europe and Latin America in Comparative Perspective*, Westview Press, Boulder, Colorado, pp. 119-146.

Ministry of Trade and Industry, Republic of Singapore (1998*), Report of the Committee on Singapore's Competitiveness*, MTI, Singapore.

Münch, C. and Wijk, R. van (2000), Restructuring of the Apparel Industry in Singapore, 1985-1998, Master's Thesis, Department of International Economics and Economic Geography, Faculty of Geographical Sciences, Utrecht University, The Netherlands.

Schmitz, H. and Knorringa, P. (1999), *Learning from Global Buyers*, IDS Working Paper 100. Brighton: Institute of Development Studies, University of Sussex.

Simon, D.F. (ed) (1995), *Corporate Strategies in the Pacific Rim. Global versus Regional Trends*, Routledge, London.

Taplin, I.M. and J. Winterton (eds) (1997), *Rethinking Global Production: A Comparative Analysis of Restructuring in the Clothing Industry*, Ashgate Publishing Ltd, Aldershot.

21 Conclusion: Institutions and Innovation in Territorial Perspective

Roger Hayter and Richard Le Heron

In this book, we have explored the nexus of the relationships underlying knowledge, industry and environment primarily from the perspective of institutional innovations as they have unfolded in a variety of territories and across different spatial scales. Our introductory comments (chapter 1) emphasized the nexus of knowledge, industry and environment, as vital to understanding the evolving economic geography of the 21st century. This final chapter briefly summarizes some lessons regarding how institutions, innovations and territory shape this nexus. If there has been variation among the chapters in their relative focus on 'knowledge', 'industry' and/or 'environment', all chapters have explicitly explored the processes of institutional innovation that are so crucial, in one way or another, to the of knowledge, industry and environment nexus. Further, apart from three framework chapters (chapters 2-4), all the chapters drew upon original empirical observations pertaining to specific territories.

In this last chapter, we first comment on the theme of institutional innovation. In this regard we do not make the claim that institutional innovation is more important than technological innovation. Rather, both institutions and technology are embedded in society, each affecting the other. Thus, given our bias on the former, we seek to confirm the importance of institutional innovation and its role, not necessarily successful, in the knowledge, industry and environment nexus. Our second theme privileges the role of institutional innovation in emphasizing the reflexivity between industry and environment, and ultimately in shaping a more knowledgeable and green economy. This priority is justified we believe because of Economic Geography's enduring difficulty in integrating economic and environmental issues and of policy makers in creating positive sum outcomes. Third, we comment on the importance and meaning of territory in our analysis. This matter may seem trite, but in an age of globalizing processes, ideas about geographic scale are readily blurred, and the forces affecting territorial development ever more complex. As part of this last segment some research priorities are mentioned, themselves drawn from suggestions by each contributor.

The Challenge of Institutional Innovation

Institutional innovation may be defined as changes in policies, organizations and attitudes that affect and govern economic and social attitudes and behaviour. Following Freeman (1974), as with technological change, institutional innovations may be incremental or minor (e.g. slight changes in work place rules), major (e.g. new national policies introducing new forms of regulation) and radical (e.g. new forms of business organization or international treaties). Further, institutional change not only occurs as explicitly stated new policies and rules but may be expressed in more intangible ways associated with new attitudes, tastes, fashions, concepts, practices, standards, benchmarks, conventions and so on.

Institutional innovations are significant to society because they often linked in their own right to productivity increases and because they are important complements to technological change. In the techno-economic model, for example, Freeman and Perez (1988) stress the idea of 'matching institutions' which refers to the new policies, organizations and attitudes required to best exploit and direct technological change.

However, a view that only emphasizes the links of institutions to technological developments is too restrictive, when considering the established desirability of shifting to a Green techno-economic paradigm (TEP). We can identify at least five reasons why the question of appropriate institutions, aligned to different goals, represents so big a challenge. First, when technology and technological change are conceived primarily in terms of products, processes and physical systems, there are major risks that wider societal goals are simply overlooked, or even worse, dismissed as not relevant. Thus, the overriding concern is recognizing that the goals towards which technological innovation might address go well beyond the purely economic.

Second, the leading target of institutional innovation in the future, we argue, is the creation of arrangements that facilitate the transition from a narrow economic and profit oriented technological focus to one that resolutely incorporates green dimensions in technological change. This needs to be seen as spanning all spheres of economic activity, from conception through to production and final consumption. The wealth of examples in *Knowledge, Industry and Environment* show how difficult this transition will be. A transition to a Green TEP requires institutional structures that will enable appropriate forces of transformation to gather momentum. This starting point partly stems from the fact that much research and policy analysis is guided by a range of conventional assumptions about how economic processes operate and therefore might be changed. These include assumptions that existing productivity measures are sufficient indicators by which to steer economic change and performance and that until economic growth is achieved there will be insufficient resources to finance and manage a greening of the economy. Yet, as many chapters in the book show, the Information and Communication TEP (ICT) is fraught with outcomes that add to rather than diminish environmental pressures. Unless sweeping and continuing new institutional forms are developed to originate and implement alternative agenda the prospects of a greener future seem dim.

Third, conventional thinking has strong elements of linearity. Time and again the findings described and analyzed in the book reveal instead the complexity, multidimensionality and uneven development of economic growth and technological change. Much theory stresses the development of institutions from technological change. We suggest that the realities in the early twenty first century put a premium on institutional innovation to provide radically different environments for the pursuit of technological innovation in the widest sense. Thus, the argument is that goals and facilitative arrangements must be brought to the fore to redirect technological change. This assertion is an inversion of the traditionally assumed relationship between technological change and institutions.

Fourth, since all institutions ultimately feature location domains from the quite local to the fully global and universal) geography is thoroughly implicated in any effort to foster institutional innovation. Again, our view is that new green-facilitative institutions are likely to have very different spatialities and temporalities to their predecessor institutions. What do we mean by this claim? In practical terms, this claim demands a new bottom line to business accountability, as existing accounting procedures are unable to accommodate the different time horizons implied by programmes to combat environmental degradation or reduce industrial pollution levels. Simultaneously, the business accounting of the future needs to fully incorporate the embryonic shift towards 'ethical' investments that explicitly contemplate how best to incorporate living wage, fair trade and human rights initiatives. Very different geographical horizons are likely to be integral to green-facilitative institutions, creating many tensions with and within institutional frameworks as transition processes are explored. We have a strong personal view that the transition into a Green TEP is an unprecedented opportunity to rework and give the ends-means relation of economic processes more holistic content.

Fifth, pressures for a Green TEP are unlikely to be confined to the developed world. *Knowledge, Industry and Environment* firmly emphasizes a range of 'divides' which are integral to the perpetuation of the existing uneven distribution of wealth, internationally and at other spatial scales. Whether as a result of ignorance or suppression, knowledge of these divisions is often missing in decision-making settings. Yet, the breadth and gravity of the issues is evident for all to see. Thus, the intensity of the public protests the Seattle World Trade Organization (WTO) meeting, rooted in growing disparities in global income, was widely documented by the media while inside the meeting Piore (2001) reports that there were 401 disputed items featured on the agenda. Subsequently, at the Bonn WTO meeting, designed to progress the Kyoto Protocol on environmental matters, various divisions among delegates were exposed, including the indicators to be used in calculating contributions to improved environmental performance. Moreover, in the flurry of international dialogue and coalition building generated by the events of September 11th 2001, the primacy of accumulation over human misery is being re-contested. A noteworthy example of this debate is the intense pressure to soften the control of multinational corporations (MNCs) over patented drugs to allow the production of generic drugs to combat HIV or Anthrax, savage killers in developing countries where people cannot afford brand names. It is all too easy to slip into cynicism over the conduct of MNCs with regard to the environment and social goals. However, as MNCs constitute the major set of investing agencies in the world, any shift in goals, strategy and behaviour of these organizations must be taken seriously. Moreover, MNCs are likely to be crucial partners in formulating and implementing alternative agendas. At the practical level, long term solutions to solve environmental and development problems need to involve MNCs.

Reflexivity of economy and environment

It is a *sine qua non* of systems theory, industrial ecology and related environmental thinking that economy and environment are interdependent, tied together by a maze of feedback loops, that are both understood and not understood, intended and not intended. It is generally recognized that industrial behaviour has modified the environment and that the modified environment has led to industrial adjustments of one kind or another. It is further widely recognized the environment has suffered greatly from industrialization, and that the fabled resilience of the environment has limits. It has also become evident during the ICT paradigm that environmental problems have become truly global. Indeed, as Barker and Soyez (1994) observe, the old admonishment to think globally and act locally' is no longer sufficient; society also has to think locally and act globally. Thus, acting locally reflects an appreciation that incremental improvements in environmental behaviour collectively have massive environment benefits. At same time, acting globally defines a need for global institutions to address problems at a global scale, most notably ozone layer depletion, climatic warming, bio-diversity loss and, perhaps less well publicized, looming water supply problems. Local industrialization is implicated in all of these global environmental problems.

The big institutional challenge is to create policies that reduce global environmental impacts without compromising development goals and reinforcing existing trends towards greater income inequality. The ideal response is for policies that facilitate positive sum games between economy and environment. Too often, in both the developed and developed world, proponents of industrialization and their environmental critics phrase the reflexivity between economy and environment in terms of trade-offs, at worst as negative sum games. Innovations in policy and attitudes, as well as in technology, are needed, if more hopeful attitudes are to prevail. In this context, we would (again) stress two points. First, developing countries will continue to aggressively seek to industrialize as they seek to catch-up with advanced countries and, second, technological change is not going to stop. The challenge is to shape the latter so that sustainability of economy and environment is achieved.

Admittedly, it has proven extremely difficult to develop frameworks that institutionalize reflexivity. At its simplist, this process speaks to the monitoring of processes and outcomes in relation to understood and agreed upon goals. Given the complexity of socio-economic life, most policy and planning processes are over-determined, making it very hard to establish which and what policy or planning dimensions provide the appropriate framework for directed or relatively autonomous action on the part of investors. In practice, reflexivity between economy and environment might be focused at a multitude of layers – households, factories and offices, firms and production systems. Meanwhile, it is important to encourage societal monitoring which is anchored in a growing understanding of economic and institutional processes. Such behaviour needs to be promoted simultaneously at a variety of geographic scales. Local, regional, national and even international monitoring that is embedded in knowledge of processes is urgently needed. The indicator movement is an important step in this direction, but more so if it is fully realigned with knowledge of how, by whom and why, particular outcomes have emerged. The innovative potential of the ICT paradigm leaves little doubt that the capacity to generate information and data is not a major issue. Where the challenge lies is in advancing behaviour and policy so that green outcomes are translated from ideas into durable social institutions at all spatial scales.

Territorial Perspective and Research Directions

The importance of the localized, territorial basis for exploring the nexus of relationships forming knowledge, industry and environment is constantly revealed throughout this book. These territories are frequently towns and cities (e.g. Dongguan, Frankfurt/Rhein-Main, Seoul, Singapore, Tokyo, Washington D.C.) or a closely allied regional context (e.g. British Columbia Gauteng, Lower Silesia, Valencia). Recognition that such territories remain valid units of investigation for Economic Geography is rooted in their differences, with respect to economic activity, technology, forms of regulation and conventions. In turn, each territory is shaped by the evolution of distinctive forms of global-local dynamics and interdependencies, in which the global and the local are broadly, simply conceived as forces originating within and beyond particular territories.

Moreover, in exploring the anatomy of particular urban and regional territories, the enduring importance of nation states is further confirmed by *Knowledge, Industry and Environment*. In contrast to much rhetoric about the power of globalization, nation states are still central to the formation of policies, perhaps increasingly so as national policies need to be articulated with international standards and considerations. It is also important to remember that as relatively open systems, cities and regions are particularly open to national influences, involving not only shared experiences, values, identity but also the relatively free movement of goods, services and people. Thus, it matters to Tong and Wang's (chapter 5) analysis of the nature of export-based industrialization in Dongguan or Schamp's (chapter 19) exploration of the idea of regional innovation systems in Frankfurt/Rhein-Main that these places are respectively in China and Germany. Other chapters make a similar point, while Stryjakiewicz's (chapter 16) study of Polish industrial transformation and Liu's (chapter 6) analysis of the Internet in China, place primary emphasis on appreciating the national perspective for understanding local variation. In this regard, part of Singapore's uniqueness is that it is both city and nation at the same time (Grunsven, chapter 20).

Around the world, local and national territories face very different institution building challenges in seeking to meet their economic and environmental goals. Indeed, in the case of the industrialization of South China and the re-industrialization of Poland, the daunting challenge is to establish or re-establish the basic institutions of capitalism in (modified) forms that guide local industrial trajectories in appropriate ways (chapters 5, 15 and 16 by Tong and Wang, Stryjakiewicz and Hardy). Seoul, Singapore, and perhaps the Valencia region of Spain have already established their (capitalist) credentials as new industrial spaces. However, even their relatively new industrial structures are already under siege, demanding new layers of institutions to carry them on the next rungs of the value added ladder of industrialization (see chapters 7, 9 and 20 by Hwang, Gil et al, and Grunsven). Elsewhere in Africa, for Rogerson (chapter 17) 'the periphery of peripheries', South Africa remains a beacon of hope, seeking to enhance its industrial base to incorporate globally competitive high tech manufacturing and services, at a time of domestic social and political transformation, while dysfunctional neighbours constitute a threatening border.

As Eraydin's chapter 4 points out, these brave attempts at industrial transformation based on technological catch-up and transfer face escalating problems as innovation and knowledge-based change itself becomes more important and qualitatively different. Moreover, the Internet explosion is part of this trend. Yet as Liu (chapter 6) observes its geographical implications for industry globally, not just in China, are barely understood. These implications are then an important research priority. Simultaneously, research on the economic geography of the knowledge economy, that has so far favoured advanced economies, needs to be extended to help new, relatively new and re-emergent industrial spaces, such as China, Poland, South Africa and Spain, as they try to improve on their own past performance (a local yardstick) and catch-up with the leaders (a global yardstick). Indeed, is it appropriate to ask which is the more appropriate yardstick, when both arenas as intimately intertwined? This research might usefully explore links between external and local, tacit and codified, and industrial and non-industrial knowledge, along with their location dynamics, in a variety of organization, sectoral and geographical circumstances.

Meanwhile, in the long established industrial centres of capitalism, such as Frankfurt/Rhein-Main and Tokyo, that are on the leading edge of innovation and knowledge creation, there are different kinds of institution building challenges. As Schamp (chapter 19) and Takuchi and Mori (chapter 18) demonstrate, many facets of production are disappearing, or at least leaving, an issue further illustrated by Alvstam (chapter 10) with respect to Sweden. Germany and Japan have long been regarded as exemplary national homes for locally coherent production systems where competition has long been tempered by nuanced forms of community spirited cooperation. Yet, in both Frankfurt/Rhein-Main and Tokyo fragmentation has become an issue. As Schamp notes, this concern extends beyond production per se to the bases of the innovation system itself, in what he calls the Triple Helix Model that links the roles of academia, business and government. He wonders whether the previously dominant (Fordist) model can be replaced by a new, coherent Triple Helix that serves the newly emerging activities of Frankfurt/Rhein-Main's economy. Meanwhile, in east Tokyo, Takeuchi and Mori note that as the vertical relationships anchored by large corporations have gone, a second generation of entrepreneurs has successfully created new small firm dominated horizontal ones. Can a third generation do the same?

Clearly, the innovation structures and inter-organizational relationships within such technologically sophisticated places as Frankfurt/Rhein-Main and Tokyo can no longer be taken for granted. The local-non-local aspects of these relationships need further investigation while there is also a need to move beyond the manufacturing sector and product innovations per se, to incorporate the service economy and public sector institutions. This observation further implies re-defining the meaning of innovation and developing new forms of measurement, an observation that connects with Alvstam's plea for mapping knowledge creation within firms and across national borders.

From South China to Lower Silesia, from Tokyo to Washington D.C., the challenge of new knowledge-based processes of industrialization for local development is further complicated by the scale and scope of environmental imperatives. If industrialization has always had profound influence on the environment, it is only been relatively recently that the latter has begun to exert a profound influence on industrialization (see chapter 2 by Hayter and Le Heron). Unfortunately, as Soyez (chapter 11) points out, Economic Geography can scarcely claim prescience in this regard as the subject has largely myopically evolved with little reference to environmental matters. But as society and the economy has turned to seriously internalize environmental values so as Economic Geography. 'Sustainability' may well have become an umbrella term that ambiguously supports different interpretations. As Park (chapter 3) urges, however, it symbolizes a shift towards a greening of the economy that has profound implications for future industrial transformation and for evolving economic geographies.

For Soyez (chapter 11), the environment is not just another 'location condition' but an institution milieu, most obviously expressed in the structures and strategies of environmental non-government organizations (ENGOs), that comprises sets of knowledge-based processes that alternatively inform and threaten corporations and governments while becoming increasingly internalized within corporate thinking. As Schulz shows (chapter 12) a substantial institutional, as well as technological, response to new environmental demands, at least within advanced country contexts, is the rapid growth of an environmental service industry, featuring new firms and sets of skills, that are widening the options available to firms and governments that seek to enhance their environmental performance. Braun (chapter 13) further reveals the emergence of an explicit environmental dimension in the evolution of corporate strategic thinking and gives particular emphasis to innovations of environmental networks, for the most part led by firms that are also innovative in other respects. We claim that these studies are on the forefront of a rapidly expanding research agenda that will systematically explore the nature of the green firm, green geographies and the Green TEP (see chapter 2 by Hayter and Le Heron). This research activity will also require a global dimension.

Indeed, if *Knowledge, Industry and Environment* has adopted traditional notions of geographic scale – the nation state, region and city – it simultaneously reveals that the processes generating change in these territories cannot be so readily constrained. Five chapters focus on the sometimes bewildering, boundary-spanning geography of four different types of learning processes in different institutional settings. In the context of the mature forest industry, Stringer (chapter 14) uses the international commodity chain as her unit of investigation to examine the shifting location dynamics of value added activity between a various sites in Japan and New Zealand, and other Pacific Rim locales, that are organized primarily by MNCs. In another mature industry, Grunsven (chapter 20) focuses on the textile commodity chain and shows how Singaporean innovations to enhance value are closely tied with various institutional arrangements and investments elsewhere in Asia, and beyond. Within these commodity changes, there are serious challenges facing policy makers to ensure 'value' stays within their jurisdictions. What policy initiatives will work best? Is the relentless pursuit of value added a game only a few can win?

In the context of high tech industry, Fromhold-Eisebeth (chapter 8) focuses on the international 'back and forth' movements of intellectual capital as embodied in (highly educated) people themselves. The 'brain drain' it appears, contrary to much popular opinion, can be turned up-side down, but the geographical consequences and the best institutional responses remain little understood. For a specific type of high tech activity, Alvstam (chapter 10) contemplates the direct and indirect creation and diffusion of intellectual capabilities within Ericsson, the Swedish telecommunications giant, within 'and between' Sweden and China. His provocative argument is that indirect R&D has become very much more important but that its geography has become increasingly invisible. He urges for new measures of this geography to better understand the spatial behaviour of MNCs and their local development consequences. Further, how does Fromhold-Eisebeth 'back and forth' migration of 'brains' relate to Alvstam's hidden movements of the know-how created directly and indirectly by intellectual capital within MNCs?

Boundary spanning activities, however, are not restricted to business organizations and their labour forces. Thus, Soyez (chapter 11) examines the proliferating growth of ENGOs and their increasingly powerful influence on industrial behaviour around the globe. As he notes, the geography of the processes underlying this influence, largely centred on the communication of information – much of it on the Internet (see Liu's chapter 6) – is largely hidden from view. An important research priority, as with Alvstam's new sources of intellectual capital, is to develop measures of this geography and of its influence. An assumption of a Green TEP, it might be noted, is that the priorities of business-driven R&D are fundamentally shaped by environmental imperatives, to respond to, and ultimately offset, the power of ENGOs.

Over the past decade or so, it has often been said that capital investment, especially that dominated by MNCs, is mobile. However, industrial capital may run but it cannot hide. Factories are physically observable. In contrast, key processes underlying the knowledge, industry and environment nexus are opaque, if not invisible, and not well understood. A key priority for Economic Geography in this regard is to render these processes more transparent and to clarify their implications for location dynamics and local and global development. With respect to policy, an important challenge is to identify institutional innovations that will use knowledge to create positive sum games between industry and environment. To contribute towards these priorities, the task for the *Organization of Economic Spaces* Commission of the IGU is to continue to be a forum for economic geographers from around the world to share ideas and experiences. If English is to remain the IGU's lingua franca, a related task is to ensure effective translation and communication of its own particular knowledge processes.

References

Barker, M. and Soyez. D. (1994), 'Think Locally, Act Globally? The Transnationalisation of Canadian land-use conflicts', *Environment*, vol. 36: 12-20, 32-6.

Freeman, C. (1974), *The Economics of Industrial Innovation*, Penguin, Harmondsworth.

Freeman, C. and Perez, C. (1988), 'Structural Crises of Adjustment: Business Cycles and Investment Behaviour', in G. Dosi, C. Freeman, R. Belson and G. Silverberg (eds), *Technological Change and Economic Theory*, Pinter Publishers, London, pp. 38-66.

Piore, A. (2001), 'From Seattle to Dora', Newsweek, 12 November, pp. 35-41

Index

acquisition 278-279
adaptation (see enterprise adaptation)
added value (see value added)
Advanced Research Projects Agency net (ARPAnet) 89
Africa 313
agenda setting (see also framing) 196
agents of change 281
agglomeration 5-7, 12, 25, 67, 84, 314, 333
amenity and protection (see environmental phase)
apparel industry (see also garment industry) 381-383, 387, 390
Asea Brown Boveri 306-308
Asia 89, 126
Asian newly industrialized economies (see newly industrialized economies)
auto industry (see car manufacturing)
AXE system 172
Bangalore 130, 132, 139
Beijing 82, 95-96, 99-100
benchmarking 276
best practice 2, 276
BioRegio contest 369
biotechnology 368-369
boundary permeating work 188, 189, 204

boundary spanning 188
brain drain 125, 126, 128, 129, 131, 135, 136, 140, 295
brain flows (see brain drain)
brain gain 132
brains 132, 136, 138, 141
branded companies 384
British Columbia 7, 199-200
build to order (BTO) 90, 91(6)
business innovation centres (BICS) 201, 202
buyer-driven chains (see also commodity chains) 255
Canada 199, 130, 200
Canadian embassy 200
capabilities 377
Cape Town 314, 318, 320, 322, 323, 324, 327, 328, 330, 331,
capital goods 151-152, 161
car manufacturing 361, 363, 364, 365, 367
catch-up 5, 49, 59, 60, 303
catching-up (see catch-up)
Central and Eastern Europe (CEE) 273, 274, 282, 283, 289, 308
ceramics 146, 152, 156, 158
chemical industry 361, 363, 364, 365, 367
China 6, 67-8, 79-80, 87, 91-101, 167, 172, 180-182, 385
China circle 70

China Internet (see Internet)
China manufacturing corporation 99
Chinanet 91, 94, 96
civil society actors 7, 195, 199, 200
civil society pressure 190, 191
clustering 55, 132, 133, 322
clusters 6, 38, 49, 52, 53, 54, 59, 108, 109, 115, 125, 126, 171, 308, 313, 356, 358, 360, 368-371
'CN' 91, 92, 93, 95
codified knowledge (see knowledge)
collective learning (see also learning) 42, 49, 50, 54, 59, 114, 118, 121, 122
commercial knowledge 168-171
commodity chain 253, 254-256, 377, 378, 382, 392
community spirit 350
compartmentlaization 297
competencies 49, 50, 51, 52, 56, 57, 377, 394
competitive advantage 49, 145, 308, 380
competitive and green 8, 227, 248
computer bandwidth 93-94, 96
computer hosts 92-94
convivial communities 24, 26
core-periphery 254-255
corporate actors 193
Cracow 307, 308
crystallization 345, 348, 349

cultural embeddedness (see also embeddedness) 274, 275-81, 286
cumulative learning 51, 60
cut-make-trim (CMT) 380-381
Darmstadt 360, 368
death of geography 88
dematerialization 16, 21, 22, 39, 40
design 153, 160
developing countries 27, 32-35, 41-44, 126, 131, 132, 136, 137
development 27
dimensions of sustainability 5
disciplinarity 187, 188
disembeddedness 285
division of labour 355, 379
Dongguan 3, 6, 67-85, 99
double dividend 233
East Asia 378
East-Central Europe (see Central East Europe)
eco-certification 23
eco-efficiency 35, 42, 44, 45
Ecofit parks 42-44
ecological dimension 230
ecological sustainability (see sustainability)
eco-modernization 187, 189, 193
economic dimenison 230
Economic Geography 32, 188, 189, 202, 400
economic imperatives 1, 2
economic sustainability (see sustainability)
economies of scale 13
economies of scope 13

e-corporation 89
education 1, 35, 58, 135, 154,
    295
electrical engineering industry
    361, 364
electronic business/commerce (e-
    com) (see Internet) 87-88,
    89, 90, 94, 98, 100, 101, 177
e-mail 94, 100, 177
embedded knowledge 58, 145,
    160, 162
embedding (see embeddedness)
embeddedness 55, 108, 110-111,
    113, 115, 117, 118, 120, 122,
    139, 146, 188, 194, 198, 201,
    269, 273, 274, 275-277
embodied knowledge 6, (9)
enabling policies 58
endogenous 146, 147, 304
Enginco 274, 277, 278-286
entrepreneurs (see also small and
    medium sized enterprises) 7,
    345
    first generation 345
    second generation (new)
    345, 349, 350
enterprise adaptation:
    deindustrialized 298, 301
    globalized 298, 299
    market oriented 298, 300
    paternalistic 298, 302
entry 278-279
environmental actors 193
environmental amenity 16
environmental audits 228
environmental business (see
    environmental services)
environmental contestation 6

environmental degradation 31,
    33, 36, 37, 40, 43
environmentalism 187, 189, 218,
    223
environmental imperatives 2, 7,
    11, 27
environmental innovations 232
environmental knowledge 7,
    192-194
environmental management 217,
    227, 228-236, 238, 244, 245,
    247, 248
environmental non-government
    organizations (ENGOs) (see
    non-governmental organiza-
    tions
environmental performance 23,
    24, 227, 228, 245, 246
environmental phase amenity
    and protection 15-17
    frontierism 14-16
    resource management 15,
    17-19
    eco-development 16, 20, 21
environmental policy 6, 41
environmental priorities 24
environmental producer services
    (see environmental service-
    providers)
environmental regulation 16, 17-
    18, 31-33, 39, 40-1, 44, 209
environmental service-providers
    209, 210, 212, 214, 216, 219,
    221, 222, 223
environmental sustainability (see
    sustainability)
Ericsson 165, 171-177, 179, 180-
    182, 184

European Community (EU) 149,
    153, 161
European Eco-Management and
    Audit Scheme (EMAS) 228,
    229, 230, 232, 233, 249
exogeneous 146, 304
export 70-72, 82-83, 378, 382,
    384
externalization 175, 183, 209,
    215-217, 223
finance 200-201, 369-371
financing environmental change
    43-4
Fletcher Challenge Forests 257,
    258, 259-61
flexibilization 209, 223
flexible production 34
flexible specialization 12, 26
footwear 146, 152, 158
Fordist 359, 360, 363, 372
Fordist techno-economic para-
    digm 13, 17
foreign direct investment (FDI)
    8, 41, 273, 274, 384, 385
foreign technology (9)
forestry 253, 257, 258, 267, 269
fragmented knowledge 171, 187
framing 187, 194, 196, 200, 201
Frankfurt/Rhein-Main 6, 356,
    360-373
frontier economics (frontierism)
    (see environmental phase)
garden cities 16
garment industry 377, 378, 379-
    381, 383, 394
Gauteng 314, 317, 318, 320, 322,
    323, 324, 327, 328, 330, 331

Germany 200, 211, 229, 356,
    361, 363, 365
Global Apparel Commodity
    Chain (GACC) 378, 379, 381,
    386, 390
global commodity chain (see
    commodity chain)
globalization 8, 9, 55, 128, 250,
    253, 355
global-local 67, 70, 126
global sourcing 383
goodwill 168
governance 8
green corporation 22-25
greenfield 275
green geographies 25-27
greening 7, 187,188, 189 190,
    199, 203, 209, 211, 222, 227
green knowledge economy 4
Greenpeace 200
green techno-economic paradigm
    2, 3, 5, 9, 16, 19-27
Guangdong 101
hardware centre 344
hidden assets 169
high technology (high tech) 67,
    165, 166, 170 (10), 315, 317-
    321, 329, 330, 331, 332, 333
Hoescht Company 364-365, 368
Hong Kong 70, 72, 75, 82-83,
    379, 380, 394
Hsinchu –Tapei 129
human capital 57
IMPIVA 156-157
incremental innovation (see
    innovation)
industrial clusters (see clusters)

industrial complex 337, 338, 343, 347, 348, 353

industrial districts 49, 54, 56, 67, 107, 108-110, 115-119, 145, 146

Industrial Ecology 38-40, 42

Industrial Geography 11, 189, 203

industrial metabolism 37, 39, 42

industrial transformation (see transformation)

informal institutions 283-285

information and communication techno-economic paradigm (ICT) 17-21, 169

information technology (IT) 90, 99, 107, 112, 117, 171, 313, 314, 315, 317, 322-328, 330, 331, 333

information technology service economy 322-327

information technology services 314, 315, 316, 322

infrastructure 12, 54, 58, 59, 358

inner Tokyo (see Tokyo)

innovation 4, 58, 148, 149, 150, 308, 400, 356, 369, 403
    national systems 356, 368
    regional systems 49, 54, 55, 57, 59, 359, 363, 365

innovation mileau 49, 50, 108-109, 122, 290, 295

innovation potential 8, 289-297

innovative synergy 114, 119, 121-122

institution bending 273

institution building strategies 4

institutional bottleneck 70, 82-4

institutional economics 1-2

institutional environment 390

institutional innovation 5, 399, 400-403

institutional matching 5, 11

institutionalist approach 4, 274

institutions 160, 391, 394

intellectual capital 166, 167, 169, 171, 181, 184

interactive learning (see learning)

internalization 24, 165, 175-177

international brain flows (see brain drain)

international migration (see migration)

international standards organization (ISO) 18

Internet 68, 87-102, 173, 176, 177, 178, 313, 324

Internet generation 100-101

Internet Protocol (IP) (see Internet)

Internet users (see Internet)

intra-corporate culture 8

intra-firm governance 279-281

invisible production 165

invisible R&D 179

Japan 25, 39, 253, 256, 258, 259, 260, 261, 263, 264, 265, 266, 267, 268

Japanese Agricultural Standard (JAS) 260

jobbing relations 216

Johannesburg 314, 319, 320, 324, 327, 328

Juken Nissho 262-264, 268

Juken Sangyo 262, 264

Jung-Chongro 108
just-in-time (JIT) 90, 91
Kangnam-Seocho 108, 115-122
knowledge (see also learning, and
    environmental knowledge)
    2, 51, 57-60, 176-177, 192-
    194, 199, 200, 210, 219, 274,
    289, 304, 308, 313, 332, 355,
    359, 360, 363, 371, 390
    codified 4, 53, 55, 56, 57,
    192, 286
    tacit 41, 49, 53, 55, 56, 192,
    220, 274, 277, 281, 283, 285,
    286
knowledge-based activities 313,
    314, 315, 316, 317, 327, 329,
    330, 332, 333
knowledge-based enterprises
    315, 331
knowledge-based industry 289,
    304, 308, 319
knowledge-based region 304,
    315
knowledge economy 2, 4, 5,
    6, 32, 50, 57, 58, 60-61, 295,
    303, 304, 307, 308, 313,
    327-328, 378, 391, 394, 395
knowledge-intensive 212
knowledge management 173-177
knowledge spillovers 56, 356
knowledge system (see also
    Triple Helix) 356, 359, 360,
    363, 367, 369, 370, 371-373
    regional knowledge system
    360, 363, 370, 371-373
Kondratiev waves 12-13
Korea 6, (7)
Kyoto protocol 18

labour 72, 74, 130
labour migrants/migration (see
    migration)
labour mobility (see migration)
large firms 347
lean production 34, 169
learning 42, 50, 51-58, 116, 117,
    195, 211, 219-221, 222, 285
learning clusters 55
learning economy 58-60
learning environment 58
learning infrastructures 59
learning networks decentralized
    55-6
learning regions 49, 50, 54
learning, spatial aspects 51-55
less developed countries (see
    developing countries)
Letchworth Garden City 16
leveraging 394
linkages 72
local knowledge system (see
    regional knowledge system)
local government 81
localization 110, 112, 118-120
lock-in 55, 57, 60, 246, 372, 380
logistic regression 236-244
London 176
Lower Silesia 274, 278
Macao 70, 73
machinery industry 338, 353
MacMillan Bloedel 200
Mapo-Yongsan 108
market pressure 190, 191
market transformation (see
    transformation)
matching institutions (see institu-
    tional matching)

mature sectors 145, 146, 161, 377

mechanical engineering (ME) 344, 345, 349, 364

metropolitan 356, 360, 361

metropolitan knowledge system (see knowledge system)

migration (see brain drain and 'brains') 126, 128, 130, 132 qualified labour 125, 126, 127-129, 130, 133-40

migration systems 126, 133, 138

milieu innovateur (see innovation mileau)

mobile Internet 173

mobile phones 88, 173-175

mobile telecommunications (see also telecommunications) 178, 181

Montreal protocol 18

Motorola 181, 308

Multi Fibre Agreement (MFA) 379

multiple regression 98

multilateral development banks

multinational corporations (MNCs) 9, 24, 38, 41-3, 68, 71-72, 135, 139, 146, 182, 290, 302, 348, 350, 352

Munich 361

NAFTA 379

Nasdaq 167

national system 338

national system of innovation (see innovation)

negative sum games 1

networking 41, 42, 67, 70, 112-113, 118, 119, 120, 122,196-197, 210, 217-219, 259, 391, 394

networks 54, 69, 71, 73, 76, 78, 79-82, 381

new economy 167

new industrial districts (see industrial districts)

new industrial spaces 109

newly industrial countries or economies (NICs or NIEs) 8, 126, 130, 132, 136, 139,140, 377, 380

New Oji 264-265

New Zealand 8, 253, 254, 256-269

Nissho Iwai 262

Nokia 171, 173, 174, 177,179

Non-governmental organizations (NGOs) (also environmental NGOs) 18-19, 23, 42,189, 191, 195, 196, 197, 199, 200, 201, 202, 204

Opel 366

Optimus Computer Firm 304-306

organizational defenses 194

organizational learning 51, 52, 188, 220, 221

original equipment manufacturers (OEM) 81, 379, 387, 388, 391, 392

Ota-ku 337-3, 347, 34, 349, 351, 352

own brand manufacturing (OBM) 379, 380, 387, 389, 391, 393

Pacific Rim 253
Pan Pacific Forest Industries
    265-266
patents 150, 151, 168, 294, 361,
    363
path dependent 50,55, 133, 134
Pearl River Delta 67, 82, 100
personal computer manufactur-
    ing (PC) 67-85
Poland 8, 273, 274, 277, 278,
    290-309
politics of leverage 197
politics of shaming 197
portfolio approach 371-373
Porterian clusters 356, 360
positive sum games 1,2, 8, 403
power of mobility 177
privatization 253, 254, 257, 258,
    274, 285, 300(16)
producer-driven chains (see also
    commodity chains) 255
production chain (see also
    commodity chain) 76, 256,
    261, 263
production systems 34, 35, 199
proximity 52
qualified labour 9, 127
qualified labour migration (see
    brain drain and migration)
radical innovations (see innova-
    tion)
recycling policies 18
reflexivity 404
regional competition 328
regional discourse styles 221
regional innovation systems (see
    innovation)

regional knowledge economies
    (see knowledge economies)
regional knowledge system (see
    knowledge system)
regulation theory 12, 211
regulatory pressure 190, 191
re-positioning 8
research and development
    (R&D) 69, 148-150,159,
    160, 161, 162, 166169, 170,
    174, 179, 182, 184, 290,
    291-297, 355, 359, 360-367,
    370, 371
    indirect 179, 180, 184
resource management see envi-
    ronmental phase
restructuring 1, 254, 256-258,
    281-283
return migration (see brain drain)
Rio Earth Summit 33
sales relations 216
second generation entrepreneurs
    (see entrepreneurs)
Seoul 108, 115, 122
service clusters 369-370
Shanghai 84, 96, 101
Shenzhen 71, 74, 76
Silicon Valley 107, 126, 129,
    137, 138, 139,
Singapore 8, 377, 378, 379, 381-
    395
small and medium-sized firms
    (SMEs) 338, 341, 342, 344,
    345, 348, 351, 353
soft capitalism 275
sogo shosha 259, 260, 261, 268
South Africa 6, 313-333,
Southern Tokyo see Tokyo

space producing lobbies 189
Spain 146, 149, 150, 151
sparring relations 216
spatial clustering (see clusters)
spatial learning (see learning)
special economic zones (SEZ)
    71, 329
state owned enterprises (SOEs)
    274, 278, 281, 282, 284
Stockholm 175, 182
strategic alliances 167, 171, 176,
    177
strategic positioning 254, 259,
    266
Stuttgart 361
subcontracting (see also
    externalization) 382, 384,
    383, 385, 386, 388, 389
success dimensions 229
Sumitomo Corporation 259
sunset industry 380, 382
supplier chains 51, 91
sustainability 5, 36-38
    environmental 34, 42
    economic 35
    networks of 42-43
sustainable development 33, 35-
    6
sustainable industrialization 36,
    38
sustainable renovation 7, 337
Sweden 9, 172, 183
Tachikawa Forest Products 266-
    267
Taiwan 70, 73-75, 82
Taiwanese PC makers 71
take-back policies 18, 24
targets 197-8

tariffs 84, 258
techno-economic paradigms
    (TEPs) 4, 5, 11-17
    and the environment 12-17
technological capability 297
technological change 12
technological clusters 139, 319,
    368-369
technological competence 152
technological dimension 232
technological frontiers 57
technological gaps 151
technological innovation 12, 386
technological institutes 145, 146,
    156-61
technology 53, 54, 146, 148,
    151, 152, 157, 158, 183
Telebit 176
telecommunications 171, 178,
    181
textiles 146, 152, 158
Tokyo 7, 337-353
toys 146, 152, 158
training 145, 154-156
transactional learning 51, 52
transformation (see techno-
    economic paradigms) 289,
    291, 297, 298, 302, 308
transnational advocacy networks
    (TANs) 195
transnational epistemic commu-
    nities 195
transnational social movement
    organizations 195
triangle manufacturing 380
Triple Helix 357, 360, 370
trust 53
ubiquities 55

United Kingdom (UK) 16, 228, 229
UN Conference on Environment and Development 33
UNIDO 33, 37, 46
universities 54, 58, 59, 87, 97, 116, 130
upgrading 140, 265, 378, 380
USA 17-18, 25, 88, 98
USA Comprehensive Environmental Response , Compensation and Liability Act 45
Valencia 6, 145, 146, 146-151, 156, 157, 162
value added 69, 145, 148, 166, 167, 168, 171, 255, 258, 259, 260, 268, 387
value-chains 172, 269
value environment 188, 198
Vehico 274, 277, 278-286
Vodafone 179

Warsaw 305, 307
Washington D.C. 200-202
Wenzhou City 98
Western Cape 314, 317, 318, 320, 322, 323, 324, 327, 328, 330, 331
Westski 279, 280
Weyerhaeuser 200
Wiesbaden 360
work practices 283-285
World Bank 191, 200, 201
World Business Council for Sustainable Develop,ment (WBCSD) 33
World Commission on Environment and Development 35
World tourism 20
WTO 69, 83-4
Yellowstone National Park 16
Youngdeungpo 108, 115, 117-119, 122